# RECOGNITION AND MANAGEMENT OF PESTICIDE POISONINGS

*Sixth Edition • 2013*

**James R. Roberts, M.D., M.P.H.**
*Professor of Pediatrics, Medical University of South Carolina*

**J. Routt Reigart, M.D.**
*Professor Emeritus, Medical University of South Carolina*

**Support for this publication was provided by:**
Office of Pesticide Programs
U.S. Environmental Protection Agency
1200 Pennsylvania Avenue, NW (7506P)
Washington, DC 20460

This manual was developed under Cooperative Agreement No. X8-83384201, awarded by the U.S. Environmental Protection Agency (EPA) to the Medical University of South Carolina. The design and printing of this manual was facilitated by the National Association of State Departments of Agriculture Research Foundation (NASDARF) under the NASDARF Cooperative Agreement with EPA, No. X8-83456201. The information in this publication does not in any way replace or supersede the restrictions, precautions, directions or other information on the pesticide label or any other regulatory requirements, nor does it necessarily reflect the position of the EPA.

## Acknowledgments

We would like to thank the Environmental Protection Agency's Office of Pesticide Programs for providing the opportunity to collaborate on this sixth edition of *Recognition and Management of Pesticide Poisonings*. We are particularly grateful to Kevin Keaney, Chief of the Pesticide Worker Safety Program, who provided the vision and support for the continuation of this manual. Elizabeth Evans, M.P.H., Environmental Protection Specialist in the Pesticide Worker Safety Program, was our project officer and provided constant oversight and assistance. We thank Khin Swe Oo, M.D., D.A.B.T., from the Toxicology and Epidemiology Branch of OPP's Health Effects Division for serving as the lead EPA final technical reviewer for all chapters. Dian D. Overbey, Environmental Protection Specialist in the Communication Services Branch, provided copy editing.

Amy K. Liebman, M.P.A., M.A., Director of Environmental and Occupational Health, Migrant Clinicians Network; Geoffrey M. Calvert, M.D., M.P.H., Senior Medical Officer with the National Institute for Occupational Safety & Health's Division of Surveillance, Hazard Evaluations, and Field Studies; and Elizabeth Evans from EPA served as co-authors on Chapter 1, *Introduction* and Chapter 2, *Making the Diagnosis*.

This edition was peer reviewed by experts in clinical toxicology. We greatly appreciate the efforts of the following reviewers:

Alvin C. Bronstein M.D., F.A.C.E.P.
Medical and Managing Director
Rocky Mountain Poison and Drug Center
Denver Health and Hospital Authority
Associate Professor
Department of Emergency Medicine
University of Colorado School of Medicine
Denver, Colorado

Catherine J. Karr, M.D., Ph.D.
Associate Professor
Departments of Pediatrics and Environmental & Occupational Health Sciences
University of Washington
Director, NW Pediatric Environmental Health Specialty Unit
Seattle, Washington

Caroline Cox, M.S.
Research Director
Center for Environmental Health
Oakland, California

Matthew C. Keifer, M.D., M.P.H.
Director, National Farm Medicine Center
Marshfield Clinic Research Foundation
Marshfield, Wisconsin

Tammi H. Schaeffer, D.O., F.A.C.M.T.
Medical Toxicologist
Rocky Mountain Poison and Drug Center
Denver Health and Hospital Authority
Assistant Professor
Department of Emergency Medicine
University of Colorado School of Medicine
Denver, Colorado

We are extremely grateful for the assistance of Katie Chamberlain, R.N., in developing this new edition. Ms. Chamberlain was instrumental in cataloguing electronic versions of all of the references from the previous edition, securing and organizing the new references, communicating with reviewers and providing editorial review. It is an understatement to say that she made this process easier than anticipated.

Sally D. O'Neal was responsible for additional editing, graphic design and formatting of this manual. Carol Black and the National Association of State Departments of Agriculture Research Foundation (NASDARF) provided financial support for the design and printing of this manual.

# Table of Contents

# Tables and Figures

**Dosage Tables** *(in alphabetical order)*



**Figures** *(in order of appearance)*



**Tables** *(in order of appearance)*

*Section I*

# GENERAL INFORMATION

CHAPTER 1

# Introduction

The purpose of this manual is to provide healthcare professionals with current consensus recommendations for treating patients with pesticide-related illnesses or injuries. The Office of Pesticide Programs of the U.S. Environmental Protection Agency has sponsored the series since 1973. The 5th edition of this manual was published in 1999; since then, much has changed with regard to the pesticide products on the market. Most indoor uses of organophosphates have been eliminated, and a combination of EPA risk mitigation actions has limited their use on food crops. Pyrethroids have largely replaced organophosphates for residential pest control. While this conversion is beneficial in that the risk to human health is lower with this relatively less acutely toxic class of pesticide, it introduces a new set of health issues for consideration. Many new pesticide products have been registered and are not necessarily widely known among health professionals. This 6th edition includes a chapter that explores potential association between low-level exposure to pesticides over time and chronic diseases.

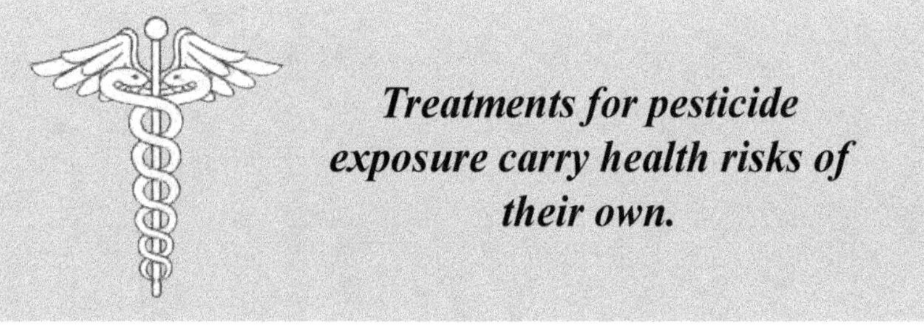


There is general agreement that *prevention* of pesticide poisoning remains a much surer path to safety and health than reliance on treatment. In addition to the inherent toxicity of pesticides, none of the medical procedures or drugs used in treating poisonings is risk free. In fact, many antidotes are toxic in their own right, and such apparently simple procedures as gastric intubation involve substantial risk. The clinician must weigh the hazards of various courses of action (including no treatment at all) against the risks of various interventions, such as gastric emptying, catharsis, administration of intravenous fluids or administration of an antidote, if available. Clinical management decisions have to be made promptly and, as often as not, on the basis of limited scientific and medical information. The complex circumstances of human poisonings rarely allow for precise comparisons of alternative management strategies. Therefore, it is important for the reader to keep in mind that the treatment recommendations in this book do not guarantee successful outcomes. They are merely consensus judgments of the best available clinical management options. Clinical toxicology is a dynamic field of medicine; new treatment methods are developed regularly, and the effectiveness of old as well as new modalities is subject to constant critical review.

## Key Principles

General methods of managing pesticide poisonings are presented in **Chapter 3** and reflect a broad base of clinical experience. Several key points deserve emphasis. The need to protect the airway from aspiration of vomitus cannot be overstated. Death has resulted from aspiration, even following ingestion of substances having relatively low toxic potential. In poisonings by agents that depress central nervous system functions or cause convulsions, airway protection by early placement of a cuffed endotracheal tube (even when this requires light general anesthesia) may be life saving. Maintenance of adequate pulmonary gas exchange is another essential element of poisoning management that deserves constant reemphasis.

The amount of pesticide absorbed is a critical factor in making treatment decisions, and estimation of dosage in many circumstances of pesticide exposure remains difficult. The terms "small amount" and "large amount" used in this book are obviously ambiguous, but the quality of exposure information obtained rarely justifies more specific terminology. Sometimes the circumstances of exposure are a rough guide to the amount absorbed. Spray drift from a pesticide properly diluted for field application is not likely to convey a large dose unless exposure has been prolonged. However, drift is the leading cause of incidents among agricultural workers reported to the Sentinel Event Notification System for Occupational Risk (SENSOR)-Pesticides.[1] Farmworkers and pesticide applicators working with pesticides on a regular basis are at risk for acute pesticide poisonings. Spills of a concentrated chemical onto the skin or clothing may well represent a large dose of pesticide unless the contamination is promptly removed. Brief dermal exposure to foliage residues of cholinesterase-inhibiting pesticides is not likely to lead to poisoning, but prolonged exposures may.

***Risk* = *Toxicity* × *Exposure***

Suicidal ingestions almost always involve "large amounts," requiring the most aggressive management. Except in children, accidental pesticide ingestions are likely to be spat out or vomited. Ingestions of pesticides by children are the most difficult to evaluate. The clinician usually must base clinical management decisions on "worst case" assumptions of dosage. Childhood poisonings are further complicated by the greater vulnerability of the very young, not only to the pesticides, but also to the drugs and treatment procedures. Children ingest a greater amount per body weight than adults. The nature of neurological development in children entails an additional level of risk that is not present in adults.

## Underreporting

*Pesticide incidents are underreported for several reasons. According to the OPP Report on Incident Information (EPA, 2007), these include:*

Lack of a universal, mandatory legal duty to report incidents

Lack of a central reporting point for all incidents

Similarity of symptoms associated with pesticide poisonings to other causes

Misdiagnosis by physicians because of a lack of familiarity with pesticide effects

Inadequate investigation of incidents to identify the pesticide that caused the effects

Difficulty in identifying and tracking chronic effects

Reluctance or inability of physicians to report incidents

Limited geographic coverage of individual poisoning databases

## Barriers to Proper Recognition and Management of Pesticide Poisonings

Pesticide-related illnesses are one example of a myriad of existing Environmental and Occupational Health (EOH) exposures of concern. For many reasons, accurate diagnosis and treatment of pesticide poisonings present a challenge to the clinician. Like many illnesses linked to environmental exposures, pesticide poisonings remain commonly under-diagnosed due in large part to barriers in seeking care and diagnosis of pesticide poisonings.

### *Seeking Care*

One important factor contributing to under-diagnosis occurs if the exposed person does not, or is unable to, seek medical attention. A pesticide applicator, for example, may not perceive the incident as significant enough to seek care, particularly if he or she has been accustomed to low-level exposure scenarios on the job. Some agricultural workers are unable to readily address a pesticide poisoning because of a complex set of socioeconomic factors including inability to take off from work, transportation problems, language and cultural barriers, lack of health insurance, scarcity of available community health services and fear of losing employment. Another scenario is the exposed person may simply not recognize his or her symptoms as pesticide related.

### *Diagnosis*

When an individual exposed to pesticides does seek care, diagnosis has its own set of challenges. Differential diagnosis is difficult because signs and symptoms of pesticide-related illnesses are often nonspecific and may be confused with common illnesses unrelated to pesticide exposure. The clinician may neglect to take an environmental and occupational exposure history,[2] a key to proper diagnosis, and thereby miss the opportunity to uncover a pesticide poisoning. Even when pesticide poisoning is suspected, few diagnostic tools are available. **Chapter 2** of this manual, entitled ***Making the Diagnosis***, is intended to guide clinicians in determining whether the patient may be experiencing symptoms of a pesticide poisoning, with an emphasis on taking an environmental and occupational exposure history.

### *Institutional*

The 1999 edition of this manual stated, "Despite recommendations by the Institute of Medicine and others urging the integration of environmental medicine into medical education, healthcare providers generally receive a very limited amount of training in occupational and environmental health, and in pesticide-related illnesses, in particular."[3] Migrant Clinicians Network surveyed clinicians in 2000 and found that more than 80% reported little or no EOH training.[4] This reality remains largely unchanged.

> ***"...environmental medicine education is largely omitted in the continuum of U.S. medical education, leaving future physicians and current practitioners without expertise in environmental medicine to provide or facilitate environmental preventative or curative patient care." (Gehel, et al., 2011)***

Few healthcare providers are adequately trained in environmental medicine despite widespread recognition of a need to better prepare the nation's frontline in public health to respond to EOH issues.[5] There is growing interest in environmental medicine among practicing clinicians[6] and medical and nursing students, but the existing education system does little to address this demand.[5] Institutional change to expand an already stressed medical curriculum has proven to be a major obstacle to inserting EOH training.

## Assessing the Relationship of Work or Environment to Disease

Pesticides and other chemical and physical hazards are often associated with nonspecific medical complaints so it is very important to link the symptoms with the timing of suspected exposure to the hazardous agent. The ***Index of Signs and Symptoms,*** beginning on page 244, provides a quick reference to symptoms and medical conditions associated with specific pesticides. Further details on the toxicology, confirmatory tests and treatment of illnesses related to pesticides are provided in each chapter of this manual. A general understanding of pesticide classes and some of the more common pesticide agents is helpful in making a pesticide-related disease diagnosis. A concurrent non-pesticide exposure can have no health effect, exacerbate an existing pesticide health effect or solely cause the health effect in a patient. In the more complicated exposure scenarios, assistance should be sought from environmental and occupational medicine (EOM) specialists.

## Common Pesticide Poisonings

Following are three pesticide incident data tables created for this manual to illustrate which pesticides are most frequently implicated in incident reports to SENSOR-Pesticides, National Poison Data System (NPDS) and California's Pesticide Illness Surveillance Program (PISP). These tables cannot be considered representative of all incidents because they only show those that were reported to these three databases. The relative frequency of cases generally reflects how widely a product is used in the environment. Organophosphate (OP) insecticides have historically topped the list of most commonly reported exposures. EPA risk mitigation measures have greatly diminished the use of organophosphates for residential, particularly indoor, use. In the United States, pyrethroids have largely replaced the OPs in terms of widespread usage. As such, they now account for the most human case reports in the United States. Although they are relatively less acutely toxic than their predecessors, some severe poisonings have similar presenting signs and symptoms as that of OP poisoning, thus complicating the process of making the correct diagnosis.

## Data Sources for Poisoning Incidents

Table 1. SENSOR-Pesticides Program

Table 2. National Poison Data System

Table 3. California Pesticide Illness and Surveillance Program

**TABLE 1**

**PESTICIDES MOST OFTEN IMPLICATED IN ACUTE OCCUPATIONAL PESTICIDE-RELATED ILLNESS AND INJURY CASES AND NUMBER OF CASES, SENSOR-PESTICIDES PROGRAM, 2005-2009 (N=9,906)**

| Rank | Pesticide Category | Number of Exposed Cases | | | | Sum of Single + Multiple Exposure Cases* (n=9,906 individuals) | |
|---|---|---|---|---|---|---|---|
| | | Exposed to Single Substance (n=6,187 individuals) | | Exposed to Multiple Substances* (n=3,719 individuals) | | | |
| | | n | % | n | % | n | % |
| 1 | Pyrethroids | 1,368 | 22.10 | 1,479 | 39.80 | 2,847 | 28.70 |
| 2 | Chlorinated compounds | 1,174 | 19.00 | 387 | 10.40 | 1,561 | 15.80 |
| 3 | Organophosphorous compounds | 600 | 9.70 | 429 | 11.50 | 1,029 | 10.40 |
| 4 | Pyrethrins | 358 | 5.80 | 620 | 16.70 | 978 | 9.90 |
| 5 | Glyphosate | 274 | 4.40 | 203 | 5.50 | 477 | 4.80 |
| 6 | Ammonium/ammonia | 32 | 0.50 | 361 | 9.70 | 393 | 4.00 |
| 7 | N-methyl carbamates | 249 | 4.00 | 112 | 3.00 | 361 | 3.60 |
| 8 | DEET | 292 | 4.70 | 59 | 1.60 | 351 | 3.50 |
| 9 | Sulfur compounds | 145 | 2.30 | 143 | 3.80 | 288 | 2.90 |
| 10 | Triazines | 168 | 2.70 | 60 | 1.60 | 228 | 2.30 |
| 11 | Fipronil | 26 | 0.40 | 135 | 3.60 | 161 | 1.60 |
| 12 | Naphthalene | 113 | 1.80 | 22 | 0.60 | 135 | 1.40 |
| 13 | Imidacloprid | 1 | 0.00 | 118 | 3.20 | 119 | 1.20 |
| 14 | Thiocarbamates/ Dithiocarbamates | 67 | 1.10 | 31 | 0.80 | 98 | 1.00 |
| 15 | Glutaraldehyde | 51 | 0.80 | 15 | 0.40 | 66 | 0.70 |
| | All other | 1,269 | 20.50 | 1,287 | 34.60 | 2,556 | 25.80 |
| **TOTAL INDIVIDUALS** | | **6,187** | **100.00** | **3,719** | **100.00** | **9,906** | **100.00** |

**Because some of the individuals exposed to multiple substances appear in the totals of more than one pesticide category, the sum of the pesticide categories exceeds the number of individuals.*

Source: Edward J. Kasner, MPH and Geoffrey M. Calvert, MD, National Institute for Occupational Safety and Health, Centers for Disease Control and Prevention.

**TABLE 2**

**PESTICIDE EXPOSURES MOST COMMONLY REPORTED TO NATIONAL POISON DATA SYSTEM ACCORDING TO THE 2010 ANNUAL REPORT[7]**

| Rank | Pesticide or Pesticide Class | | Child <5 years | 6-12 years | 13-19 years | ≥ 20 years | Unknown age | Total |
|---|---|---|---|---|---|---|---|---|
| 1 | Pyrethrins and pyrethroids | | 7,717 | 1,672 | 1,222 | 14,800 | 2,706 | 28,117 |
| 2 | Disinfectants | Hypochlorite disinfectants | 5,024 | 563 | 837 | 5,471 | 1,355 | 13,250 |
| | | Other disinfectants (*e.g.*, pine oil and phenols) | 6,994 | 619 | 433 | 2,435 | 537 | 11,018 |
| 3 | Rodenticides | Anticoagulant rodenticides | 9,176 | 204 | 95 | 796 | 225 | 10,496 |
| | | Other rodenticides | 1,785 | 89 | 67 | 250 | 183 | 2,374 |
| 4 | Insect repellents | DEET | 3,194 | 685 | 251 | 934 | 189 | 5,253 |
| | | Others (*e.g.*, naphthalene moth repellent) | 3,178 | 328 | 130 | 1,338 | 491 | 5,465 |
| 5 | Herbicides (*e.g.*, glyphosate, chlorophenoxy herbicides) | | 2,019 | 362 | 246 | 4,593 | 817 | 8,037 |
| 6 | Borates and boric acid pesticides | | 4,270 | 92 | 62 | 466 | 110 | 5,000 |
| 7 | Organophosphates | OPs alone | 722 | 171 | 107 | 1,331 | 321 | 2,652 |
| | | OP + carbamate and OP + non-carbamate insecticides | 158 | 47 | 49 | 495 | 83 | 832 |
| 8 | Carbamate insecticides | | 804 | 119 | 83 | 1,027 | 221 | 2,254 |
| 9 | Fungicides | | 171 | 25 | 21 | 414 | 73 | 704 |
| 10 | Organochlorine insecticides | | 182 | 30 | 15 | 245 | 58 | 530 |
| 11 | Fumigants | | 48 | 19 | 14 | 213 | 56 | 350 |
| | All other insecticides (including unknown) | | 5,526 | 615 | 387 | 5,264 | 1,371 | 13,163 |
| **TOTAL PESTICIDES/DISINFECTANTS** | | | **50,968** | **5,640** | **4,019** | **40,072** | **8,796** | **109,495** |

The pesticides most commonly reported to Poison Control Centers, according to the 2010 Annual Report data from the American Association of Poison Control Centers' (AAPCC) National Poison Data System (NPDS) are listed in Table 2, above. Cases listed as organophosphates (and the other categories as well) may also include other insecticides such as carbamates and organochlorines in a single product. Asymptomatic cases are included in Table 2 only.

| TABLE 3 | | | | |
|---|---|---|---|---|
| **SUMMARY OF PESTICIDE EXPOSURES AMONG CASES IDENTIFIED BY THE CALIFORNIA PESTICIDE ILLNESS SURVEILLANCE PROGRAM FROM 2005–2009 AND EVALUATED, AFTER INVESTIGATION, AS DEFINITELY, PROBABLY OR POSSIBLY RELATED TO PESTICIDE EXPOSURE, BY PESTICIDE CATEGORY** | | | | |
| **Pesticide category** | **Occupational** | | **Non-Occupational** | |
| | **Only pesticide implicated** | **Two or more pesticides involved** | **Only pesticide implicated** | **Two or more pesticides involved** |
| **Antimicrobials** | | | | |
| Hypochlorite | 422 | 69 | 98 | 81 |
| Quaternary Ammonium | 227 | 106 | 15 | 14 |
| Glutaraldehyde | 69 | 3 | 0 | 0 |
| Other/Unknown | 197 | 297 | 92 | 88 |
| **Insecticides/ Miticides/Insect Growth Regulators** | | | | |
| Organophosphates | 162 | 227 | 52 | 91 |
| Carbamates | 13 | 16 | 12 | 4 |
| Pyrethrins/ Pyrethroids | 56 | 425 | 134 | 294 |
| Organochlorines | 0 | 1 | 0 | 2 |
| Other/Unknown | 61 | 612 | 124 | 136 |
| **Herbicides/Defoliants** | 80 | 184 | 28 | 44 |
| **Fungicides** | 81 | 548 | 29 | 62 |
| **Fumigants** | 228 | 106 | 366 | 134 |
| **Other/unknown*** | 41 | 568 | 83 | 97 |
| **TOTAL EXPOSURES** | **1,637** | **3,162** | **1,033** | **1,047** |

**The majority of other/unknown pesticides are adjuvants, which are registered in California but not necessarily identified by active ingredients. Additionally, this category includes a molluscicide, a nematicide and several pheromones, plant growth regulators, preservatives, repellents, rodenticides, synergists, pesticides with multiple functions and products that never were identified.*

Table 3 shows the numbers of occupational and non-occupational exposures from 2005–2009 that the California Pesticide Illness Surveillance Program associated with various categories of pesticides. All exposures that occurred while the affected person was at work are considered occupational. Occupational exposures probably continue to be more fully reported than non-occupational exposures. A case represents one individual's exposure to pesticide(s). Cases in which only one exposure was credibly implicated are distinguished from those to which any or all of two or more pesticides may have contributed. This table illustrates exposures; when more than one pesticide active ingredient is implicated, an exposure is counted for each person/pesticide combination. Multiple pesticide active ingredients were implicated in the cases of 2,657 people exposed occupationally and 432 exposed non-occupationally. These cases are counted in each pesticide category for which they qualify, for totals of 3,162 occupational exposures and 1,047 non-occupational exposures.

## Special Populations and Environmental Justice

Environmental justice strives to ensure that no population is forced to shoulder a disproportionate burden of the negative human health and environmental impacts of pollution or other environmental hazards.[8] EPA seeks to ensure the fair treatment and meaningful involvement of all people regardless of race, color, national origin, educational level or income with respect to the development, implementation and enforcement of environmental laws, regulations and policies.[9]

With regard to pesticide exposure and environmental justice, the farmworker population is of particular concern. The majority of farmworkers and their family members in the United States are Latinos living in poverty. Farmworkers are the population most often affected by pesticide overexposure. Children represent another population of concern as they may be at greater risk from pesticide exposures because they are growing and developing. Women of reproductive age and pregnant and nursing women may also be more vulnerable because of the effects of pesticide exposures on fetuses and infants. These three populations face higher risk of harmful pesticide exposure because of occupation or developmental susceptibility, or combination thereof. Each is discussed in more detail below.

### *Agricultural Workers*

In the United States, between 1 million and 2.5 million hired farmworkers earn their living from agriculture.[10,11] Farmworkers are the working population most often affected by pesticide overexposure, especially Latino farmworkers.[12] Farmworker patients should be considered to be at high risk for pesticide exposure; their screening or exposure history should include specific questions about any agricultural work being done. For example:

- Are pesticides being used at home or at work?
- Do you mix or apply pesticides?
- Are the fields or orchards wet when you pick, prune or harvest?
- Was spraying taking place in or near the fields or orchards while you were working?
- Do you get sick during or after working in the fields or orchards?
- Do you use agricultural pesticides in your home?
- Did you learn about adverse health effects of pesticides and how to protect yourself from exposure while using pesticides?

Farmworkers often reside in agricultural communities where they and their family members may be further exposed in their homes because of pesticide drift from spraying of nearby fields or orchards and drinking contaminated water. Para-occupational exposure factors such as pesticide residue on workers and their clothing, shoes and vehicles and lack of adequate facilities to clean pesticide-contaminated work clothes may increase the risk of pesticide exposure for other household members as well.

### *Children*

Children face particular risks from pesticides, as their physical makeup, behavior and physiology may make them more susceptible than adults.[13,14,15] As such, it is important to assess pesticide exposures by asking about where pediatric patients live, the occupation of their parents and whether pesticides are used in the home, childcare facility, school and play areas. It is also important to remind parents to store pesticides out of the reach of children.

Children from agricultural families and those living in close proximity to agricultural areas are exposed to higher levels of pesticides than those whose parents do not work in agriculture and who do not live close to farms.[16,17,18] The higher pesticide levels may result from parents' tracking pesticides from the workplace into the home or by pesticide drift.[19,20]

Adolescents working in agriculture are also at risk of exposure to pesticides.[21,22] The incidence rate of acute occupational pesticide-related illness in adolescents is significantly higher compared to adolescents not working in agriculture.[23] This is a particular concern for young farmworkers since adolescents are permitted to work in agriculture at younger ages than in other industries. While the research examining the impact of neurotoxicants on the central nervous system of adolescents is limited,[24,25,26] there is strong evidence of neural remodeling and brain development during adolescence.[25,26,27,28] Dose responses, metabolic rates and routes of exposure may vary by age, gender and maturation.[21,22,28] Extra caution is merited as consideration is given to acute and chronic pesticide exposures of adolescents.[21,22]

### *Women of Reproductive Age and Pregnant Women*

Pesticides may cause the most damage in humans during periods of rapid development, especially *in utero* through transplacental absorption.[29,30] Even prior to fetal periods of increased sensitivity, studies have found that *preconception* exposure of either the mother or father may have an effect on reproductive outcome and offspring.[31,32,33,34] Maternal exposure to pesticides should be minimized during pregnancy and during the preconception period. The period of maximal sensitivity to a teratogen varies depending on the birth defect, but is almost always within the first 10 weeks of the pregnancy. However, the central nervous system, eyes, teeth and external genitalia may be susceptible to teratogenic exposures throughout the pregnancy.[35] Although no pesticides have been proven to be human teratogens, several studies have shown associations between pesticide exposures and reproductive toxicity in humans. For example, *in utero* exposure to organophosphates has been associated with low birth weight, mental and motor delay, attention deficit hyperactivity disorder (ADHD), and reduced IQ.[36,37] Women who are pregnant or planning a pregnancy, especially those currently engaging in agricultural activities, should be informed of the implications of exposure before conception and during the pre- and peri-natal periods, and assisted in making decisions that are appropriate for their individual work and home situations.[38] See **Chapter 21, *Chronic Effects*,** for further information and examples.

## References

1. Calvert GM, Karnik J, Mehler L, Beckman J, Morrissey B, Sievert J, Barrett R, Lackovic M, Mabee L, Schwartz A, Mitchell Y, Moraga-McHaley S. Acute pesticide poisoning among agricultural workers in the United States, 1998-2005. *Am J Ind Med.* 2008;883-898.

2. Trasande, et al. Pediatrician Attitudes, Clinical Activities, and Knowledge of Environmental Health in Wisconsin. *Wisconsin Med J.* 2006;105(2).

3. Institute of Medicine. *Role of the Primary Care Physician in Occupational and Environmental Medicine,* Washington, DC: Institute of Medicine, 1988.

4. Liebman A., Harper S. *Environmental Health Perceptions Among Clinicians and Administrators Caring for Migrants.* MCN Streamline. 2001;7(1).

5. Gehle, et al. Integrating Environmental Health Into Medical Education. *Am J Prev Med.* 2011;41(4S3):S296-S301.

6. Trasande, et al. Pediatrician Attitudes, Clinical Activities, and Knowledge of Environmental Health in Wisconsin. *Wisconsin Med J.* 2006;105(2).

7. 2010 Annual *Report of the American Association of Poison Control Centers' National Poison Data System (NPDS): 28th Annual Report.* Table A. "Demographic profile of Single Substance Nonpharmaceuticals exposure cases by generic category."

8. U.S. Department of Health and Human Services. Subcommittee on Environmental Justice, Environmental Health Policy Committee. Strategic elements for environmental justice. *Environ Health Perspect.* 1995 Sep; 103(9):796- 801.

9. Environmental Protection Agency. Environmental Justice. http://www.epa.gov/environmentaljustice/index.html.

10. Kandel W. Profile of Hired Farmworkers, A 2008 Update. Economic Research Report No. 60. Economic Research Service, U.S. Department of Agriculture. 2008.

11. Martin P. Immigration reform: implications for agriculture. Agricultural and Resource Economics Update. Davis, CA: University of California, Giannini Foundation. 2006.

12. Calvert GM, Karnik J, Mehler L et al. Acute pesticide poisoning among agricultural workers in the United States, 1998-2005. *Am J Ind Med.* 2008;51(12):883-98.

13. Landrigan, P. Pesticides and PCBs: Does the evidence show that they threaten children's health? *Contemp Pediatr.* 2001;18(2):110-124.

14. Faustman EM, Silbernagel SM, Fenske RA, Burbacher TM, Ponce RA. Mechanisms underlying children's susceptibility to environmental toxicants. *Environ Health Perspect.* 2001;108 suppl 1:13-21.

15. Reigart JR, Roberts JR. Pesticides in children. *Pediatr Clin North Am.* 2001 Oct;48(5):1185-98, ix.

16. Simcox NJ, Fenske RA, Wolz SA, Lee IC, Kalman DA. Pesticides in household dust and soil: exposure pathways for children in agricultural families. *Environ Health Perspect.* 1995;103(12):1126-34.

17. Fenske RA, Kissel JC, Lu C, Kalman DA, Simcox NJ, Allen EH, Keifer MC. Biologically based pesticide dose estimates for children in an agricultural community. *Environ Health Perspect.* 2000;108(6):515-20.

18. Curl C, Fenske RA, Kissel JC, Shirai JH, Moate TF, Griffith W. Evaluation of take-home organophosphorus pesticide exposure among agricultural workers and their children. *Environ Health Perspect.* 2002;110:A787-A792.

19. Thompson B, Coronado GD, Grossman JE, Puschel K, Solomon CC, Islas I, Curl CL, Shirai JH, Kissel JC. Pesticide take-home pathway among children of agricultural workers: Study design, methods, and baseline findings. *J Occup Environ Med.* 2003;45:43-53.

20. Eskenai B, Bradman A, Castorina R. Exposures of children to organophosphate pesticides and their potential adverse health effects. *Environ Health Perspect.* 1999;107 Suppl 3:409-19.

21. Rohlman DS, Nuwayhid I, Ismail A, Saddik B. Using epidemiology and neurotoxicology to reduce risks to young workers. *Neurotoxicology*. 2012 Aug;33(4):817-22.

22. Rohlman DS, Lasarev M, Anger WK, Scherer J, Stupfel J, McCauley L. Neurobehavioral performance of adult and adolescent agricultural workers. *Neurotoxicology*. 2007 Mar;28(2):374-80.

23. Calvert GM, Mehler LN, Rosales R, Baum L, Thomsen C, Male D, Shafey O, Das R, Lackovic M, Arvizu E. Acute pesticide-related illnesses among working youths, 1988-1999. *Am J Public Health*. 2003 Apr;93(4):605-10.

24. Adams J, Barone S Jr, LaMantia A, Philen R, Rice DC, Spear L, Susser E. Workshop to identify critical windows of exposure for children's health: neurobehavioral work group summary. *Environ Health Perspect.* 2000 Jun;108 Suppl 3:535-44.

25. Brown SA, Tapert SF, Granholm E, Delis DC. Neurocognitvie functioning of adolescents: effects of protracted alcohol use. *Alcohol Clin Exp Res.* 2000;24:164–71.

26. Spear LP. Alcohol's effect on adolescents. *Alcohol Res Health.* 2002;26: 287–91.

27. Andersen SL. Trajectories of brain development: point of vulnerability or window of opportunity. *Neurosci Biobehav Rev.* 2003;27:3–18.

28. Spear LP. Assessment of adolescent neurotoxicity: rationale and methodological considerations. *Neurotoxicol Teratol.* 2007 Jan-Feb;29(1):1-9.

29. Jurewicz J, Hanke W, Johansson C, Lundquist C, Ceccatelli S, Van Den Hazel P, Saunders M, Zetterström R. Adverse health effects of children's exposure to pesticides: What do we really know and what can be done about it. *Acta Pædiatrica.* 2006;95 Suppl 453:71.

30. Committee on Pesticides in the Diets of Infants and Children: Pesticides in the Diets of Infants and Children. National Academy Press, Washington, DC, 1993. 408 pp.

31. Arbuckle TE, Lin Z, Mery LS. An exploratory analysis of the effect of pesticide exposure on the risk of spontaneous abortion in an Ontario farm population. *Environ Health Perspect.* 2001 Aug;109(8):851-7. PubMed PMID: 11564623; PubMed Central PMCID: PMC1240415.

32. Vinson F, Merhi M, Baldi I, Raynal H, Gamet-Payrastre L. Exposure to pesticides and risk of childhood cancer: a meta-analysis of recent epidemiological studies. *Occup Environ Med.* 2011 Sep;68(9):694-702. Epub 2011 May 23. PubMed PMID: 21606468.

33. Abadi-Korek I, Stark B, Zaizov R, Shaham J. Parental occupational exposure and the risk of acute lymphoblastic leukemia in offspring in Israel. *J Occup Environ Med.* 2006 Feb;48(2):165-74. PubMed PMID: 16474265.

34. Murphy LE, Gollenberg AL, Buck Louis GM, Kostyniak PJ, Sundaram R. Maternal serum preconception polychlorinated biphenyl concentrations and infant birth weight. *Environ Health Perspect.* 2010 Feb;118(2):297-302. PubMed PMID: 20123616; PubMed Central PMCID: PMC2831933.

35. Moore KL, Persaud TVN. *The developing human: clinically oriented embryology.* 7th edition. Sauders, Philadelphia, Pennsylvania. 2003. 544 pp.

36. Rauh V, Arunajadai S, Horton M, Perera F, Hoepner L, Barr DB, Whyatt R. Seven-year neurodevelopmental scores and prenatal exposure to chlorpyrifos, a common agricultural pesticide. *Environ Health Perspect.* 2001 Aug;119(8):1196-1201.

37. Bouchard MF, Chevrier J, Harley KG, Kogut K, Vedar M, Calderon N, Trujillo C, Johnson C, Bradman A, Barr DB, Eskanazi B. Prenatal exposure to organophosphate pesticides and IQ in 7-year old children. *Environ Health Perspect.* 2011 Aug;119(8):1189-1195.

38. McDiarmid MA, Gehle K: Preconception Brief: Occupational/Environmental Exposures. *Maternal and Child Health J.* 2006;10:S123-S128.

CHAPTER 2

# Making the Diagnosis

## *Tools for Clinicians to Ascertain Environmental and Occupational Health Exposures*

## OVERVIEW

Accurate identification of the patient's exposure can lead to improved diagnostic, therapeutic and rehabilitative decisions by the clinician and result in improved patient outcomes. Without an accurate diagnosis, the clinician may decide upon a symptom-based treatment that may be less effective.

Once identified, a pesticide exposure incident should be considered a potential sentinel health event that may require follow-up efforts to locate the source and any additional cases. By identifying the source of exposure, the clinician can avert further exposure in the initial patient and other exposed individuals. Post-diagnostic activities are important to support a systems approach to pesticide exposure cases, including reporting the incident, filing a workers' compensation claim, and conducting specialty care referrals. The clinician must also be aware of several ethical and public health considerations. Lastly, there are key resources available to assist clinicians and patients in dealing with pesticide-related illnesses or injuries.

## TAKE INITIAL SCREENING FOR PESTICIDE EXPOSURE

Asking the patient a few initial screening questions is critical for making an accurate diagnosis and may flag the need to take a more extensive exposure history. Given that time constraints in a primary care setting compete with the need to identify a patient's potential EOH exposures, it is **highly recommended that a few short screening questions be incorporated into the routine patient intake procedure** in order to identify relevant EOH exposures.[1] See the *Sample Screening Questions* in the sidebar.

## OBTAIN DETAILED EXPOSURE HISTORY

If the initial screening suggests a potential EOH exposure concern, a detailed exposure interview is often needed. An extensive exposure history can take up to an hour and provides a more complete picture of pertinent exposure factors. The detailed interview includes questions on occupational exposure, environmental exposure, symptoms and medical conditions. Data collection guidelines specific to patients with confirmed acute pesticide illnesses or injuries is provided at the end of this chapter, on pages 26-27. Although the focus is on pesticide exposures and related health effects, concurrent non-pesticide exposures need to be considered in the overall patient health assessment.

Questions typical of a detailed EOH history are provided in **Appendix A, *Detailed Occupational and Environmental Exposure History Questions,*** on page 240. For further information on taking a history for all types of occupational and environmental hazards, consult a general occupational and environmental medicine reference text[2] or Agency for Toxic Substances and Disease Registry's *Case Study in Environmental Medicine: Taking an Exposure History.*[3]

### SAMPLE SCREENING QUESTIONS

***For an adult patient***

*After establishing the chief complaint and history of present illness:*

What kind of work do you do?

(*If unemployed*) Do you think your health problems are related to your home or other location?

(*If employed*) Do you think your health problems are related to your work? Are your symptoms better or worse when you are at home or at work?

Are you now or have you previously been exposed to pesticides, solvents or other chemicals, dusts, fumes, radiation or loud noise?

***For a pediatric patient***

*Questions asked of parent or guardian*

Do you think the patient's health problems are related to the home, child care setting, school or other location?

Has there been any exposure to pesticides, solvents or other chemicals, dusts, fumes, radiation or loud noise?

In what kind of work are the parents and other household members engaged?

## DEALING WITH A SUSPECTED PESTICIDE EXPOSURE

After conducting exposure screening and possibly a detailed exposure history, the clinician should take the following steps once they suspect a pesticide poisoning.[4] It should be noted that each pesticide incident is a unique situation with varying levels of severity and urgency; therefore, these steps are not always achieved in the order they are presented. It is, however, crucial to obtain and preserve any evidence of the exposure as soon as possible.

### 1. Collect Information on the Pesticide

When you suspect a pesticide poisoning, try to get as much information about the pesticide(s) as possible, including: the name of the pesticide used, the EPA pesticide registration number, and the pesticide label and/or the Material Safety Data Sheet (MSDS) for the pesticide(s). If this is a case involving agricultural workers or residents in an agricultural area, try to talk directly to the farm manager, safety coordinator or the pesticide applicator to get this information in addition to a description of the incident itself. Often application records will be made available if requested. Under EPA's Worker Protection Standard (40 CFR 170), agricultural employers are required to make the name of the pesticide and the label available to healthcare providers and workers if it is requested. Refer to the material entitled *Data Collection on an Acutely Pesticide Exposed Patient* found on pages 26-27 at the end of this chapter.

### *2. Follow Decontamination Procedures*

Follow the decontamination procedures as outlined in **Chapter 3, *General Principles,*** beginning on page 29.

### *3. Collect Evidence of Contamination*

Obtain an unlaundered sample of clothing that the patient was wearing at the time of the incident, if available. Put it in a plastic bag to prevent further exposure and to preserve the specimens for subsequent analysis; freezing is optimal. It can be difficult to find appropriate clothing to sample if the worker has been instructed to go home and thoroughly wash his/her clothing. If most clothing has been washed or is not available, it is likely the patient's hat or shoes would still be contaminated and could be analyzed.

### *4. Obtain a Urine Sample*

If an exposure seems likely, either based on the history or the clinical exam, obtain a urine sample and freeze it. If more than one patient is exposed, obtain a urine sample for each patient. Freezing the urine allows you extra time to determine if the sample needs to be analyzed and to which laboratory it should be sent.

### *5. Order Laboratory Tests*

The National Pesticide Information Center (NPIC) provides a list of pesticides that can be analyzed by clinical laboratories. This list and a list of accredited laboratories can be accessed at: *http://npic.orst.edu/mcapro/PesticidesTestingForExposure.pdf.*

If the patient appears to have been exposed to an organophosphate or N-methyl carbamate insecticide, order cholinesterase blood tests, both plasma and red blood cell, to determine the clinical level of cholinesterase activity. Some experts recommend blood testing if a clinician believes any significant exposure has occurred regardless of a baseline test. Unless a dramatic depression is present, the results of post-exposure testing are likely to be difficult to interpret in the absence of baseline cholinesterase testing. In this instance, it is advisable to conduct periodic re-tests, until it appears that the cholinesterase level has returned to normal. A "negative" cholinesterase (*i.e.*, results within the "reference range") does not rule out the possibility that the patient's symptoms are due to pesticides if the patient was reacting to a pesticide other than an organophosphate or N-methyl carbamate or if s/he was reacting to other ingredients in the organophosphate or carbamate formulation (*e.g.*, the solvents, propellents and carriers in the pesticide product formulation). However, negative results could be misinterpreted by an employer or insurer to mean that no exposure occurred. Post-exposure cholinesterase tests need to be compared to baseline pre-exposure test results or re-testing of cholinesterase several weeks post-exposure. The recovery rate for depressed cholinesterase can be estimated to be 0.8% per day for red blood cells and 1.2% per day for plasma.

### *6. Consult with the Appropriate Specialists*

You may need to consult with others, such as toxicologists, occupational and environmental medicine specialists, and industrial hygienists, who have expertise in dealing with chemical exposures. Pesticide Information Resources including the Association of Occupational and Environmental Clinics (p. 25) are listed later in this chapter, beginning on page 23.

### *7. Schedule/Conduct Patient Follow-up*

Make arrangements with the patient(s) for follow-up appointments and for reporting test results. Once the patient has been cared for, inform everyone else who needs to know about the incident – the workers' compensation case manager and the employer, in particular. The healthcare provider must obtain the employee's permission before notifying the employer.

While a diagnosis can be based on a group exposure for the purpose of treatment, workers' compensation systems generally deal with workers one at a time. Therefore the clinician must collect the information needed to document the exposure, symptomatology and confirmatory data for each individual involved in a multiple-patient poisoning. While illness consistent with other members in a clearly sick group may be sufficient for the clinician facing an outbreak, it may not be sufficient objective information to establish causality for a worker compensation claim.

### *8. Report the Pesticide Incident*

**a. Contact the Appropriate State Health Agency**

Pesticide exposures are reportable as health incidents and occupational incidents may also be reportable as a violation of the Agricultural Worker Protection Standard. Both of these important reporting requirements are discussed here.

If a healthcare professional suspects that a patient has a pesticide-related illness, the clinician should report it to the appropriate state health agency. If the healthcare professional is in one of the 30 states that mandate these reports, than s/he should send the report to the appropriate state health agency.

More information about state-specific reporting requirements can be found at *http://www.migrantclinician.org/exposurereportingmap.* The healthcare professional can notify the local poison control center (PCC) by calling (800) 222-1222.

The National Institute for Occupational Safety and Health (NIOSH), Centers for Disease Control and Prevention (CDC) and EPA support surveillance for pesticide-related illness and injury through the SENSOR-Pesticides program that aggregates pesticide incident data from 11 states (California, Florida, Iowa, Louisiana, Michigan, New Mexico, New York, North Carolina, Oregon, Texas and Washington) and has an occupational focus. The California Department of Pesticide Regulation (DPR) maintains the

Pesticide Incident Surveillance System (PISP). These surveillance systems collect case reports on pesticide-related illnesses and injuries from clinicians and other sources (*e.g.*, poison control centers, workers' compensation agencies and state agencies that regulate pesticides); conduct selected interviews, field investigations and research projects; and function as a resource for pesticide information within their state.

The impacts of these surveillance programs extend beyond the participant states by identifying emerging pesticide exposure issues that steer intervention efforts to prevent future incidents with similar exposure scenarios nationwide. However, there remains a need for systematic reporting of pesticide poisonings in all states into a central agency in order to compile accurate statistics on the frequency and circumstances of poisoning and facilitate efforts to limit these occurrences.

**b. Contact Pesticide State Lead Agency**

Before sending the patient(s) home, call the appropriate EPA-recognized pesticide State Lead Agency (SLA), which can investigate pesticide poisoning incidents. To find your SLA contact, go to *http://aapco.org/officials.html.*

The SLA will help determine if there was any violation of the Agricultural Worker Protection Standard. It can also tell you if additional action or information is needed.

### *9. Discuss Workers' Compensation with the Patient*

If the case involves an occupational exposure, each patient's chart should document it as such. A workers' compensation report must be completed for each exposed worker.

To achieve a successful workers' compensation claim, the healthcare provider must document evidence of the exposure and the illness and conclude that it is more likely than not that the illness was caused or aggravated by a workplace pesticide exposure. The legal standard for a workers' compensation case is that **there must be a "preponderance of evidence" that the disease is work related**. A preponderance of evidence is defined as meaning that it is more likely than not (*i.e.*, greater than 50% probability) that the poisoning was caused or aggravated by a workplace pesticide exposure.

Workers' compensation laws exist in all states, but benefit levels vary across states and not all states require coverage for agricultural workers. In the realm of workers' compensation, the worker is responsible for proving that his/her disease is occupational in origin. It is not the employer's responsibility. Workers' compensation claims for minor ailments or for injuries that are obviously work related are rarely contested by the compensation insurance companies.[5] This tends to be true for many acute pesticide poisoning cases where the illness is consistent with the known toxicology of the pesticide, where there is objective evidence that the patient experienced a pesticide exposure, and where the dose was sufficient to produce illness. Costly claims, such as death claims or claims involving permanent total disability are often contested by the workers' compensation insurance company. The proportion of workers' compensation claims for acute pesticide poisoning that are contested is not known. However, in those cases with little or no objective evidence that a pesticide exposure occurred (*i.e.*, lack of biological or residue evidence of exposure), especially when the poisoning signs and symptoms resemble a common respiratory or gastrointestinal illness, achieving a successful workers' compensation claim may be difficult. Finally, clinicians should be aware that reporting a workers' compensation case can have substantial deleterious implications for the worker being evaluated (*e.g.*, job loss or disciplinary action).

## SPECIAL CONCERNS

### Ethical Considerations

Ethical guidelines and codes of conduct have been established that can guide healthcare professionals who are dealing with dilemmas involving pesticide poisoning.[6,7] Three fundamental values underpin these guidelines and codes of conduct: (1) it is the duty of the healthcare professional to do good for the patient and to place the patient's interests above those of the healthcare professional, (2) the individual is the best judge of his or her own best interests and (3) social justice promotion of a fair and equitable distribution of finite health resources. Among the codes of ethics most relevant to the realm of pesticide poisoning is the need to keep confidential all individual medical information, only releasing such information "with proper authorization when required by law, for overriding public health considerations, to other healthcare professionals according to accepted medical practice, to others at the request of the individual, or when there is reasonable concern about potential endangerment of third parties."[6]

Investigation of a suspected occupational pesticide illness may necessitate obtaining further information from the worksite manager or owner. Any contact with the worksite should be taken in consultation with the patient because of the potential for retaliatory actions against the patient (such as job loss or other disciplinary action). Similarly, a request for a workplace visit or more information about pesticide exposure at the workplace should occur only after gaining the patient's permission. Even when investigating non-occupational pesticide illnesses, the patient's permission should be obtained before calling the patient's neighbors or others potentially responsible for the pesticide exposure. The discovery of pesticide contamination in a residence, school, childcare setting, food product or other environmental site or product can have public health, financial and legal consequences for the patient and other individuals (*e.g.*, building owner, school district, food producer). It is prudent to discuss these potential adverse consequences and follow-up options with the patient before pursuing an investigation.

In situations where the pesticide hazard is substantial and many individuals might be affected, a request can be made to the state health department to obtain the assistance needed for a disease outbreak investigation. If an outbreak investigation demands more resources than the state health department can provide, the state health department can request assistance from the Centers for Disease Control and Prevention. In such a situation, even if the initial case patient objects to disclosing the pesticide hazard to public health authorities, state reporting requirements and overriding public health considerations may require this notification.

### Public Health Considerations

Healthcare providers must recognize and diagnose cases of pesticide poisoning to ensure that pesticides are not producing unreasonable harm to human health. Cases of suspected pesticide poisoning can lead to detection of new pesticide hazards. Healthcare professionals are often the first to see a poisoned patient who may represent evidence of a new or re-emerging pesticide hazard. Such patients may also represent a full-blown disease outbreak.

A disease outbreak is defined as a statistically elevated rate of disease among a well-defined population as compared to a standard population. For example, in 2010, two workers were diagnosed with methyl bromide poisoning after being exposed to methyl bromide over several months while inspecting produce in a California cold storage facility. Methyl bromide was being used to fumigate grapes imported from Chile. Both workers had profound neurologic symptoms and elevated serum

### Steps in Investigating a Disease Outbreak

Confirm diagnosis of initial case reports (the "index" cases)

Identify other unrecognized cases

Establish a case definition

Characterize cases by person, place, and time characteristics (*e.g.*, age, race, ethnicity, gender and location within a company or a neighborhood, timeline of exposure and health events)

Create plot of case incidence by time (an epidemic curve)

Determine if a dose-response relationship exists (*i.e.*, more severe clinical case presentation for individuals with higher exposures)

Derive an attack rate and determine if statistical significance is achieved (divide number of incident cases by number of exposed individuals and multiply by 100 to obtain attack rate percentage)

### Items Contained in a Material Safety Data Sheet (MSDS)

Material identification

Ingredients and occupational exposure limits

Physical data

Fire and explosion data

Reactivity data

Health hazard data

Spill, leak and disposal procedures

Special protection data

Special precautions and comments

bromide levels. The physician for one of these workers notified the local poison control center, which notified the California DPR. The California DPR conducted an investigation and found that methyl bromide reached unsafe concentrations in enclosed areas during the transportation and storage of fumigated grapes. Stakeholders (*e.g.,* commodity groups, warehouse operators, USDA, EPA and the Chilean produce industry) were notified of these findings, and measures were adopted to reduce methyl bromide exposures.[8]

Disease outbreak investigations are conducted for many types of exposures and health events, not only those in the occupational and environmental areas. Usually, assistance from government or university experts is needed because the investigation may require access to information, expertise and resources beyond those available to the average clinician. The steps involved in such an investigation and the types of information typically gathered in the preliminary clinical stages are outlined in the Steps in Investigating a Disease Outbreak list in the margin on the previous page. The clinician must be aware that an outbreak investigation may be needed when severe and widespread exposure and disease scenarios exist. For more information on disease outbreak investigations, consult the literature.[9,10]

Clinicians are typically prohibited from sharing identifiable health data without the consent of the patient. However, an exception is made when the clinician disclosure is for public health purposes. The Health Insurance Portability and Accountability (HIPAA) Privacy Rule balances the protection of individual privacy with the need to protect public health. This privacy rule permits identifiable health data disclosures without patient consent to public health authorities authorized by law to collect or receive the information for the purpose of preventing or controlling disease, injury or disability [45 CFR 164.512(b)].[11] In other words, when state public health authorities need identifiable health data to address a public health need, this need overrides the HIPAA privacy rule requirements for patient consent before sharing.

## RESOURCES

### Material Safety Data Sheets and Pesticide Labels

In addition to the patient history, it is often helpful to obtain further information on suspect pesticide products. Two documents are useful starting points in the identification and evaluation of the pesticide exposure: the Material Safety Data Sheet (MSDS) and the pesticide label.

#### *Material Safety Data Sheet (MSDS)*

Under OSHA's Hazard Communications Standard (29 CFR 1910.1200), all chemical manufacturers are required to provide an MSDS for each hazardous chemical they produce or import. Employers are required to keep copies of the MSDS for all chemicals used at the workplace and make them available to the workers. The items contained in an MSDS are shown in the margin.

These documents tend to provide very limited information on health effects, and some of the chemical ingredients may be omitted because of trade secret considerations. One cannot rely solely on an MSDS when making medical determinations.

#### *Pesticide Label*

EPA requires that all pesticide products bear labels that provide certain information. This information can help in evaluating pesticide health effects and necessary precautions. Pesticide labels must include the information listed on the next page. The general organization of a pesticide label is illustrated in the front panel schematic below and the back panel schematic on the following page.

Note that for some products with multiple uses (typically agricultural products) or products with very small containers, EPA allows some information, such as directions for use or worker protection requirements, to be contained in an accompanying booklet rather than affixed on the container. The booklet is part of the legal label, which is reviewed and approved by EPA. The most important safety-related elements of the label, such as the signal word, ingredients, hazard statements, treatment statement and EPA registration number, must be on the container itself.

The EPA registration number is very useful when contacting EPA for information or when calling the National Pesticide Information Center hotline (see page 24). Pesticide product labels may differ from one state to another based on marketing or other area-specific considerations. Also, different formulations of the same active ingredients may result in different label information. The pesticide label generally lists information only for active ingredients (not for inert/other components) and rarely

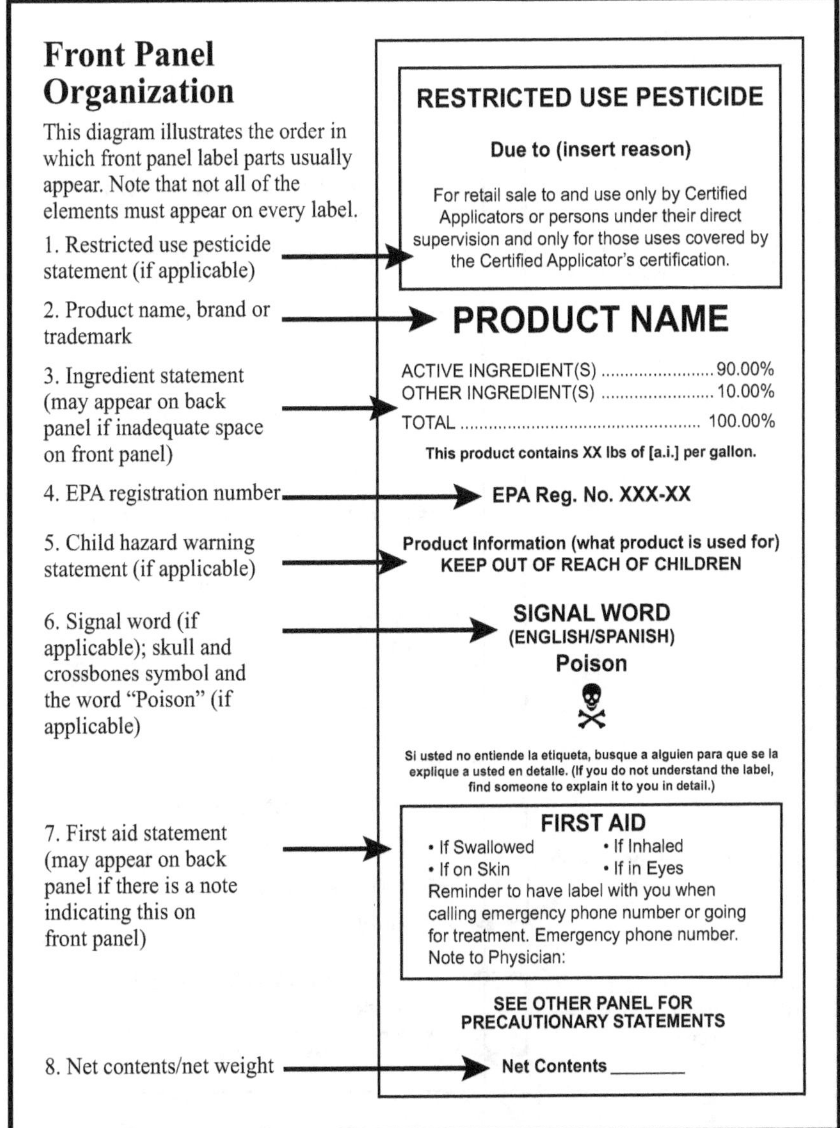


## Items Required on Pesticide Labels

Product name

Manufacturer name and address

EPA registration number

Active ingredients

Precautionary statements:

- Human hazard signal words "Danger" (most hazardous), "Warning," and "Caution" (least hazardous)
- "Poison" and symbol, if applicable
- Child hazard warning
- Statement of practical treatment (signs and symptoms of poisoning, first aid, antidotes and note to physicians in the event of a poisoning)
- Hazards to humans and domestic animals
- Environmental hazards
- Physical or chemical hazards

Directions for use

Net contents

EPA establishment number

Worker Protection Standard (WPS) designation, including restricted entry interval and personal protection equipment required (agricultural products only) (see WPS description on page 21) contains information on chronic health effects (*e.g.,* cancer and neurologic, reproductive and respiratory diseases). Although further information is often needed, pesticide labels and labeling should be considered as the first step in identifying and understanding the health effects of a given pesticide. The Agricultural Worker Protection Standard provides the legal basis for the healthcare provider(s) to obtain from the employer the name of the pesticide product to which the patient was exposed. When requesting this information, the clinician should keep the patient's name confidential whenever possible.

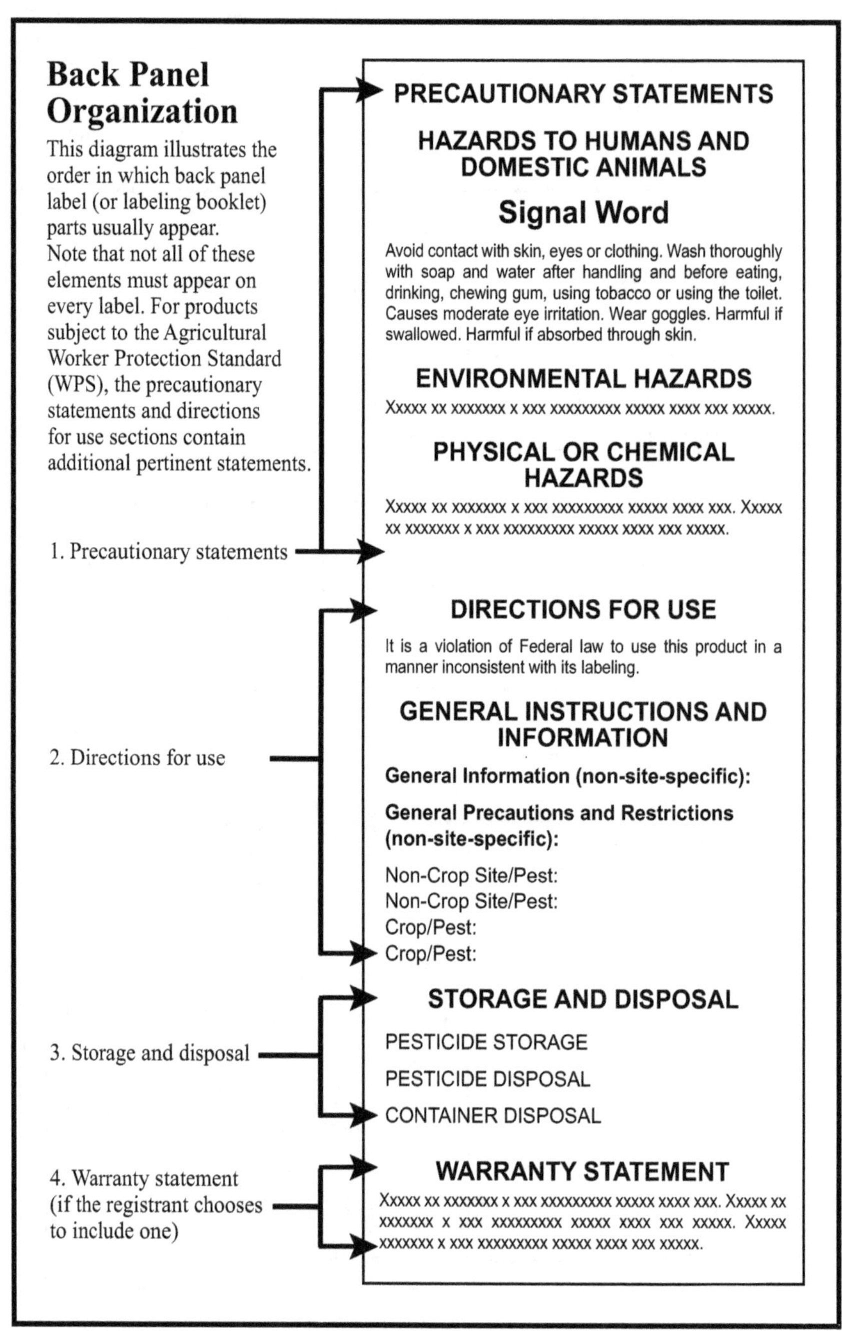

## Federal Regulatory Agencies

### *U.S. Environmental Protection Agency*

#### *a. Office of Pesticide Programs*

Since its formation in 1970, EPA has been the lead agency for the regulation of pesticide use under the Federal Insecticide, Fungicide, and Rodenticide Act. EPA's mandates include the registration of all pesticides used in the United States, setting restricted entry intervals (*i.e.,* the time interval during which individuals should not enter or be present in a pesticide-treated area, unless the individual is using appropriate personal protective equipment), specification and approval of label information and setting acceptable food and water tolerance (*i.e.,* residue) levels. In addition, EPA works in partnership with state, territorial, and tribal agencies to implement two field programs. First, the certification and training program for pesticide applicators sets national standards for those who apply restricted use pesticides, currently just under 1 million people. Second, the Agricultural Worker Protection Standard protects agricultural workers and pesticide handlers from pesticide exposures through training, field posting, requirements for protective equipment and decontamination protocols.

The authority to enforce EPA pesticide regulations is delegated to the states. Concerns about non-compliance with these regulations can typically be directed to your pesticide State Lead Agency (SLA). The EPA-recognized pesticide SLA is typically the state agriculture department but in some states and territories it can be another state agency (*e.g.,* the state environmental protection agency). To identify the pesticide SLA in your state, visit the Association of American Pesticide Control Officials (AAPCO) website at *http://aapco.org/.* If a worker would like to report a pesticide violation to the SLA but fears possible retaliatory action by management (*e.g.* , job loss or disciplinary action), the worker can make an anonymous call to the SLA. Note that not all state departments of agriculture have identical regulations. For instance, only California and Washington State require employers to obtain cholinesterase testing of agricultural pesticide handlers who apply pesticides containing cholinesterase-inhibiting compounds.

For pesticide contamination in water, EPA sets enforceable maximum containment levels. EPA also works jointly with the Food and Drug Administration (FDA) and the U.S. Department of Agriculture (USDA) to monitor and regulate pesticide residues and their metabolites in food and drugs. Tolerance limits are established by EPA for pesticides and their metabolites in raw agricultural commodities.

#### *b. Agricultural Worker Protection Standard (WPS)*

Recognizing that agricultural employees needed increased protection from pesticide exposures, EPA promulgated 40 CFR 170, the Agricultural Worker Protection Standard (WPS). The intent of the regulation is to protect agricultural employees by eliminating or reducing pesticide exposure, mitigating exposures that occur and informing agricultural employees about the hazards of pesticides. The WPS applies to two types of employees in the farm, greenhouse, nursery and forest industries: (1) agricultural pesticide handlers (mixer, loader, applicator, equipment cleaner or repair person, and flagger) and (2) field workers performing hand labor tasks (cultivator or harvester). The regulation does not cover agricultural employees in livestock production. The WPS includes requirements that agricultural employers notify employees about pesticide applications in advance, offer basic pesticide safety training, provide necessary personal protective equipment for direct work with pesticides and observe restricted entry interval (REI) times. Of special interest to healthcare providers, the WPS also requires agricultural employers to:

- Post an emergency medical facility address and phone number in a central location.
- Arrange immediate transport from the agricultural establishment to a medical facility for field workers or pesticide handlers who become ill or injured after an acute work-related pesticide exposure.
- Provide the exposed worker or handler and medical personnel with the pesticide product name, EPA registration number, active ingredient(s), medical information from the label, a description of how the pesticide was used, and any other relevant exposure information.

### *Occupational Safety and Health Administration*

The Occupational Safety and Health Administration (OSHA) plays a less substantial role than EPA in pesticide regulation. Whereas EPA has authority over pesticides in the home, environment and workplace, OSHA has authority only in the workplace. Like EPA, OSHA allows states to enforce federal OSHA regulations or their own adaption of the federal regulations (which must be approved by federal OSHA and be at least as stringent as the federal regulations). A total of 25 states, Puerto Rico and the Virgin Islands have such OSHA-approved state plans. In the other 25 states, regulations are enforced by federal OSHA.

OSHA has fewer responsibilities in agricultural workplaces compared to non-agricultural workplaces. For example, small farms (employing 10 or fewer non-family workers and having no temporary labor camps within the last 12 months) are exempt from enforcement of all OSHA rules, regulations and standards *(http://www.osha.gov/pls/oshaweb/owadisp.show_document?p_table=DIRECTIVES&p_id=1519).* The only exceptions to this are in California, Oregon and Washington, where the OSHA-approved state plans enforce OSHA rules, regulations and standards on farms of all sizes. OSHA is authorized to inspect farms with 11 or more employees but generally defers to EPA-delegated state agencies for enforcement of all pesticide-related activities in crop-based agriculture. The pesticide enforcement activities deferred to these EPA-delegated state agencies include the Worker Protection Standard, compliance with language on the pesticide label and compliance with pesticide registration, classification and labeling requirements.

In the non-agricultural setting, OSHA has greater jurisdiction over workplace pesticide exposures. All workers involved in pesticide manufacturing are covered by OSHA, which has established permissible exposure levels for selected pesticides (*e.g.*, captan, carbaryl, carbofuran, chlorpyrifos, chloropicrin, 2,4-D, diazinon, propoxur and pyrethrum). Similar to the option of anonymous reporting of suspected pesticide exposures or violations in agriculture to EPA or the State Lead Agency, a worker in a non-agricultural setting who fears possible retaliatory action can anonymously report a suspected pesticide violation to OSHA.

## Pesticide Information Resources

### *EPA Office of Pesticide Programs*

EPA's Office of Pesticide Programs (OPP) is responsible for registering pesticide products and regulating their use.

***a. Pesticide Worker Safety Program.*** Within OPP, the Pesticide Worker Safety Program conducts a variety of regulatory and outreach activities aimed at protecting the pesticide workforce, including agricultural workers, handlers and pesticide applicators. EPA/OPP also leads the National Strategies for Health Care Providers: Pesticides Initiative with the goal of improving the training of healthcare providers in the recognition, diagnosis, treatment and prevention of pesticide poisonings among those who work with pesticides. See **Appendix B, *Key Competencies for Clinicians*** to learn more. Pesticide safety materials developed through the Pesticide Worker Safety Program, including this manual, can be ordered online at no charge from the National Agricultural Center at:

http://www.epa.gov/agriculture/awor.html

Further information on Pesticide Worker Safety Program activities is available at:

http://www2.epa.gov/pesticide-worker-safety

***b. Pesticide Chemical Search.*** Pesticide Chemical Search was created by EPA/OPP to allow users to easily find information such as Reregistration Eligibility Decisions (REDs), factsheets, science reviews and regulatory actions on the chemical of interest. The site is searchable by chemical name or active ingredient (CAS number or pc code) and is located on the EPA Pesticides website at http://iaspub.epa.gov/apex/pesticides/f?p=chemicalsearch:1 or click on the Chemical Search icon on the EPA Pesticides homepage.

### *National Institute for Occupational Safety and Health (NIOSH) Centers for Disease Control and Prevention*

NIOSH is the federal agency responsible for conducting research on occupational disease and injury. NIOSH investigates potentially hazardous working conditions upon request, makes recommendations on preventing workplace disease and injury, and provides training to occupational safety and health professionals.

(800) 356-4674 or http://www.cdc.gov/niosh/homepage.html

***a. Centers for Agricultural Disease and Injury Research, Education, and Prevention.*** NIOSH has funded eight Agricultural Health and Safety Centers throughout the country. These centers conduct research and develop intervention programs aimed at preventing occupational disease and injury of agricultural workers and their families.

http://www.cdc.gov/niosh/agctrhom.html

***b. Sentinel Event Notification System for Occupational Risk (SENSOR)-Pesticides.*** Surveillance for pesticide-related illness and injury is designed to protect the public by determining the magnitude and underlying causes of over-exposure to pesticides. Surveillance also serves as an early warning system of any harmful effects not detected by manufacturer testing of pesticides. The NIOSH Centers for Disease Control and Prevention (CDC) and EPA support surveillance for pesticide-related illness and injury through the SENSOR-Pesticides program. In 2012, 11 states were participating in the SENSOR-Pesticides program. The success of these state-based pesticide poisoning surveillance systems relies on healthcare providers to report cases of suspected pesticide poisoning. Further information about SENSOR-Pesticides is available at the website.

http://www.cdc.gov/niosh/topics/pesticides/

### National Pesticide Information Center

The National Pesticide Information Center (NPIC) is based at Oregon State University and is cooperatively sponsored by the university and EPA. NPIC serves as a source of objective, science-based pesticide information on a wide range of pesticide-related topics, such as recognition and management of pesticide poisonings, safety information, health and environmental effects, referrals for investigation of pesticide incidents, emergency treatment for both humans and animals and cleanup and disposal procedures. NPIC also provides a rapid response in the form of skilled technical assistance to persons suspected of being adversely affected by pesticide exposures. Highly qualified pesticide specialists and a physician with extensive experience in pesticide toxicology provide and deliver appropriate information to all inquiries. A toll-free telephone service provides pesticide information in both English and Spanish to callers in the continental United States, Puerto Rico and the Virgin Islands. Additionally, pesticide questions and comments can be sent to an email address. The website (in both English and Spanish) has links to other sites and databases for further information.

**(800) 858-7378**
*(Hotline hours of operation: 6:30 am – 3:30 pm PST, Monday through Friday, except holidays)*

**http://www.npic.orst.edu**

### Migrant Clinicians Network

The Migrant Clinicians Network strengthens healthcare services and infrastructure for migrants and other mobile poor through training and technical assistance to clinicians and communities. As a partner to EPA's Health Care Provider Initiative, MCN assists primary care providers in recognizing, managing and preventing pesticide exposures and provides critically needed referral to occupational and environmental specialists. MCN's pesticide website provides clinical tools and resources and patient educational materials as part of its comprehensive pesticide exposure prevention and response efforts.

**(512) 327-2017** or **http://www.migrantclinician.org**

**http://www.migrantclinician.org/clinical_topics/pesticides.html**

**http://www.migrantclinician.org/clinical_topics/environmental-and-occupational-health.html**

### American Association of Poison Control Centers

The American Association of Poison Control Centers (AAPCC) is a non-profit, national organization founded in 1958. AAPCC represents the poison control centers of the United States and the interests of poison prevention and treatment of poisoning.

Emergencies
Local Poison Control
1-800-222-1222

**(800) 222-1212**
*(local Poison Control Center access)*

**http://www.aapcc.org**

### *Association of Occupational and Environmental Clinics*

The Association of Occupational and Environmental Clinics (AOEC) is a network of more than 60 clinics and more than 250 specialists that facilitates the prevention and treatment of occupational and environmental illnesses and injuries.

**(202) 347-4976** or **http://www.aoec.org**

### *Farmworker Justice*

The Farmworker Justice Fund can provide an appropriate referral to a network of legal services and nonprofit groups which represent farmworkers for free.

**(202) 776-1757** or **http://www.farmworkerjustice.org**

## Pesticide Information Databases

### *California Department of Pesticide Regulation Pesticide Illness Surveillance Program*

Since 1971, California law has required doctors to report any disease or condition that they know or have reason to believe resulted from pesticide exposure. The California Department of Pesticide Regulation (DPR) collects these reports in its Pesticide Illness Surveillance Program. To supplement physician reporting, DPR cooperates with the California Department of Public Health and California Department of Industrial Relations to search workers' compensation documents for pesticide-related disability. More recently, DPR has contracted with the California Poison Control System to help doctors fulfill their responsibility to report. As of 2011, a law requires clinical laboratories to send DPR the results of cholinesterase tests done to evaluate pesticide exposure. County agricultural commissioners (CACs) investigate every case identified and send reports of their findings to DPR. Scientists of the Pesticide Illness Surveillance Program review, evaluate and abstract all reports received from CACs and are working to integrate cholinesterase reports. Data from this program and others (including pesticide use, product label, enforcement, school IPM and more) can be retrieved from the website.

**http://www.cdpr.ca.gov/dprdatabase.htm**

### *National Pesticide Information Retrieval Service (NPIRS)*

The National Pesticide Information Retrieval System (NPIRS) receives funding from EPA to maintain a pesticide information database. NPIRS provides publicly available registration information on approximately 90,000 EPA-registered pesticides. The data include: product number and name, company number and name, registration date, cancellation date and reason, existing stocks date and product manager name and phone number. NPIRS is administered by the Center for Environmental and Regulatory Information Systems at Purdue University in West Lafayette, Indiana.

**http://ppis.ceris.purdue.edu/**

### *Agency for Toxic Substances and Disease Registry*

The Agency for Toxic Substances and Disease Registry (ATSDR), part of the Department of Human Health and Services, publishes fact sheets and information on pesticides and other toxic substances.

**http://www.atsdr.cdc.gov/**

## Data Collection on an Acute Pesticide Exposed Patient

When patients present with an identified pesticide poisoning, the following data collection format has been recommended to guide the clinician on the appropriate information to obtain as well as an evaluation of appropriate samples and other materials.

1. PT ID: Name/Age/Sex/Occupation
2. Initial and subsequent symptoms and signs*
3. Name of pesticide product and active ingredients, their concentration and EPA registration number
4. Date and time when exposure occurred
5. How the pesticide was applied, when applied and on what crop or for what use
6. Route(s) of exposure: dermal, ocular, oral, respiratory
7. How much of the product was ingested, if ingested
8. Circumstances of exposure – intentional or accidental, occupational or non-occupational
9. A detailed description of how the exposure happened
10. Treatment already received
    a. Skin exposure:
        i. Was affected area washed? If so, when? If not, proceed with skin decontamination procedures
        ii. Was any clothing contaminated?
        iii. If so did they change clothes?
    b. Ocular exposure:
        i. Were the eyes irrigated?
        ii. If so, with what and for how long?
    c. GI exposure:
        i. Were any emetics used?
        ii. Were any absorbents used?
        iii. Were any home remedies (*e.g.,* water, milk, lemon juice) used?
        iv. Was there any emesis before arrival?

### *Data Collection on an Acute Pesticide Exposed Patient, continued*

**Materials to be Gathered:**

1. A copy of the pesticide label and/or a copy of the Material Safety Data Sheet (MSDS).
2. A copy of the pesticide application record (tank mix, concentration, etc.) if applicable. This should be available from the pesticide applicator or the grower.
3. 10 cc whole blood, anticoagulated with sodium heparin (refrigerate).
4. 5 cc plasma anticoagulated with sodium heparin (refrigerate).
5. A fresh urine sample (label and freeze).
6. Any contaminated clothing, hats, foliage from the site. Place in clean sealable plastic bag; label, seal and freeze.
7. Other options:
   a. Fingernail residue. If the worker handled the pesticide or materials with pesticide residue, some pesticide may be lodged under the fingernails. Clean under the nails. Place in clean sealable plastic bag, label, seal and freeze.
   b. Saliva sample. Some pesticides can be detected in saliva. Have the patient spit repeatedly into a clean glass or plastic container. Seal the container, label and freeze.
   c. Hair sample, if the head was exposed. Place in clean sealable plastic bag, label, seal and freeze.
   d. A skin wipe with ethanol-impregnated swab
      i. Wipe skin that was contaminated if possible. Use a newly opened alcohol wipe. Wipe an area of skin and if possible estimate the size of the area wiped and record this on the sample label. Try to focus on an area that is likely to have been contaminated in the exposure.
      ii. Place wipe in clean sealable plastic bag, label, seal and freeze.

**For the pediatric patient, note parents' occupations and child's appearance compared to his/her usual baseline. It is important to ask if the child is acting normally, if there is an abnormal gait, stumbling or ataxia; and if the child has experienced excessive sleepiness, irritability or other personality changes.*

*Developed by Matthew C. Keifer MD, MPH*
*National Farm Medicine Center*

## References

1. Liebman AK, Rowland MM. To ask or not ask: The critical role of the primary care provider in screening for occupational injuries and exposures. *J Public Health Manag Pract.* 2009:15(2):173-5.

2. Levy BS, Wegman DH (eds). *Occupational and Environmental Health,* 5th ed. Lippincott Williams and Wilkins, Phila. 2006. 847 pp.

3. Agency for Toxic Substances and Disease Registry. *Case studies in environmental medicine: taking an exposure history. 2008, 2011.* Accessed 10/12/12: http://www.atsdr.cdc.gov/csem/exphistory/docs/exposure_history.pdf.

4. Rowland MM, Liebman AK, Sudakin DL, Keifer MC. Learning Opportunities from the Reported Incident of Pesticide Poisoning. *Streamline.* 2006, 12(5):6.

5. Ashford NA. Workers' compensation. In: Rom WN, editor. *Environmental and Occupational Medicine.* 4th ed. Philadelphia: Wolters Kluwer/Lippincott Williams and Wilkins. 2007. pp 1712-19.

6. ACOEM. The seven ethical principals of occupational and environmental medicine. ACOEM: Elk Grove Village, IL. 2010. Available at http://www.acoem.org/codeofconduct.aspx.

7. Blank L, Kimball H, McDonald W, Merino J; ABIM Foundation; ACP Foundation; European Federation of Internal Medicine. Medical professionalism in the new millennium: a physician charter 15 months later. *Ann Intern Med.* 2003;138:839-41.

8. Centers for Disease Control and Prevention. Illnesses associated with exposure to methyl bromide-fumigated produce — California, 2010. MMWR. 2011;60:923-26.

9. Brooks SM, Gochfield M, Herzstein J, et al. *Environmental Medicine.* St. Louis, MO: Mosby Yearbook. 1995.

10. Steenland K. *Case Studies in Occupational Epidemiology.* New York: Oxford University Press. 1993.

11. Centers for Disease Control and Prevention. HIPAA Privacy Rule and Public Health: Guidance from the CDC and the U.S. Department of Health and Human Services. Available at http://www.cdc.gov/mmwr/preview/mmwrhtml/m2e411a1.htm.

## Other References

Blondell, J. Epidemiology of pesticide poisonings in the United States, with special reference to occupational cases. *Occup Med-C.* 1997; 12(2):209-20.

McCauley LA, Lazarev MR, Higgins G, Rothlein J, Muniz J, Ebbert C, et al. Work characteristics and pesticide exposures among migrant agricultural families: A community based research approach. *Environ Health Perspect.* 2001;109:533-538.

Stanbury M, Anderson H, Rogers P, Bonauto D, Davis L, Materna B, Rosenman K. Guidelines for Minimum and Comprehensive State-Based Public Health Activities in Occupational Safety and Health, DHHS Publication No. 2008-148. National Institute for Occupational Safety and Health, Cincinnati, OH. 2008. Online at http://www.cdc.gov/niosh/docs/2008-148.

CHAPTER 3

# General Principles in the Management of Acute Pesticide Poisonings

## Introduction

This chapter describes basic management techniques applicable to most acute pesticide exposures. Where special considerations and treatments are required for a particular pesticide, they are addressed separately in the appropriate chapter.

Remember: Treat the patient, not the poison. Symptomatic and supportive care is the mainstay of therapy. Severe poisoning should be treated in an intensive care unit setting, preferably with toxicological consultation, if available. Consultation with the regional poison control center is highly advisable. Its staff can assist with treatment recommendations or advise when no treatment is needed, helping to avoid unnecessary and possibly harmful interventions.

The American Association of Poison Control Centers (AAPCC) maintains the National Poison Data System (NPDS). NPDS records data from the 57 U.S. poison centers in near real-time. In 2010, 2.4 million human exposures were reported to NPDS. Of these, 90,037 (3.8%) were exposed to some type of pesticide. The chart below demonstrates the seasonal variation for 2000–2010, with peak exposures in July of each year.

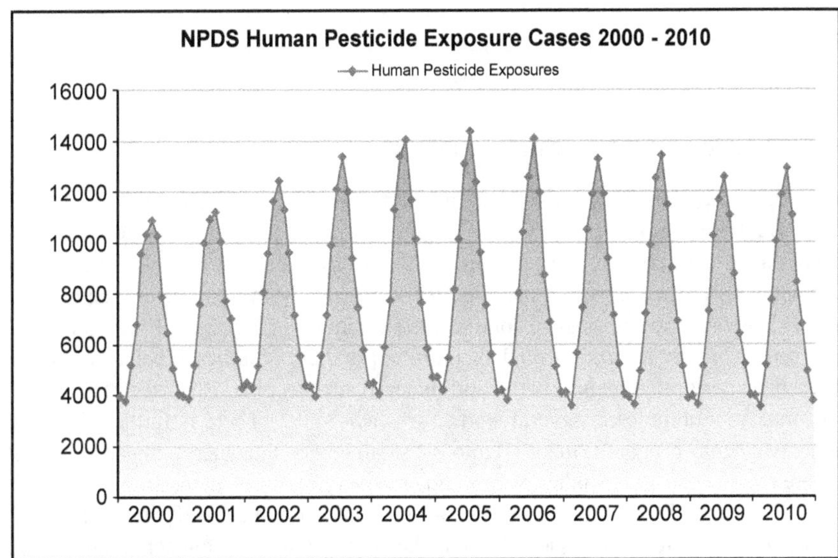


## Skin Decontamination

Decontamination must proceed concurrently with whatever resuscitative and antidotal measures are necessary to preserve life. Be careful not to expose yourself or other care providers to potentially contaminating substances. Wear protective gear (gloves, gown and goggles) and wash exposed areas promptly. Persons attending the victim should avoid direct contact with heavily contaminated clothing and bodily fluids.

Place all contaminated clothing and personal effects in an appropriate container. While no glove will provide complete protection to all possible chemical contamination, butyl rubber gloves generally provide the best protection compared to latex and other surgical or precautionary gloves. If butyl rubber gloves are not available, nitrile gloves may be an option. A double layer of gloves will increase protection, but will decrease manual dexterity.[1]

Flush exposed areas with copious amounts of water. Wash carefully behind ears, under nails and in skin folds. Use soap and shampoo for oily substances. If the patient exhibits any signs of weakness, ataxia or other neurologic impairment, clothing should be removed and a complete bath and shampoo given while the victim is recumbent.

## Eye Decontamination

Ocular exposures should be treated by irrigating the exposed eyes with copious amounts of clean water for at least 15 minutes. Remove contact lenses if present prior to irrigation. If irritation persists after irrigation, patients should be referred to a healthcare facility for an ophthalmic exam.

## Airway Protection

Support airway, breathing and circulation. Suction any oral secretions using a large bore suction device if necessary. Intubate and ventilate as needed, especially if the patient has respiratory depression or if the patient appears obtunded or otherwise neurologically impaired. Administer oxygen as necessary to maintain adequate tissue perfusion. In severe poisonings, it may be necessary to mechanically support pulmonary ventilation for several days.

There are a couple of special considerations with regard to certain pesticides. In **organophosphate** and **carbamate** poisoning, adequate tissue oxygenation is essential prior to administering atropine. In **paraquat** and **diquat** poisoning, oxygen is **contraindicated** early in the poisoning because of progressive oxygen toxicity to the lung tissue. See specific chapters for more details.

## Gastrointestinal Decontamination

Control seizures before attempting any method of GI decontamination.[2]

**Gastric lavage** should NOT be routinely used in pesticide exposure management and is contraindicated in poisonings due to hydrocarbon ingestion. Lavage is indicated only when a patient has ingested a potentially life-threatening amount of poison and the procedure can be done within 60 minutes of ingestion. Even then, clinical benefit has not been confirmed in controlled studies.[2,3] Studies of poison recovery have been performed mainly with solid material such as pills. Reported recovery of material at 60 minutes in several studies was 8%-32%.[4,5] There is further evidence that lavage may propel the material into the small bowel, thus increasing absorption.[6] There are no controlled studies of pesticide recovery by these methods.

For gastric lavage, a large bore (36-40 French for adult, 24-28 French for children) orogastric tube is passed through the mouth into the stomach followed by administration of small volumes (200-300 mL adults, 10mL/kg child) warmed saline or water (avoid water in children, use saline instead), which is then allowed to drain back out with the hope of removing poisons in the stomach. Patient must be able to maintain airway or be intubated prior to lavage. Do not attempt to lavage a patient with ingestion of poisons that may cause seizures or rapid CNS depression, unless intubated. Measure the patient for the correct placement of tube; place in left-lateral decubitus position. Place on cardiac monitor and pulse oximetry. Have suction equipment nearby. Continue lavage process until returns are clear. Volume of fluid returned should be the same as the amount instilled to avoid fluid and electrolyte imbalance. Negative or poor lavage does not rule out significant ingestion.

Complications of gastric lavage may include aspiration, fluid and electrolyte imbalance, mechanical injury to the throat-esophagus-stomach and hypoxia. Lavage is contraindicated in hydrocarbon ingestion, a common solvent used in many pesticide formulations. Therefore, for most pesticide exposures, gastric lavage should not be performed. Contraindications to gastric lavage are listed in the adjacent table.

| GASTRIC LAVAGE CONTRAINDICATIONS[2] | |
|---|---|
| 1 | Patients with unprotected airway |
| 2 | Patients with decreased level of consciousness without intubation |
| 3 | Patients who have ingested drugs that may cause abrupt CNS depression or seizures and who have not been intubated |
| 4 | Patients who have Ingested corrosive substances: acid or alkali |
| 5 | Patients who have ingested hydrocarbon and have high risk of aspiration |
| 6 | Patients at risk of bleeding or GI perforation because of recent surgery or medical conditions such as coagulopathy |

**Cathartics** have NO role in management of poisoned patients and are NOT recommended as a way to decontaminate the GI tract. Repeat doses of cathartics may result in fluid and electrolyte imbalances, particularly in children. [7]

Saline cathartics include magnesium citrate, magnesium sulfate, sodium sulfate and magnesium hydroxide. Osmotic cathartics increase the water content and weight of the stool. Sorbitol is a sugar alcohol that functions as an osmotic cathartic and is slowly metabolized in humans. Sorbitol is often combined with charcoal to improve the taste and mask the grittiness of charcoal. Previously given along with charcoal, cathartics were intended to decrease the absorption of poisons by speeding movement of the charcoal-poison complex through the gut resulting in bowel evacuation. The use of sorbitol is not recommended in poisonings with organophosphates, carbamates or arsenicals, which generally result in profuse diarrhea, or in poisonings with diquat or paraquat, which may result in an ileus.

Contraindications to cathartic use include absent bowel sounds, abdominal trauma or surgery, or intestinal perforation or obstruction. Cathartics are also contraindicated in volume depletion, hypotension, electrolyte imbalance or the ingestion of a corrosive substance.[7] A 2004 revision of a 1997 position paper on cathartics determined that there was no new evidence that required a change in the 1997 conclusions.[8]

**Activated charcoal** is an effective adsorbent for many poisonings. Volunteer studies suggest that it reduces the amount of poison absorbed if given within 60 minutes of ingestion.[9] There are insufficient data to support or exclude its use if time from ingestion is prolonged, although some poisons that are less soluble may be adsorbed beyond 60 minutes.

Nearly all clinical trials with charcoal have been conducted with poisons other than pesticides. There is evidence that paraquat is well adsorbed by activated charcoal.[10,11,12,13] *In vitro* data demonstrated that boric acid is well adsorbed by charcoal.[14] Charcoal has been anecdotally successful in cases of poisoning from other pesticides. There are *in vitro* data that evaluated the effect of the herbicide 2,4-D, although the purpose of the study was to evaluate charcoal for environmental adsorption. It was not simulated in a gastric environment, so the data do not strictly reflect an effect in human poisoning.[15]

### Dosage of Activated Charcoal

It is difficult to determine the precise dosage, as clinical studies are either conducted in animals or in humans with a known quantity of ingestant. The following dosages are recommended.[16]

- ***Infants up to 1 year of age: 10–25 g or 0.5–1.0 g/kg***
- ***Children 1 to 12 years of age: 25–50 g or 0.5–1.0 g/kg***
- ***Adolescents and adults: 25–100 g***

Administer charcoal as an aqueous slurry. Encourage the victim to swallow the adsorbent. Antiemetic therapy may help control vomiting in adults or older children. As an alternative, activated charcoal may be administered through an orogastric tube or diluted with water and administered slowly through a nasogastric tube. Repeated administration of charcoal or other absorbent every 2-4 hours may be beneficial in both children and adults. Repeated doses of activated charcoal should not be administered if the gut is atonic. The use of charcoal without airway protection is contraindicated in the neurologically impaired patient.

Charcoal should be used with caution in cases of poisoning from organophosphates, carbamates and organochlorines if they are prepared in a hydrocarbon solution as this will increase the risk for aspiration.

Single-dose activated charcoal should not be used routinely in the management of poisoned patients. Charcoal appears to be most effective within 60 minutes of ingestion and may be considered for use for this time period. Although it may be considered 60 minutes after ingestion, there is insufficient evidence to support or exclude its use for this time period. Despite improved binding of poisons within 60 minutes, only one study exists to suggest that there is improved clinical outcome.[17]

Activated charcoal is contraindicated in an unprotected airway, a GI tract not anatomically intact, and when charcoal therapy may increase the risk of aspiration, such as when a hydrocarbon-based pesticide has been ingested.[9] A 2004 position paper by the American Academy of Clinical Toxicology (AACT) reviewed data since its 1997 statement was published and essentially reiterated the position of those guidelines.[8] A randomized controlled trial of multiple dose charcoal conducted in Sri Lanka was published after the 2004 AACT position paper. This study did not find a difference in mortality between the two groups, and the researchers concluded that routine use of multiple-dose activated charcoal could not be recommended in rural Asia Pacific.[18]

**Syrup of ipecac** was historically given to patients to induce emesis, both prior to emergency department referral and on emergency department arrival. Ipecac syrup was used as an intervention in order to prevent healthcare facility referral in minor ingestions. Ipecac has been used as an emetic since the 1950s. In a pediatric study, administration of syrup of ipecac resulted in emesis within 30 minutes in 88% of children.[19] Most clinical trials involve the use of pill form ingestants, such as aspirin,[5,20] acetaminophen,[21] ampicillin[22] and multiple types of tablets.[23] No clinical trials have been done with pesticides. In 2010, the National Poison Data System of the American Association of Poison Control Centers (AAPCC) reported more than 2.4 million human exposures, and syrup of ipecac was administered in only 359 (0.02%).[24] This was a significant decrease from 2009, when syrup of ipecac was administered for decontamination in only 658 (0.03%) of all human exposures.[25]

In 1993, the American Academy of Clinical Toxicology (AACT) advised that ipecac syrup should not be routinely administered to poison patients in a healthcare setting. In the 1997 AACT guidelines, syrup of ipecac was not considered first-line therapy. The guidelines acknowledged that clinical studies have demonstrated no benefit from its use.[26] A subsequent revised position statement in 2004 acknowledged that no changes to the 1997 statement recommendations were required.[27]

In 2003, the American Academy of Pediatrics recommended that ipecac syrup not be used as home treatment in a child who ingested any toxic substance. The policy statement also recommended that existing ipecac in homes should be disposed of safely.[28] This recommendation was reinforced in 2005 when the AAPCC issued guidelines on syrup of ipecac use. The AAPCC concluded that there were only rare circumstances in which use of ipecac syrup should be considered. All of the following would have to be true:

1. syrup of ipecac is not contraindicated,
2. the poisoning in question will give substantial risk of toxicity to the patient,
3. there is no available alternative gastrointestinal decontamination therapy,
4. there will be a delay of greater than 1 hour to get to an emergency medical facility, and
5. ipecac syrup would not adversely affect definitive treatment available at the hospital or other medical facility.[29]

It is unlikely that syrup of ipecac would ever be indicated for home or pre-hospital gastric decontamination.

## Seizure Management

Lorazepam is increasingly being recognized as the benzodiazepine of choice for toxicological induced single or multiple seizures, although there are few reports of its use with certain pesticides. With any benzodiazepine or other seizure control medication, one must be prepared to assist ventilation.

### Dosage of Lorazepam for Seizure Control

- ***Adults: 2-4 mg/dose given IV over 2-5 minutes. Repeat if necessary to a maximum of 8 mg in a 12-hour period.***
- ***Adolescents: Same as adult dose, except maximum dose is 4 mg.***
- ***Children under 12 years: 0.05-0.10 mg/kg IV over 2-5 minutes. Repeat if necessary 0.05 mg/kg 10-15 minutes after first dose, with a maximum dose of 4 mg.***

*CAUTION:* ***Be prepared to assist pulmonary ventilation mechanically and endotracheally intubate the patient if laryngospasm or respiratory depression occurs and hypoxia is possible. Monitor for hypotension and cardiac dysrhythmias. Also remember to evaluate for hypoglycemia, electrolyte disturbances and hypoxia.***

For organochlorine compounds, use of lorazepam has not been reported in the literature. Diazepam is often used for this and other pesticide poisonings.

**Dosage of Diazepam**

- ***Adults: 5-10 mg IV and repeat every 5-10 minutes to maximum of 30 mg.***
- ***Children: 0.2 to 0.5 mg/kg IV every 5 minutes to maximum of 10 mg in children over 5 years and 5 mg in children under 5 years.***

Phenobarbital is an additional treatment option for seizure control.

**Dosage of Phenobarbital**

- ***Infants, children and adults: 15-20 mg/kg as an IV loading dose. Give an additional 5 mg/kg IV every 15-30 minutes to a maximum of 30 mg/kg. The drug should be pushed no faster than 1 mg/kg/minute.***

For seizure management, most patients respond well to usual management consisting of benzodiazepines and phenobarbital.

### *Management of Refractory Seizures*

These patients require intensive care management and should be referred to a tertiary center.

Consider an infusion of propofol in patients who continue to experience seizures despite adequate benzodiazepine and/or phenobarbital dosing. Monitor closely for propofol infusion syndrome, cardiac failure, rhabdomyolysis, metabolic acidosis and renal failure, which may be fatal.[30]

**Dosage of Propofol**

- ***Infants, children, and adults: Start with a bolus dose of 1 to 2 mg/kg IV. Follow with an infusion of 2 mg/kg/hour IV and titrate up as needed for sedation and control of seizures.***

## References

1. OSHA. Hospital-Based Receivers of Victims from Mass Casualty Incidents Involving the Release of Hazardous Substances. 2005; http://www.osha.gov/dts/osta/bestpractices/first-receivers_hospital.pdf.

2. Vale JA, Kulig K. Position paper: gastric lavage. *J Toxicol Clin Toxicol.* 2004;42(7):933-943.

3. Vale JA. Position statement: gastric lavage. American Academy of Clinical Toxicology; European Association of Poisons Centres and Clinical Toxicologists. *J Toxicol Clin Toxicol.* 1997;35(7):711-719.

4. Tenenbein M, Cohen S, Sitar DS. Efficacy of ipecac-induced emesis, orogastric lavage, and activated charcoal for acute drug overdose. *Ann Emerg Med.* Aug 1987;16(8):838-841.

5. Danel V, Henry JA, Glucksman E. Activated charcoal, emesis, and gastric lavage in aspirin overdose. *Br Med J (Clin Res Ed).* May 28 1988;296(6635):1507.

6. Saetta JP, March S, Gaunt ME, Quinton DN. Gastric emptying procedures in the self-poisoned patient: are we forcing gastric content beyond the pylorus? *J R Soc Med.* May 1991;84(5):274-276.

7. Barceloux D, McGuigan M, Hartigan-Go K. Position statement: cathartics. American Academy of Clinical Toxicology; European Association of Poisons Centres and Clinical Toxicologists. *J Toxicol Clin Toxicol.* 1997;35(7):743-752.

8. Position paper: cathartics. *J Toxicol Clin Toxicol.* 2004;42(3):243-253.

9. Chyka PA, Seger D. Position statement: single-dose activated charcoal. American Academy of Clinical Toxicology; European Association of Poisons Centres and Clinical Toxicologists. *J Toxicol Clin Toxicol.* 1997;35(7):721-741.

10. Idid SZ, Lee CY. Effects of Fuller's Earth and activated charcoal on oral absorption of paraquat in rabbits. *Clin Exp Pharmacol Physiol.* Aug 1996;23(8):679-681.

11. Nakamura T, Kawasaki N, Tamura T, Tanada S. *In vitro* adsorption characteristics of paraquat and diquat with activated carbon varying in particle size. *Bull Environ Contam Toxicol.* Mar 2000;64(3):377-382.

12. Gaudreault P, Friedman PA, Lovejoy FH, Jr. Efficacy of activated charcoal and magnesium citrate in the treatment of oral paraquat intoxication. *Ann Emerg Med.* Feb 1985;14(2):123-125.

13. Terada H, Miyoshi T, Imaki M, Nakamura T, Tanada S. Studies on *in vitro* paraquat and diquat removal by activated carbon. *Tokushima J Exp Med.* Jun 1994;41(1-2):31-40.

14. Oderda GM, Klein-Schwartz W, Insley BM. *In vitro* study of boric acid and activated charcoal. *J Toxicol Clin Toxicol.* 1987;25(1-2):13-19.

15. Belmouden M, Assabbane A, Ichou YA. Adsorption characteristics of a phenoxy acetic acid herbicide on activated carbon. *J Environ Monit.* Jun 2000;2(3):257-260.

16. Chyka PA, Seger D, Krenzelok EP, Vale JA. Position paper: Single-dose activated charcoal. *Clin Toxicol (Phila).* 2005;43(2):61-87.

17. Merigian KS, Woodard M, Hedges JR, Roberts JR, Stuebing R, Rashkin MC. Prospective evaluation of gastric emptying in the self-poisoned patient. *Am J Emerg Med.* Nov 1990;8(6):479-483.

18. Eddleston M, Juszczak E, Buckley NA, et al. Multiple-dose activated charcoal in acute self-poisoning: a randomised controlled trial. *Lancet.* Feb 16 2008;371(9612):579-587.

19. Robertson WO. Syrup of ipecac--a slow or fast emetic? *Am J Dis Child.* Feb 1962;103:136-139.

20. Curtis RA, Barone J, Giacona N. Efficacy of ipecac and activated charcoal/cathartic. Prevention of salicylate absorption in a simulated overdose. *Arch Intern Med.* Jan 1984;144(1):48-52.

21. McNamara RM, Aaron CK, Gemborys M, Davidheiser S. Efficacy of charcoal cathartic versus ipecac in reducing serum acetaminophen in a simulated overdose. *Ann Emerg Med.* Sep 1989;18(9):934-938.

22. Tenenbein M, Cohen S, and Sitar DS. Efficacy of ipecac-induced emesis, orogastric lavage, and activated charcoal for acute drug overdose. *Ann Emerg Med.* 1987;16:838-41.

23. Neuvonen PJ, Vartiainen M, Tokola O. Comparison of activated charcoal and ipecac syrup in prevention of drug absorption. *Eur J Clin Pharmacol.* 1983;24(4):557-562.

24. Bronstein AC, Spyker DA, Cantilena LR, Jr., Green JL, Rumack BH, Dart RC. 2010 Annual Report of the American Association of Poison Control Centers' National Poison Data System (NPDS): 28th Annual Report. *Clin Toxicol (Phila).* Dec 2011; 49(10):910-941.

25. Bronstein AC, Spyker DA, Cantilena LR, Jr., Green JL, Rumack BH, Giffin SL. 2009 Annual Report of the American Association of Poison Control Centers' National Poison Data System (NPDS): 27th Annual Report. *Clin Toxicol (Phila).* Dec 2010;48(10):979-1178.

26. Krenzelok EP, McGuigan M, Lheur P. Position statement: ipecac syrup. American Academy of Clinical Toxicology; European Association of Poisons Centres and Clinical Toxicologists. *J Toxicol Clin Toxicol.* 1997;35(7):699-709.

27. Position paper: Ipecac syrup. *J Toxicol Clin Toxicol.* 2004;42(2):133-143.

28. Poison treatment in the home. American Academy of Pediatrics Committee on Injury, Violence, and Poison Prevention. *Pediatrics.* Nov 2003;112(5):1182-1185.

29. Manoguerra AS, Cobaugh DJ. Guideline on the use of ipecac syrup in the out-of-hospital management of ingested poisons. *Clin Toxicol (Phila).* 2005;43(1):1-10.

30. Abend NS, Dlugos DJ. Treatment of refractory status epilepticus: literature review and a proposed protocol. *Pediatr Neurol.* Jun 2008;38(6):377-390.

*Section II*

# INSECTICIDES

*Pyrethrins*
**HIGHLIGHTS**

Strongly lipophilic

Crude pyrethrum is a dermal & respiratory allergen

Easily absorbed by GI tract & pulmonary membranes

Relatively low mammalian toxicity

**SIGNS & SYMPTOMS**

Contact dermatitis

Rhinitis, asthma

**TREATMENT**

Antihistamines

Epinephrine for anaphylaxis as required

Topical corticosteroid for contact dermatitis

Flush eyes as necessary

Consider gastric emptying or charcoal adsorption

CHAPTER 4

# Pyrethrins and Pyrethroids

## PYRETHRINS

Pyrethrum is the oleoresin extract of dried chrysanthemum flowers. The extract contains about 50% active insecticidal ingredients known as pyrethrins. The ketoalcoholic esters of chrysanthemic and pyrethroic acids are known as **pyrethrins, cinerins** and **jasmolins**. These strongly lipophilic esters rapidly penetrate many insects and paralyze their nervous systems. Both crude pyrethrum extract and purified pyrethrins are contained in various commercial products, commonly dissolved in petroleum distillates. Some are packaged in pressurized containers ("bug bombs"), usually in combination with the synergists piperonyl butoxide and n-octyl bicycloheptene dicarboximide. The synergists retard enzymatic degradation of pyrethrins. Pyrethrum and pyrethrin products are used mainly for indoor pest control. They are not sufficiently stable in light and heat to remain as active residues on crops. The synthetic insecticides known as pyrethroids (chemically similar to pyrethrins) have the stability needed for agricultural applications. Pyrethroids are discussed separately below.

### Toxicology

**Crude pyrethrum** is a dermal and respiratory allergen, probably due mainly to non-insecticidal ingredients. Contact dermatitis and allergic respiratory reactions (rhinitis and asthma) have occurred following exposures.[1,2] Single cases exhibiting anaphylactic[3] and pneumonitic manifestations[4] have also been reported. Pulmonary symptoms may be due to inhalation of the hydrocarbon vehicle(s) of the insecticides. The refined pyrethrins are probably less allergenic but appear to retain some irritant and/or sensitizing properties.

**Pyrethrins** are absorbed across the gastrointestinal tract and pulmonary membranes, but only slightly across intact skin. They are very effectively hydrolyzed to inert products by mammalian liver enzymes. This rapid degradation, combined with relatively poor bioavailability, probably accounts in large part for their relatively low mammalian toxicity. Dogs fed extraordinary doses exhibit tremor, ataxia, labored breathing and salivation. Similar neurotoxicity has been rarely observed in humans, even in individuals who have had extensive contact from using pyrethrins for body lice control or have ingested pyrethrum as an anthelmintic.

In cases of human exposure to commercial products, the possible role of other toxicants in the products should be kept in mind. The synergists piperonyl butoxide and n-octyl bicycloheptene dicarboximide have low toxic potential in humans, which is further discussed in **Chapter 19, *Miscellaneous Pesticides, Solvents and Adjuvants.*** However, the hydrocarbon vehicle(s) may have significant toxicity. Pyrethrins themselves do not inhibit the cholinesterase enzymes.

### Confirmation of Poisoning

No practical tests for pyrethrin metabolites or pyrethrin effects on human enzymes or tissues are currently available.

## Treatment of Pyrethrin or Pyrethrum Toxicosis

1. Use antihistamines, which are effective in controlling most allergic reactions. Severe asthmatic reactions, particularly in predisposed persons, may require administration of inhaled β-agonists and/or systemic corticosteroids. Inhalation exposure should be carefully avoided in the future.
2. For anaphylaxis-type reactions, use subcutaneous epinephrine, epinephrine and respiratory support as necessary.[3]
3. In cases of contact dermatitis, administer topical corticosteroid preparations for an extended period, as necessary, under the supervision of a physician. Future contact with the allergen must be avoided.
4. Remove eye contamination by flushing the eye with copious amounts of clean water or saline. Specialized ophthalmologic care should be obtained if irritation persists.
5. Treat toxic manifestations caused by other ingredients according to their respective toxic actions, independent of pyrethrin-related effects.
6. Even though most ingestions of pyrethrin products present little risk, if a large amount of pyrethrin-containing material has been ingested and the patient is seen within 1 hour, consider gastric emptying. If seen later, or if gastric emptying is performed, consider administration of activated charcoal as described in **Chapter 3, *General Principles*.**

*Pyrethrins*
**COMMERCIAL PRODUCTS**

Aquacide
Black Flag
Chemsico
Evercide
Hot Shot Flea Killer
Prentox
Purge
Pyrocide Fogging Concentrate
Raid Ant & Roach Killer
Raid Fogger
Supra-Quick Flea & Tick Mist

## PYRETHROIDS

These modern synthetic insecticides are similar chemically to natural pyrethrins, but pyrethroids are modified to increase stability in the natural environment. They are now widely used in agriculture, in homes and gardens, and for treatment of ectoparasitic disease. There has been increasing use of these agents as use of organophosphate pesticides becomes more restricted.[5]

## Toxicology

Although certain **pyrethroids** exhibit striking neurotoxicity in laboratory animals when administered by intravenous injection and some are toxic by the oral route, systemic toxicity by inhalation and dermal absorption is low. While limited absorption may account for the low toxicity of some pyrethroids, rapid biodegradation by mammalian liver enzymes (ester hydrolysis and oxidation) is probably the major factor responsible for this phenomenon.[6,7] Neonatal rats have been demonstrated to have decreased ability to metabolize and excrete pyrethroids. The $LD_{50}$ for weanling rats with deltamethrin has been reported at 12 mg/kg, while the adult $LD_{50}$ is about 80 mg/kg. At these doses, the brain levels of deltamethrin at death are equivalent in both weanling and adult rats.[8] Most pyrethroid metabolites are promptly excreted, at least in part, by the kidneys.

Multiple mechanisms and targets have been evaluated for mammalian toxicity. At concentrations as low as $10^{-10}$ M in *in vitro* systems, pyrethroids alter sodium and chloride channels and result in norepinephrine release. At concentrations around $10^{-7}$ M, membrane depolarization and apoptosis occur, as well as other cellular effects. In laboratory animal studies, this results in a state of hyperexcitability at lower expo-

*Pyrethroids*
## HIGHLIGHTS

Low systemic toxicity via inhalation and dermal route

Sites of action: sodium & chloride channels; GABA, nicotinic acetylcholine, peripheral benzodiazepine receptors

Type I (*e.g.*, permethrin) usually do not contain a cyano group

Type II (*e.g.*, cypermethrin, fenvalerate) always contain a cyano group

Type II acute poisonings are generally more severe

## SIGNS & SYMPTOMS

Type I: fine tremor, reflex hyperexcitability

Type II: severe salivation, hyperexcitability, choreoathetosis

May include dizziness, headache, fatigue, vomiting, diarrhea

Stinging, burning, itching, tingling, numb skin may be reported

Severe cases: pulmonary edema, seizures, coma

## TREATMENT

Decontaminate skin, eyes

Consider GI decontamination

Treat seizures as necessary

sures, followed by depolarization, conduction block and cell death at very high levels of exposure.[7] In addition to the calcium and sodium channel sites of action, multiple other sites described include GABA receptors (for Type II effects, see following), nicotinic acetylcholine receptors and peripheral benzodiazepine receptors. They have also been shown to alter mitochondrial electron transport.[9]

These discrete effects at differing levels and the relative resistance of mammals to these agents explain the typical syndromes of human poisoning. However, the possibility of neuronal death with prenatal exposure or with repeated dosing in adults has been raised.[7] The potential decreased ability of the fetus to metabolize these agents could result in higher levels in the developing brain, with resulting neurotoxicity.

Pyrethroids have been divided into two types based on clinical findings with overdosing. Type I pyrethroids, such as **permethrin**, usually do not contain a cyano group, while most Type II pyrethroids, such as **cypermethrin** and **fenvalerate**, always do.[10] Both of these types show marked stimulus of catecholamine release from the adrenals with overdosing. This release of epinephrine and norepinephrine results in marked sympathetic symptoms.

There have been recent reports of illnesses due to these agents. A report of 466 episodes of illnesses and injuries related to total release foggers notes that eight of the ten most commonly reported active ingredients in these episodes are pyrethroid compounds, representing 86% of all reported episodes.[11] In these cases, 18% were reported as moderate severity and 2% were classified as high severity.

## Signs and Symptoms of Poisoning

Type II acute poisonings are generally more severe than Type I.[10] Type I poisoning has been described as characterized by fine tremor and reflex hyperexcitability. Type II poisoning has typically shown severe salivation, hyperexcitability and choreoathetosis. Other signs and symptoms of toxicity include abnormal facial sensation, dizziness, headache, fatigue, vomiting, diarrhea and irritability to sound and touch. In more severe cases, pulmonary edema, muscle fasiculations, seizures and coma can develop. A large ingestion (200 to 500 mL) of concentrated formulations may cause coma and seizures within 20 minutes. Initial symptoms following ingestion include gastrointestinal events (*i.e.,* abdominal pain, vomiting and diarrhea) generally within 10 to 60 minutes. Of 573 cases reviewed in China, 51 included disturbed consciousness and 34 included seizures. Of those 85 symptomatic cases, only five were from occupational exposure.[12]

A report of illnesses in 27 farmworkers and 4 emergency responders was related to pesticide drift of the pyrethroid **cyfluthrin**.[13] In this episode, the most commonly reported symptoms were headache (96%), nausea (89%), eye irritation (70%), muscle weakness (70%), anxiety (67%) and shortness of breath (64%).[13]

Apart from central nervous system toxicity, some pyrethroids do cause distressing paresthesias when liquid or volatilized materials contact human skin. These symptoms are more common with exposure to the Type II pyrethroids than the Type 1.[6] Sensations are described as stinging, burning, itching and tingling, progressing to numbness.[12,14,15] The skin of the face seems to be most commonly affected, but the hands, forearms and neck are sometimes involved. Sweating, exposure to sun or heat and application of water enhance the disagreeable sensations. Sometimes the paresthetic effect is noted within minutes of exposure, but a 1-2 hour delay in appearance of symptoms is more common.[14,16] Sensations rarely persist more than 24 hours.[7] Little or no inflammatory reaction is apparent where the paresthesias are reported; the effect is presumed to result from pyrethroid contact with sensory nerve endings in the skin. The paresthetic reaction is not allergic in nature, though sensitization and allergic responses have been reported as an independent phenomenon with pyrethroid exposure. However, allergic responses are less likely with pyrethroids than with pyrethrins. Race, skin type and disposition to allergic disease do not affect the likelihood or severity of the reaction.

Persons treated with permethrin for lice or flea infestations sometimes experience itching and burning at the site of application, but this is chiefly an exacerbation of sensations caused by the parasites themselves and is not typical of the paresthetic reaction described above.

Pyrethroids are not cholinesterase inhibitors. However, there have been some cases in which pyrethroid poisoning has been misdiagnosed as organophosphate poisoning, due to similar presenting signs.[7,12] There are also reports of mixed poisoning where the initial diagnosis of organophosphate poisoning had to be reconsidered when the response to atropine was more prompt and complete than expected.[16]

## Confirmation of Poisoning

Pyrethroid metabolites can be measured in the urine; however, this is not routinely available for the acutely poisoned patient. The following metabolites have been detected in occupationally exposed workers: cis- and trans-3-(2,2-dichlorovinyl)-2,2-dimethylcyclopropanecarboxylic acid, cis-3-(2,2-dibromovinyl)-2,2-dimethylcyclopropanecarboxylic acid, 3-phenoxybenzoic acid and 4-fluoro-3-phenoxybenzoic acid.[17]

## Treatment of Pyrethroid Toxicosis

Decontaminate the skin promptly with soap and water as outlined in **Chapter 3, *General Principles***. If irritant or paresthetic effects occur, obtain treatment by a physician. Because volatilization of pyrethroids apparently accounts for paresthesia affecting the face, strenuous measures should be taken (ventilation, protective face mask and hood) to avoid vapor contact with the face and eyes. Vitamin E oil preparations (dl-alpha tocopheryl acetate) are uniquely effective in preventing and stopping the paresthetic reaction.[14,18] They are safe for application to the skin under field conditions. Corn oil is somewhat effective, but possible side effects with continued use make it less suitable. Vaseline is less effective than corn oil, and zinc oxide actually worsens the reaction.

Treat eye contamination immediately by prolonged flushing with copious amounts of clean water or saline. Some pyrethroid compounds can be very corrosive to the eyes, so extraordinary measures should be taken to avoid eye contamination. If irritation persists, professional ophthalmologic care should be obtained.

If large amounts of pyrethroids, especially the cyano-pyrethroids, have been ingested and the patient is seen soon after exposure, consider gastrointestinal decontamination as outlined in **Chapter 3**. Based on observations in laboratory animals[6] and humans,[12] large ingestions of allethrin, cismethrin, fluvalinate, fenvalerate or deltamethrin would be the most likely to generate neurotoxic manifestations.

If only small amounts of pyrethroid have been ingested, or if treatment has been delayed, administer activated charcoal and a cathartic orally as this probably represents optimal management. Do not give cathartic if patient has diarrhea or an ileus.

***Pyrethroids***

**COMMERCIAL PRODUCTS**

*This list includes the names of several pyrethroids that are not currently in commercial production. These are included because they may be marketed in the future, if not in the United States, then possibly in other countries.*

allethrin (Pynamin)

barthrin*

bioallethrin (D-trans)

bioresmethrin*

biopermethrin*

cismethrin*

cyfluthrin (Baythroid)

cypermethrin (Ammo, Barricade, CCN52, Cymbush, Cymperator, Cynoff, Cyperkill, Demon, Folcord, KafilSuper, NRDC 149, Polytrin, Siperin, Ripcord, Flectron, Ustadd, Cyrux, many others)

deltamethrin (DeltaDust, DeltaGard, Suspend, Deltex, Decis)

dimethrin

fenothrin*

fenpropanate*

fenpropathrin (Danitol, Herald, Meothrin, Rody)

fenvalerate (Belmark, Sumicidin, Fenkill)

flucythrinate (Cybolt, Payoff, Fluent)

fluvalinate*

furethrin*

*continued next page*

*Pyrethroids*
**COMMERCIAL PRODUCTS**
*continued*

permethrin (Ambush, Dragnet, Eksmin, Kafil, Permasect, Perthrine, Pounce, Pramex, Outflank, Talcord and others)

phthalthrin*

resmethrin (Benzofuroline, Chrysron, Pynosect)

tetramethrin (Neopynamin)

tralomethrin (Tralex, SAGA)

Nix and Elimite – permethrin creams applied to control human ectoparasites

**Not registered in the U.S.*

Treat seizures as outlined in **Chapter 3**. Several drugs are effective in relieving the pyrethroid neurotoxic manifestations observed in deliberately poisoned laboratory animals, but none has been tested in human poisonings. Therefore, neither efficacy nor safety under these circumstances is known. Furthermore, moderate neurotoxic symptoms and signs are likely to resolve spontaneously.

## References


1. Moretto A. Indoor spraying with the pyrethroid insecticide lambda-cyhalothrin: effects on spraymen and inhabitants of sprayed houses. *Bull World Health Organ.* 1991;69(5):591-594.
2. Newton JG, Breslin AB. Asthmatic reactions to a commonly used aerosol insect killer. *Med J Aust.* Apr 16 1983;1(8):378-380.
3. Culver CA, Malina JJ, Talbert RL. Probable anaphylactoid reaction to a pyrethrin pediculocide shampoo. *Clin Pharm.* Nov 1988;7(11):846-849.
4. Carlson JE, Villaveces JW. Hypersensitivity pneumonitis due to pyrethrum. Report of a case. *JAMA.* Apr 18 1977;237(16):1718-1719.
5. Williams MK, Rundle A, Holmes D, et al. Changes in pest infestation levels, self-reported pesticide use, and permethrin exposure during pregnancy after the 2000-2001 U.S. Environmental Protection Agency restriction of organophosphates. *Environ Health Perspect.* Dec 2008;116(12):1681-1688.
6. Dorman DC, Beasley VR. Neurotoxicology of pyrethrin and the pyrethroid insecticides. *Vet Hum Toxicol.* Jun 1991;33(3):238-243.
7. Ray DE, Fry JR. A reassessment of the neurotoxicity of pyrethroid insecticides. *Pharmacol Ther.* Jul 2006;111(1):174-193.
8. Sheets LP, Doherty JD, Law MW, Reiter LW, Crofton KM. Age-dependent differences in the susceptibility of rats to deltamethrin. *Toxicol Appl Pharmacol.* May 1994;126(1):186-190.
9. Soderlund DM, Clark JM, Sheets LP, et al. Mechanisms of pyrethroid neurotoxicity: implications for cumulative risk assessment. *Toxicology.* Feb 1 2002;171(1):3-59.
10. Ray DE, Forshaw PJ. Pyrethroid insecticides: poisoning syndromes, synergies, and therapy. *J Toxicol Clin Toxicol.* 2000;38(2):95-101.
11. Illnesses and injuries related to total release foggers--eight states, 2001-2006. *MMWR Morb Mortal Wkly Rep.* Oct 17 2008;57(41):1125-1129. Available on-line: http://www.cdc.gov/mmwr/preview/mmwrhtml/mm5741a3.htm, accessed 12-27-12.
12. He F, Wang S, Liu L, Chen S, Zhang Z, Sun J. Clinical manifestations and diagnosis of acute pyrethroid poisoning. *Arch Toxicol.* 1989;63(1):54-58.
13. Worker illness related to ground application of pesticide--Kern County, California, 2005. *MMWR Morb Mortal Wkly Rep.* May 5 2006;55(17):486-488.
14. Flannigan SA, Tucker SB, Key MM, et al. Synthetic pyrethroid insecticides: a dermatological evaluation. *Br J Ind Med.* Jun 1985;42(6):363-372.
15. Tucker SB, Flannigan SA. Cutaneous effects from occupational exposure to fenvalerate. *Arch Toxicol.* Nov 1983;54(3):195-202.
16. Tripathi M, Pandey R, Ambesh SP, Pandey M. A mixture of organophosphate and pyrethroid intoxication requiring intensive care unit admission: a diagnostic dilemma and therapeutic approach. *Anesth Analg.* Aug 2006;103(2):410-412, table of contents.
17. Leng G, Kuhn KH, Idel H. Biological monitoring of pyrethroid metabolites in urine of pest control operators. *Toxicol Lett.* Nov 1996;88(1-3):215-220.
18. Tucker SB, Flannigan SA, Ross CE. Inhibition of cutaneous paresthesia resulting from synthetic pyrethroid exposure. *Int J Dermatol.* Dec 1984;23(10):686-689.

CHAPTER 5

# Organophosphate Insecticides

Organophosphates (OPs) are a class of insecticides, several of which are highly toxic. Until the 21st century, they were among the most widely used insecticides available. Thirty-six of them are presently registered for use in the United States, and all can potentially cause acute and subacute toxicity. Organophosphates are used in agriculture, homes, gardens and veterinary practices; however, in the past decade, several notable OPs have been discontinued for use, including **parathion**, which is no longer registered for any use, and **chlorpyrifos**, which is no longer registered for home use. All share a common mechanism of cholinesterase inhibition and can cause similar symptoms, although there are some differences within the class. Since they share this mechanism, exposure to the same organophosphate by multiple routes or to multiple organophosphates by multiple routes may lead to serious additive toxicity. It is important to understand, however, that there is a wide range of toxicity in these agents and wide variation in dermal absorption, making specific identification of the agent and individualized management quite important.

## HIGHLIGHTS

Acts through phosphorylation of the acetylcholinesterase enzyme at nerve endings

Efficiently absorbed by inhalation and ingestion

Dermal penetration/ absorption varies

Muscarinic, nicotinic, CNS effects

## SIGNS & SYMPTOMS

Headache, hypersecretion, muscle twitching, nausea, diarrhea, vomiting

Tachycardia/bradycardia, bronchospasm/ bronchorrhea

Respiratory depression, seizures (esp. pediatric), loss of consciousness

Miosis is often a helpful diagnostic sign

Depressed RBC AChE and/ or butyrylcholinesterase levels

## TREATMENT

Ensure a clear airway

Administer atropine sulfate or glycopyrolate

Pralidoxime may be indicated

Decontaminate concurrently

## CONTRAINDICATED

Morphine, succinylcholine, theophylline, phenothiazines, reserpine

## Toxicology

**Organophosphates** poison insects and other animals, including birds, amphibians and mammals, primarily by phosphorylation of the acetylcholinesterase enzyme (AChE) at nerve endings. The result is a loss of available AChE so that the effector organ becomes overstimulated by the excess acetylcholine (ACh, the impulse-transmitting substance) in the nerve ending. The enzyme is critical to normal control of nerve impulse transmission from nerve fibers to smooth and skeletal muscle cells, secretory cells and autonomic ganglia, and within the central nervous system (CNS). Once a critical proportion of the tissue enzyme mass is inactivated by phosphorylation, symptoms and signs of cholinergic poisoning become manifest.

At sufficient dosage, loss of enzyme function allows accumulation of ACh peripherally at cholinergic neuroeffector junctions (muscarinic effects), skeletal nerve-muscle junctions and autonomic ganglia (nicotinic effects), as well as centrally. At cholinergic nerve junctions with smooth muscle and secretory cells, high ACh concentration causes muscle contraction and secretion, respectively. At skeletal muscle junctions, excess ACh may be excitatory (cause muscle twitching) but may also weaken or paralyze the cell by depolarizing the end plate. Impairment of the diaphragm and thoracic skeletal muscles can cause respiratory paralysis. In the CNS, high ACh concentrations cause sensory and behavioral disturbances, incoordination, depressed motor function and respiratory depression. Increased pulmonary secretions coupled with respiratory failure are the usual causes of death from organophosphate poisoning. Recovery depends ultimately on generation of new enzyme in critical tissues.

Organophosphates are efficiently absorbed by inhalation and ingestion. Dermal penetration and subsequent systemic absorption varies with the specific agents. There is considerable variation in the relative absorption by these various routes. For instance, the oral $LD_{50}$ of **parathion** in rats is between 3-8 mg/kg, which is quite toxic,[1,2] and essentially equivalent to dermal absorption with an $LD_{50}$ of 8 mg/kg.[2] On the other hand, the toxicity of **phosalone** is much lower from the dermal route than the oral route, with rat $LD_{50}$s of 1,500 mg/kg and 120 mg/kg, respectively.[2] In general, the highly toxic agents are more likely to have higher-order dermal toxicity

## COMMERCIAL PRODUCTS

### *Highly Toxic*[1]

azinphos-methyl (Guthion, Gusathion)

bomyl[2] (Swat)

carbophenothion (Trithion)

chlorfenvinphos (Apachlor, Birlane)

chlormephos[2] (Dotan)

chlorthiophos[2] (Celathion)

coumaphos (Co-Ral, Asuntol)

cyanofenphos[2] (Surecide)

demeton[3] (Syntox)

dialifor (Torak)

dicrotophos (Bidrin)

dimefos[2] (Hanane, Pestox XIV)

dioxathion (Delnav)

disulfoton[3] (Disyston)

endothion[2]

EPN

ethyl parathion (E605, Parathion, thiophos)

famphur[2] (Famfos, Bo-Ana, Bash)

fenamiphos (Nemacur)

fenophosphon[2] (trichloronate, Agritox)

*continued next page*

[1]*Compounds are listed approximately in order of descending toxicity. "Highly toxic" organophosphates have listed oral LD50 values (rat) less than 50 mg/kg; "moderately toxic" agents have LD50 values in excess of 50 mg/kg and less than 500 mg/kg.*

[2]*Products no longer registered in the United States.*

[3]*These organophosphates are systemic: they are taken up by the plant and translocated into foliage and sometimes into the fruit.*

than the moderately toxic agents. To a degree, the occurrence of poisoning depends on the rate at which the pesticide is absorbed. Breakdown occurs chiefly by hydrolysis in the liver, and rates of hydrolysis vary widely from one compound to another. In those organophosphates for which breakdown is relatively slow, significant temporary storage in body fat may occur. Some organophosphates, such as **diazinon**, **fenthion** and **methyl parathion**, have significant lipid solubility, allowing fat storage with delayed toxicity due to late release.[3,4] Delayed toxicity may also occur atypically with other organophosphates, specifically **dichlorofenthion** and **demton-methyl**.[5] Many organothiophosphates readily undergo conversion from thions (P=S) to oxons (P=O). Conversion occurs in the environment under the influence of oxygen and light and, in the body, chiefly by the action of liver microsomal enzymes. Oxons are much more toxic than thions, but oxons break down more readily than thions. Ultimately, both thions and oxons are hydrolyzed at the ester linkage, yielding alkyl phosphates and leaving groups, both of which are of relatively low toxicity. They are either excreted or further transformed in the body before excretion.

After the initial exposure of the effector junction and the organophosphate, the enzyme-phosphoryl bond is strengthened by loss of one alkyl group from the phosphoryl adduct. This process is known as aging. The bond is then essentially permanent. Time of aging varies by agent and can occur within minutes to days. Depending on the time of aging of the agent, some phosphorylated acetylcholinesterase enzyme can be de-phosphorylated (reactivated) by a compound known as an oxime. The only currently FDA-approved oxime in the United States is pralidoxime. Other oximes include obidoxime and HI-6, which have been used in Europe and Asia. Depending on the agent, pralidoxime reactivation may be no longer possible after a couple of days,[6] although in some cases, improvement has still been seen with pralidoxime administration days after exposure.[7] Oximes have been used for OP poisoning for more than 50 years.[8] However, controversy remains as to the effectiveness of oximes because of conflicting and limited evidence of efficacy.[4,9,10]

Rarely, certain organophosphates have caused a different kind of neurotoxicity consisting of damage to the afferent fibers of peripheral and central nerves and associated with inhibition of "neuropathy target esterase" (NTE). Certain organophosphates are exceptionally prone to storage in fat tissue, prolonging the need for antidote for several days as stored pesticide is released back into the circulation.[3,4,11] This delayed syndrome has been termed organophosphate-induced delayed neuropathy (OPIDN) and is manifested chiefly by weakness or paralysis and paresthesia of the extremities.[12] OPIDN predominantly affects the legs and may persist for weeks to years. Only a few of the many organophosphates used as pesticides have been implicated as causes of delayed neuropathy in humans. EPA guidelines require that organophosphate and carbamate compounds that are candidate pesticides be tested in susceptible animal species for this neurotoxic property.

In addition to acute poisoning episodes and OPIDN, an intermediate syndrome has been described. This syndrome occurs after resolution of the acute cholinergic crisis, generally 24–96 hours after exposure. It is characterized by acute respiratory paresis and muscular weakness, primarily in the facial, neck and proximal limb muscles. In addition, it is often accompanied by cranial nerve palsies and depressed tendon reflexes. Like OPIDN, this syndrome lacks muscarinic symptoms and appears to result from a combined pre- and post-synaptic dysfunction of neuromuscular transmission. Symptoms do not respond well to atropine and oximes; therefore, treatment is mainly supportive.[13,14] The most common compounds involved in this syndrome are methyl parathion, fenthion and **dimethoate**, although one case with **ethyl-parathion** was also observed.[4,13]

Other specific properties of individual organophosphates may render them more hazardous than basic toxicity data suggest. Certain organophosphates are exceptionally prone to storage in fat tissue, prolonging the need for antidote for several days as stored pesticide is released back into the circulation. *In vitro* and animal studies have demonstrated potentiation or additive effects when two or more organophosphates are absorbed simultaneously, thereby creating a cumulative effect.[15,16] Animal studies have also demonstrated additive effects when orgranophosphates are combined with other pesticides including herbicides, carbamates and pyrethroids.[17,18,19] Animal studies have demonstrated a protective effect of toxicity from phenobarbital, which induces hepatic degradation of the pesticide.[1] Degradation of some compounds to a trimethyl phosphate can cause restrictive lung disease.[20]

In the late 1950s and 1960s, several reports appeared suggesting that long-term effects have occurred following acute and often massive exposures. Symptoms that are consistently reported from exposed persons include depression, memory and concentration problems, irritability, persistent headaches and motor weakness.[21,22,23] In these rare, anecdotal cases, symptoms have persisted for months to years. These hypothesis-generating cases eventually led to larger epidemiological studies with an exposed group and a control group that also supported the hypothesis that a proportion of patients acutely poisoned from any organophosphate can experience some long-term neuropsychiatric sequelae. The findings included significantly impaired performance on a battery of neuro-behavioral tests and compound-specific peripheral neuropathy, in some cases. Specific functions included impaired memory and concentration, depressed mood and peripheral neuropathy. These findings were subtle and, in some cases, picked up only on neuropsychologic testing rather than during neurologic exam.[24,25,26] For information on chronic and long-term effects from OPs, including subacute effects and long-term exposure without acute poisoning, see **Chapter 21, *Chronic Effects*.**

## Signs and Symptoms of Poisoning

Symptoms of acute organophosphate poisoning develop during or after exposure, within minutes to hours, depending on method of exposure. Exposure by inhalation results in the fastest appearance of toxic symptoms, followed by the oral route and finally the dermal route. All signs and symptoms are cholinergic in nature and affect muscarinic, nicotinic and central nervous system receptors.[6] The critical symptoms in initial management are the respiratory symptoms. Sufficient muscular fasciculations and weakness are often observed and require respiratory support, as respiratory arrest can occur suddenly. Bronchospasm and bronchorrhea can occur, producing chest tightness, wheezing, productive cough and pulmonary edema. These can impede efforts at adequate oxygenation of the patient. A life-threatening severity of poisoning is signified by loss of consciousness, incontinence, seizures and respiratory depression. The primary cause of death is respiratory failure.

There usually is a secondary cardiovascular component to the respiratory symptoms. The classic cardiovascular sign is bradycardia, which can progress to sinus arrest. However, this may be superseded by tachycardia and hypertension from nicotinic (sympathetic ganglia) stimulation.[27] Toxic cardiomyopathy has been reported after severe poisoning due to sarin, a weaponized organophosphate compound structurally similar to the insecticides.[28]

Some of the most commonly reported early symptoms include headache, nausea, dizziness and hypersecretion, the latter of which is manifested by sweating, salivation, lacrimation and rhinorrhea. Muscle twitching, weakness, tremor, incoordination, vomiting, abdominal cramps and diarrhea all signal worsening of the poisoned state.

***Highly Toxic Commercial Products***
*continued*

fensulfothion (Dasanit)

fonofos (Dyfonate, N-2790)

fosthietan (Nem-A-Tak)

isofenphos (Amaze, Oftanol)

mephosfolan[2,3] (Cytrolane)

methamidophos (Monitor)

methidathion (Supracide, Ultracide)

methyl parathion (E601, Penncap-M)

mevinphos (Phosdrin, Duraphos)

mipafox[2] (Isopestox, Pestox XV)

monocrotophos (Azodrin)

phorate (Thimet, Rampart, AASTAR)

phosfolan[2,3] (Cyolane, Cylan)

phosphamidon (Dimecron)

prothoate[2,3] (Fac)

schradan[2] (OMPA)

sulfotep (Thiotepp, Bladafum, Dithione)

terbufos (Counter, Contraven)

tetraethyl pyrophosphate[2] (TEPP)

***Moderately Toxic***[1]

acephate (Orthene)

bensulide (Betasan, Prefar)

bromophos-ethyl[2] (Nexagan)

bromophos[2] (Nexion)

chlorphoxim[2] (Baythion-C)

chlorpyrifos (Dursban, Lorsban, Brodan)

*continued next page*

***Moderately Toxic Commercial Products***
*continued*

crotoxyphos (Ciodrin, Cypona)
crufomate[2] (Ruelene)
cyanophos[2] (Cyanox)
cythioate[2] (Proban, Cyflee)
DEF (De-Green, E-Z-Off D)
demeton-S-methyl[3] (Duratox, Metasystox-R)
diazinon (Spectracide)
dichlofenthion (VC-13 Nemacide)
dichlorvos (DDVP, Vapona)
edifenphos[2]
EPBP[2] (S-Seven)
ethion (Ethanox)
ethoprop (Mocap)
etrimfos[2] (Ekamet)
fenitrothion (Accothion, Agrothion, Sumithion)
fenthion (mercaptophos, Entex, Baytex, Tiguvon)
formothion[2] (Anthio)
heptenophos[2] (Hostaquick)
IBP (Kitazin)
iodofenphos[2] (Nuvanol-N)
isoxathion[2] (E-48, Karphos)
leptophos[2] (Phosvel)

*continued next page*

*[1]Compounds are listed approximately in order of descending toxicity. "Highly toxic" organophosphates have listed oral LD50 values (rat) less than 50 mg/kg; "moderately toxic" agents have LD50 values in excess of 50 mg/kg and less than 500 mg/kg.*

*[2]Products no longer registered in the United States.*

*[3]These organophosphates are systemic: they are taken up by the plant and translocated into foliage and sometimes into the fruit.*

Miosis is often a helpful diagnostic sign, and the patient may report blurred and/or dark vision. Anxiety and restlessness are prominent. There are a few reports of choreiform movements.[29,30] Psychiatric symptoms including depression, memory loss and confusion have been reported. Toxic psychosis, manifested as confusion or bizarre behavior, has been misdiagnosed as alcohol intoxication.

Children often present with a slightly different clinical picture from adults. Four series have been published that describe children with poisoning from cholinesterase-inhibiting insecticides. Some of the more typical cholinergic signs of bradycardia, muscular fasciculations, lacrimation and sweating were less common. Seizures (range 8%-39%) and mental status changes, including lethargy and coma (range 55%-100%) were common in children.[31,32,33,34] In comparison, only around 2%-3% of adults present with seizures.[35,36] Other common presenting signs in children include flaccid muscle weakness, miosis and excessive salivation. In one of the studies, 80% of all cases were transferred with the wrong preliminary diagnosis.[33] In another study, 88% of parents initially denied a history of organophosphate exposure.[32] See the preceding section for information regarding the features of the intermediate syndrome and OPIDN.

## Confirmation of Poisoning

> CAUTION: ***If strong clinical indications of acute organophosphate poisoning are present, treat patient immediately. Do not wait for laboratory confirmation, which can take days. Initial medical care should be based on clinical presentations.***

Blood samples can measure plasma butyrylcholinesterase (pseudocholinesterase) and red blood cell (RBC) AChE levels.[37] Depressions of plasma pseudocholinesterase and/or RBC acetylcholinersterase enzyme activities are generally available biochemical indicators of excessive organophosphate absorption. Rarely, there have been reports of cases of symptomatic organophosphate toxicity in which the initial red blood cell cholinesterase levels were not depressed. Subsequent testing eventually demonstrated depressed cholinesterase levels. Certain organophosphates may selectively inhibit either plasma pseudocholinesterase or RBC acetylcholinesterase.[38] A minimum amount of organophosphate must be absorbed to depress blood cholinesterase activities, but enzyme activities, especially plasma pseudocholinesterase, may be lowered by dosages considerably less than are required to cause symptomatic poisoning. A 20%-30% depression of AChE may indicate a significant OP poisoning that, even without symptoms, needs antidotal treatment. In severe cases, the enzyme is usually depressed by 80%-90% of normal levels. The latter group typically requires significantly high doses of atropine.[4,37] Enzyme depression is usually apparent within a few minutes or hours of significant absorption of organophosphate. Depression of the plasma enzyme generally persists several days to a few weeks; the RBC enzyme activity may not reach its minimum for several days, and usually remains depressed longer, sometimes 1-3 months, until new enzyme replaces that inactivated by organophosphate. Lower limits of cholinesterase levels vary among laboratories and methods, so clinicians should interpret levels based on the given reference ranges. Patients with clinical signs of toxicity and accompanied by AChE levels depressed by 20%-50% should be managed as outlined in the treatment section.

In certain conditions, the activities of plasma and RBC cholinesterase are depressed in the absence of chemical inhibition. About 3% of individuals have a genetically determined low level of plasma pseudocholinesterase. These persons are particularly vulnerable to the action of the muscle-paralyzing drug succinylcholine, often administered to surgical patients, but not organophosphates. Patients with hepatitis, cirrhosis, malnutrition, chronic alcoholism and dermatomyositis exhibit low plasma cholinesterase activities. A number of toxicants, notably cocaine, carbon disulfide, benzalkonium salts, organic mercury compounds, ciguatoxins and solanines may reduce plasma pseudocholinesterase activity. Early pregnancy, oral contraception and metoclopramide may also cause some depression. The RBC acetylcholinesterase is less likely than the plasma enzyme to be affected by factors other than organophosphates. It is reduced, however, in certain rare conditions that damage the red cell membrane, such as hemolytic anemia.

The alkyl phosphates and phenols to which organophosphates are hydrolyzed in the body can often be detected in the urine during pesticide absorption and up to about 48 hours thereafter. These analyses are sometimes useful in identifying and quantifying the actual pesticide to which workers have been exposed. Urinary alkyl phosphate and phenol analyses can demonstrate organophosphate absorption at lower dosages than those required to depress cholinesterase activities and at much lower dosages than those required to produce symptoms and signs. Their presence may simply be a result of organophosphates in the food chain. These metabolites are among the numerous chemical metabolites measured in a U.S. sample via the National Health and Nutrition Education Survey (NHANES) and can be found in CDC's National Report on Human Exposure to Environmental Chemicals.[39]

Detection of intact organophosphates in the blood usually is not possible except during or soon after absorption of a substantial amount. In general, organophosphates do not remain unhydrolyzed in the blood more than a few minutes or hours, unless the quantity absorbed is large or the hydrolyzing liver enzymes are inhibited. Blood should be obtained for cholinesterase testing as described above, but it is not feasible or practical to attempt to test for specific compounds. It may be useful to obtain a urine sample from the poisoned patient and send it for metabolite detection as discussed in the preceding paragraph. For a patient with an unknown poisoning, a frozen sample of urine for later testing may be useful.

## Treatment of Organophosphate Toxicosis

*CAUTION:* ***Persons attending the victim should avoid direct contact with heavily contaminated clothing and vomitus. All caregivers should have appropriate protective gear when in contact with a patient poisoned by organophosphates. Wear rubber gloves while washing pesticide from skin and hair.***

1. Ensure that a clear airway exists. Intubate the patient and aspirate the secretions with a large bore suction device if necessary. Administer oxygen by mechanically assisted pulmonary ventilation if respiration is depressed and keep patient on a high $FiO_2$. In severe poisonings, patients should be treated in an intensive care unit setting.
2. Administer atropine sulfate intravenously, or intramuscularly if intravenous injection is not possible. Remember that atropine can be administered through an endotracheal tube if initial IV access is difficult to obtain. Depending on the severity of poisoning, doses of atropine ranging from very low to as high as 300 mg per day or more may be required,[40] or even continuous infusion.[41, 42] (See dosage on following page.)



***Moderately Toxic Commercial Products***
*continued*

malathion (Cythion)

merphos (Folex, Easy Off-D)

methyl trithion[2], dimethoate (Cygon, DeFend)

naled (Dibrom)

oxydemeton-methyl[3] (Metasystox-R)

oxydeprofos[2,3] (Metasystox-S)

phencapton[2] (G 28029)

phenthoate[2] (dimephenthoate, Phenthoate)

phosalone (Zolone)

phosmet (Imidan, Prolate)

phoxim[2] (Baythion)

pirimiphos-ethyl[2] (Primicid)

pirimiphos-methyl (Actellic)

profenofos (Curacron)

propetamphos (Safrotin)

propyl thiopyrophosphate[2] (Aspon)

pyrazophos[2] (Afugan, Curamil)

pyridaphenthion[2] (Ofunack)

quinalphos[2] (Bayrusil)

ronnel (Fenchlorphos, Korlan)

sulprofos[2] (Bolstar, Helothion)

temephos (Abate, Abathion).

tetrachlorvinphos (Gardona, Apex, Stirofos)

thiometon[2] (Ekatin)

triazophos[2] (Hostathion)

trichlorfon (Dylox, Dipterex, Proxol, Neguvon)

The objective of atropine antidotal therapy is to antagonize the effects of excessive concentrations of acetylcholine at end-organs having muscarinic receptors. Atropine does not reactivate the cholinesterase enzyme or accelerate disposition of organophosphate. Recrudescence of poisoning may occur if tissue concentrations of organophosphate remain high when the effect of atropine wears off, and multiple doses will be required. Atropine is effective against muscarinic manifestations, but it is ineffective against nicotinic actions, specifically muscle weakness and twitching, and respiratory depression. Despite these limitations, atropine is often a life-saving agent in organophosphate poisonings. Favorable response to a test dose of atropine can help differentiate poisoning by anticholinesterase agents from other conditions.

### Test Dosage of Atropine

- ***Adults**: 1 mg*
- ***Children under 12 years: 0.01 mg/kg***

Note, however, that lack of response with no evidence of atropinization (atropine refractoriness), may also indicate a more severe poisoning. The adjunctive use of nebulized atropine has been reported to improve respiratory distress, decrease bronchial secretions and increase oxygenation.[43]

### Dosage of Atropine

In ***moderately severe poisoning*** (hypersecretion and other end-organ manifestations without central nervous system depression), the following dosage schedules have been used.

- ***Adults and children over 12 years: Initial dose 1-3 mg IV. Repeat in 3-5 minutes if no change in clinical symptoms. Dose may be doubled with each administration until the patient is atropinized. Once adequate atropinization has been achieved, the patient can be maintained on an atropine continuous infusion at about 10%-20% of the loading dose and titrated to effect.[4,44,45,46]***

- ***Children under 12 years: There is less agreement regarding pediatric dosing. Recent studies recommend beginning with 0.02 mg/kg body weight, and doubling the dose every 5 minutes until atropinization is achieved.[4,44] Patients seen in a pediatric ICU setting were given 0.05 mg/kg every 15 minutes.[31] Since children sometimes present differently than adults and have more CNS findings, aggressive atropinization should proceed when there are muscarinic signs such as bradycardia, salivation, diarrhea and miosis that can be observed to change with adequate atropine.[31]***

Clear breath sounds and absent pulmonary secretions are the primary end-point. Other signs of atropinization may occur, including flushing, dry mouth, dilated pupils and tachycardia (pulse of 140 per minute). Early in therapy, monitor for improving blood pressure and heart rate (above 80 beats/minute), normal pupil size and drying of the skin and axillae.[4,45]

*WARNING:* ***In cases of ingestion of liquid concentrates of organophosphate pesticides, hydrocarbon aspiration may complicate these poisonings. Pulmonary edema and poor oxygenation in these cases will not respond to atropine and should be treated as a case of acute respiratory distress syndrome.***

Maintain atropinization by repeated doses based on recurrence of symptoms for 2–12 hours or longer depending on severity of poisoning. Crackles in the lung bases usually indicate inadequate atropinization. Pulmonary improvement may not parallel other signs of atropinization. Continuation of or return of cholinergic signs indicates the need for more atropine.

Maintain atropinization with repeated dosing as indicated by clinical status. When symptoms are stable for as much as 6 hours, the dosing may be decreased.

Severely poisoned individuals may exhibit remarkable tolerance to atropine; two or more times the dosages suggested above may be needed. The dose of atropine may be increased and the dosing interval decreased as needed to control symptoms. Continuous intravenous infusion of atropine may be necessary when atropine requirements are massive. The desired end-point is the reversal of muscarinic symptoms, most predominantly drying of secretions, and signs of improvement in pulmonary status and oxygenation, without an arbitrary dose limit. Preservative-free atropine products should be used whenever possible.

*NOTE:* ***Persons not poisoned or only slightly poisoned by organophosphates may develop signs of atropine toxicity from large doses. Fever, muscle fibrillations and delirium are the main signs of atropine toxicity. If these appear while the patient is fully atropinized, atropine administration should be discontinued, at least temporarily while the severity of poisoning is reevaluated.***

3. Consider administering glycopyrrolate. Glycopyrrolate has been studied as an alternative to atropine and found to have similar outcomes using continuous infusion. Ampules of 7.5 mg of glycopyrrolate were added to 200 mL of saline, and this infusion was titrated to the desired effects of dry mucous membranes, heart rate above 60 beats/minute and absent muscle fasciculations. During this study, atropine was used as a bolus for a heart rate less than 60 beats/minute. The other apparent advantage to this regimen was a decreased number of respiratory infections. This may represent an alternative when there is a concern for respiratory infection due to excessive and difficult-to-control secretions, and in the presence of altered level of consciousness where distinction between atropine toxicity or relapse of organophosphate poisoning is unclear.[47]

4. Draw a blood sample (heparinized) for cholinesterase analysis before administration of pralidoxime, which tends to reverse the cholinesterase depression.

5. Consider administering pralidoxime (Protopam, 2-PAM), a cholinesterase reactivator, in cases of moderate-to-severe OP poisoning in which respiratory depression, muscle weakness and/or twitching are severe. Pralidoxime works by reacti-

vating the cholinesterase and also by slowing the "aging" process, in which there is a loss of an akyl group. The AChE can no longer be reactivated. It is important to administer it early in the poisoning, preferably within 48 hours; however, this varies by the OP that is ingested. Some OPs will age much faster than others, (*e.g.*, parathion ages within 20 minutes, while diethyl-OPs tend to require >48 hours). Pralidoxime given after the aging process will be ineffective. Delayed treatment appears to be one factor in previous studies with oximes and OP poisoning that did not show a beneficial effect. [48,49]

As noted previously, there are limited data supporting the efficacy of oximes in OP poisoning, particularly from randomized controlled trials (RCTs), although they have been used for over 50 years.[37,50,51] One recent RCT demonstrated that pralidoxime substantially and moderately reactivated red cell AChE activity in patients poisoned by diethyl and dimethyl compounds, respectively, when given as a continuous infusion of 500 mg/hour. Though mortality was higher in the group receiving pralidoxime, the difference was not statistically significant.[9]

Another well-designed RCT compared two different dosing regimens after all patients first received a 2-gram loading dose of pralidoxime. The authors found that a continuous infusion of 1 gram of pralidoxime per hour was superior to what had been previously considered a standard bolus dosing of 1 gram pralidoxime every 4 hours. Mortality and morbidity, as measured by atropine requirements, need for intubation and duration of ventilator support, were all reduced in the group receiving continuous infusion. In this study, both groups appear to have received appropriate intensive care management that would closely match care provided in a U.S. hospital.[10] While further study is needed, particularly as to optimal dose and delivery time with respect to type of OP ingested, pralidoxime continues to be recommended in the United States for moderate-to-severe OP poisoning. Unfortunately, all studies have been performed on adults, so there are no adequate or updated data regarding proper dosing for children.

NOTE: ***Pralidoxime is of limited value, and may be hazardous, in poisonings by the cholinesterase-inhibiting carbamate compounds (see* Chapter 6*).***

### Dosage of Pralidoxime

#### Loading Dose

- ***Adults and children over 12 years: 2.0 gm by intravenous infusion over a 30-minute period.[10]***
- ***Children under 12 years: 20-50 mg/kg body weight given intravenously (depending on severity of poisoning), mixed in 100 mL of normal saline and infused over 30 minutes.***

#### Subsequent Dose

- ***1 gram per hour as a continuous infusion over a 48-hour period. Subsequent doses, if required, should be given every 4 hours, infused over an hour. Alternatively, dosage of pralidoxime may be repeated in 1-2 hours, then at 4-hour intervals if needed.***

Repeated doses of pralidoxime are usually required. Dosing should continue while ventilator support is required. In cases that involve continuing absorption of organophosphate (as after ingestion of large amount) or continuing transfer of highly lipophilic organophosphate from fat into blood, it may be necessary to continue administration of pralidoxime for several days beyond the 48-hour post-exposure interval usually cited as the limit of its effectiveness.

Blood pressure should be monitored during administration because of the occasional occurrence of hypertensive crisis. Administration should be slowed or stopped if blood pressure rises to hazardous levels. Be prepared to assist pulmonary ventilation mechanically if respiration is depressed during or after pralidoxime administration.

If intravenous injection is not possible, the bolus regimen of pralidoxime may be given by deep intramuscular injection.

6. Decontaminate skin, clothing, hair and/or eyes of patients who have been poisoned by organophosphates, concurrently with whatever resuscitative and antidotal measures are necessary to preserve life. Decontaminate eyes by flushing with copious amounts of clean water. If no symptoms are evident in a patient who remains alert and physically stable, a prompt shower and shampoo may be appropriate, provided the patient is carefully observed to ensure against sudden appearance of poisoning symptoms. If there are any indications of weakness, ataxia or other neurologic impairment, clothing should be removed and a complete bath and shampoo, using copious amounts of soap and water, should be given while the victim is recumbent. Attendants should wear rubber gloves, as latex or polyvinyl chloride provides no protection against skin absorption.[52,53] Even nitrile butadiene rubber gloves exhibited some defects following exposure to chlorpyrifos and diazinon, although these appeared much later (24-48 hours after exposure) compared to some almost immediate defects appearing in PVC gloves.[52] The possibility of pesticide sequestered under fingernails or in skin folds should not be overlooked. Contaminated clothing should be promptly bagged and not returned until it has been thoroughly laundered. Contaminated leather shoes should be discarded. Pesticide may have contaminated the inside surfaces of gloves, boots and/or headgear as well.

7. Consider gastrointestinal decontamination if organophosphate has been ingested in quantity sufficient to cause poisoning, if the patient receives care within 30 minutes of the exposure and if there is sufficient airway protection. If the patient has already vomited, which is most likely in serious exposures, further efforts at GI decontamination may not be indicated. In significant ingestions, diarrhea and/or vomiting are so constant that charcoal adsorption and catharsis are not indicated.

   A. Take rigorous precautions to protect the airway from aspiration of regurgitated gastric contents. If a victim is unconscious, obtunded, has an altered mental status or any respiratory compromise, orotracheal intubation should be performed prior to gastric aspiration.

   B. Save a sample of emesis or initial gastric aspirate for chemical analysis.

8. Observe patient closely for at least 72 hours after atropinization has been withdrawn to ensure that symptoms (sweating, visual disturbances, vomiting, diarrhea, chest and abdominal distress, and sometimes pulmonary edema) do not recur. In very severe poisonings by ingested organophosphates, particularly the more lipophilic and slowly hydrolyzed compounds, metabolic disposition of toxi-

cant may require as many as 5-14 days. In some cases, this slow elimination may combine with profound cholinesterase inhibition to require atropinization for several days or even weeks. As dosage is reduced, the lung bases should be checked frequently for crackles. If crackles are heard, or if there is a return of miosis, bradycardia, sweating or other cholinergic signs, atropinization must be reestablished promptly.

9. Monitor pulmonary status carefully even after apparent recovery from muscarinic symptoms, particularly in poisonings by large ingested doses of organophosphate. In some cases, respiratory failure has developed several days following organophosphate ingestion, and has persisted for days to weeks.

10. Monitor cardiac status in severely poisoned patients by continuous ECG recording. Some organophosphates have significant cardiac toxicity.

11. Do not use the following drugs: morphine, succinylcholine, theophylline, phenothiazines and reserpine. They are contraindicated in nearly all organophosphate poisoning cases. Adrenergic amines should be given only if there is a specific indication, such as marked hypotension.

12. If seizures occur despite therapy with atropine and pralidoxime, ensure that causes unrelated to pesticide toxicity are not responsible: head trauma, cerebral anoxia or mixed poisoning. Seizures occur rarely in severe organophosphate poisonings. Drugs useful in controlling seizures are discussed in **Chapter 3, *General Principles.*** The benzodiazepines – diazepam or lorazepam – are the agents of choice as initial therapy.

13. Warn persons who have been clinically poisoned by organophosphate pesticides to avoid re-exposure to cholinesterase-inhibiting chemicals until symptoms and signs have resolved completely and blood cholinesterase activities have returned to at least 80% of pre-poisoning levels. If blood cholinesterase was not measured prior to poisoning, blood enzyme activities should reach at least minimum normal levels before the patient is returned to a pesticide-contaminated environment.

14. Treat ingestion of liquid concentrates of organophosphate pesticides like a case of acute respiratory distress syndrome. Hydrocarbon aspiration may complicate these poisonings. Pulmonary edema and poor oxygenation in these cases will not respond to atropine.

15. Do not administer atropine or pralidoxime prophylactically to workers exposed to organophosphate pesticides. Prophylactic dosage with either atropine or pralidoxime may mask early signs and symptoms of organophosphate poisoning and thus allow the worker to continue exposure and possibly progress to more severe poisoning. Atropine itself may enhance the health hazards of the agricultural work setting, impairing heat loss (due to reduced sweating) and impairing the ability to operate mechanical equipment (due to blurred vision caused by mydriasis).

## References

1. DuBois KP. The toxicity of organophosphorus compounds to mammals. *Bull World Health Organ.* 1971;44(1):233-240.
2. Pasquet J, Mazuret A, Fournel J, Koenig FH. Acute oral and percutaneous toxicity of phosalone in the rat, in comparison with azinphosmethyl and parathion. *Toxicol Appl Pharmacol.* Jul 1976;37(1):85-92.
3. Garcia-Repetto R, Martinez D, Repetto M. Coefficient of distribution of some organophosphorus pesticides in rat tissue. *Vet Hum Toxicol.* 1995;37:226-229.
4. Roberts DM, Aaron CK. Management of acute organophosphorus pesticide poisoning. *BMJ.* Mar 24 2007;334(7594):629-634.
5. Gallo MA, Lawryk NJ. Organic phosphorus pesticides. In: Haves WJ, Laws ER, eds. *Handbook of pesticide toxicology.* Vol 2. San Diego: Academic Press Inc.; 1991.
6. Taylor P. Anticholinesterase agents. In: Gilman AG, Goodman LS, eds. *The pharmacological basis of therapeutics.* New York: Macmillian Publishing Co., Inc; 1985:110-128.
7. de Kort WL, Kiestra SH, Sangster B. The use of atropine and oximes in organophosphate intoxications: a modified approach. *J Toxicol Clin Toxicol.* 1988;26(3-4):199-208.
8. Antonijevic B, Stojiljkovic MP. Unequal efficacy of pyridinium oximes in acute organophosphate poisoning. *Clin Med Res.* Mar 2007;5(1):71-82.
9. Eddleston M, Eyer P, Worek F, et al. Pralidoxime in acute organophosphorus insecticide poisoning--a randomised controlled trial. *PLoS Med.* Jun 30 2009;6(6):e1000104.
10. Pawar KS, Bhoite RR, Pillay CP, Chavan SC, Malshikare DS, Garad SG. Continuous pralidoxime infusion versus repeated bolus injection to treat organophosphorus pesticide poisoning: a randomised controlled trial. *Lancet.* Dec 16 2006;368(9553):2136-2141.
11. Akgur S, Ozturk P, Yemiscigil A, Ege B. Rapid communication: postmortem distribution of organophosphate insecticides in human autopsy tissues following suicide. *J Toxicol Environ Health A.* Dec 12 2003;66(23):2187-2191.
12. Jamal GA. Neurological syndromes of organophosphorus compounds. *Adverse Drug React Toxicol Rev.* Aug 1997;16(3):133-170.
13. De Bleecker J, Van den Neucker K, Colardyn F. Intermediate syndrome in organophosphorus poisoning: a prospective study. *Crit Care Med.* Nov 1993;21(11):1706-1711.
14. De Bleecker J, Willems J, Van Den Neucker K, De Reuck J, Vogelaers D. Prolonged toxicity with intermediate syndrome after combined parathion and methyl parathion poisoning. *J Toxicol Clin Toxicol.* 1992;30(3):333-345; discussion 347-339.
15. Scholz NL, Truelove NK, Labenia JS, Baldwin DH, Collier TK. Dose-additive inhibition of chinook salmon acetylcholinesterase activity by mixtures of organophosphate and carbamate insecticides. *Environ Toxicol Chem.* May 2006;25(5):1200-1207.
16. Axelrad JC, Howard CV, McLean WG. Interactions between pesticides and components of pesticide formulations in an *in vitro* neurotoxicity test. *Toxicology.* 2002;173(3):259-269.
17. Costa LG, Murphy SD. Unidirectional cross-tolerance between the carbamate insecticide propoxur and the organophosphate disulfoton in mice. *Fundam Appl Toxicol.* Sep-Oct 1983;3(5):483-488.
18. Ahmad M. Potentiation/Antagonism of pyrethroids with organophosphate insecticides in *Bemisia tabaci* (Homoptera: Aleyrodidae). *J Econ Entomol.* Jun 2007;100(3):886-893.
19. Trimble AJ, Lydy MJ. Effects of triazine herbicides on organophosphate insecticide toxicity in *Hyalella azteca*. *Archives of environmental contamination and toxicology.* Jul 2006;51(1):29-34.
20. Aldridge WN, Nemery B. Toxicology of trialkylphosphorothioates with particular reference to lung toxicity. *Fundam Appl Toxicol.* Apr 1984;4(2 Pt 2):S215-223.

21. Gershon S, Shaw FH. Psychiatric sequelae of chronic exposure to organophosphorus insecticides. *Lancet.* Jun 24 1961;1(7191):1371-1374.

22. Holmes JH, Gaon MD. Observations on acute and multiple exposure to anticholinesterase agents. *Trans Am Clin Climatol Assoc.* 1957;68:86-103.

23. Metcalf DR, Holmes JH. VII. Toxicology and physiology. EEG, psychological, and neurological alterations in humans with organophosphorus exposure. *Ann N Y Acad Sci.* 1969;160(1):357-365.

24. Rosenstock L, Keifer M, Daniell WE, McConnell R, Claypoole K. Chronic central nervous system effects of acute organophosphate pesticide intoxication. The Pesticide Health Effects Study Group. *Lancet.* Jul 27 1991;338(8761):223-227.

25. Savage EP, Keefe TJ, Mounce LM, Heaton RK, Lewis JA, Burcar PJ. Chronic neurological sequelae of acute organophosphate pesticide poisoning. *Arch Environ Health.* Jan-Feb 1988;43(1):38-45.

26. Steenland K, Jenkins B, Ames RG, O'Malley M, Chrislip D, Russo J. Chronic neurological sequelae to organophosphate pesticide poisoning. *Am J Public Health.* May 1994;84(5):731-736.

27. Bardin PG, van Eeden SF, Moolman JA, Foden AP, Joubert JR. Organophosphate and carbamate poisoning. *Arch Intern Med.* Jul 11 1994;154(13):1433-1441.

28. Okudera H. Clinical features on nerve gas terrorism in Matsumoto. *J Clin Neurosci.* Jan 2002;9(1):17-21.

29. Joubert J, Joubert PH. Chorea and psychiatric changes in organophosphate poisoning. A report of 2 further cases. *S Afr Med J.* Jul 2 1988;74(1):32-34.

30. Joubert J, Joubert PH, van der Spuy M, van Graan E. Acute organophosphate poisoning presenting with choreo-athetosis. *J Toxicol Clin Toxicol.* 1984;22(2):187-191.

31. Lifshitz M, Shahak E, Sofer S. Carbamate and organophosphate poisoning in young children. *Pediatr Emerg Care.* Apr 1999;15(2):102-103.

32. Sofer S, Tal A, Shahak E. Carbamate and organophosphate poisoning in early childhood. *Pediatr Emerg Care.* Dec 1989;5(4):222-225.

33. Zwiener RJ, Ginsburg CM. Organophosphate and carbamate poisoning in infants and children. *Pediatrics.* Jan 1988;81(1):121-126.

34. Levy-Khademi F, Tenenbaum AN, Wexler ID, Amitai Y. Unintentional organophosphate intoxication in children. *Pediatr Emerg Care.* Oct 2007;23(10):716-718.

35. Hayes MM, van der Westhuizen NG, Gelfand M. Organophosphate poisoning in Rhodesia. A study of the clinical features and management of 105 patients. *S Afr Med J.* Aug 5 1978;54(6):230-234.

36. Jamil H, Kundi A, Akhtar S, Sultana N. Organo-phosphorus insecticide poisoning—review of 53 cases. *J Pak Med Assoc.* Jul 1977;27(7):361-363.

37. Eddleston M, Buckley NA, Eyer P, Dawson AH. Management of acute organophosphorus pesticide poisoning. *Lancet.* Feb 16 2008;371(9612):597-607.

38. Sullivan JB, Blose J. Organophosphate and carbamate insecticides. In: Sullivan JB, Krieger GR, eds. *Hazardous materials toxicology*. Baltimore: Williams and Wilkins; 1992:1015-1026.

39. Prevention CfDCa. National report on human exposure to environmental chemicals. 2010; http://www.cdc.gov/exposurereport/. Accessed March 24, 2010.

40. Goswamy R, Chaudhuri A, Mahashur AA. Study of respiratory failure in organophosphate and carbamate poisoning. *Heart Lung.* Nov-Dec 1994;23(6):466-472.

41. du Toit PW, Muller FO, van Tonder WM, Ungerer MJ. Experience with the intensive care management of organophosphate insecticide poisoning. *S Afr Med J.* Aug 8 1981;60(6):227-229.

**42.** LeBlanc FN, Benson BE, Gilg AD. A severe organophosphate poisoning requiring the use of an atropine drip. *J Toxicol Clin Toxicol.* 1986;24(1):69-76.

**43.** Shockley LW. The use of inhaled nebulized atropine for the treatment of malathion poisoning. *J Toxicol Clin Toxicol.* 1989;27(3):183-192.

**44.** Eddleston M, Buckley NA, Checketts H, et al. Speed of initial atropinisation in significant organophosphorus pesticide poisoning--a systematic comparison of recommended regimens. *J Toxicol Clin Toxicol.* 2004;42(6):865-875.

**45.** Eddleston M, Dawson A, Karalliedde L, et al. Early management after self-poisoning with an organophosphorus or carbamate pesticide - a treatment protocol for junior doctors. *Crit Care.* Dec 2004;8(6):R391-397.

**46.** Perera PM, Shahmy S, Gawarammana I, Dawson AH. Comparison of two commonly practiced atropinization regimens in acute organophosphorus and carbamate poisoning, doubling doses vs. ad hoc: a prospective observational study. *Hum Exp Toxicol.* Jun 2008;27(6):513-518.

**47.** Bardin PG, Van Eeden SF. Organophosphate poisoning: grading the severity and comparing treatment between atropine and glycopyrrolate. *Crit Care Med.* Sep 1990;18(9):956-960.

**48.** Cherian AM, Peter JV, Jaydevan R, et al. Effectiveness of P2AM (PAM - pralidoxime) in the treatment of organophosphorus poisoning (OPP) a randomized, double blind placebo controlled clinical trial. *JAPI.* 1997;45(1):22-24.

**49.** Johnson S, Peter JV, Thomas K, Jeyaseelan L, Cherian AM. Evaluation of two treatment regimens of pralidoxime (1 gm single bolus dose vs 12 gm infusion) in the management of organophosphorus poisoning. *J Assoc Physicians India.* Aug 1996;44(8):529-531.

**50.** Freeman G, Epstein MA. Therapeutic factors in survival after lethal cholinesterase inhibition by phosphorus insecticides. *N Engl J Med.* Aug 18 1955;253(7):266-271.

**51.** Namba T, Hiraki K. PAM (pyridine-2-aldoxime methiodide) therapy for alkyl-phosphate poisoning. *J Am Med Assoc.* Apr 12 1958;166(15):1834-1839.

**52.** Canning KM, McQuillan P, Jablonski W. Laboratory simulation of splashes and spills of organophosphate insecticides on chemically protective gloves used in agriculture. *Ann Agric Environ Med.* 1998;5(2):155-167.

**53.** Geller RJ, Singleton KL, Tarantino ML, Drenzek CL, Toomey KE. Nosocomial poisoning associated with emergency department treatment of organophosphate toxicity--Georgia, 2000. *J Toxicol Clin Toxicol.* 2001;39(1):109-111.

**HIGHLIGHTS**

Muscarinic, nicotinic, CNS effects

Absorbed by inhalation, ingestion, skin

Lipophilic

Poisonings tend to be of shorter duration than OPs

**SIGNS & SYMPTOMS**

Malaise, muscle weakness, dizziness, sweating

May include blurred vision, incoordination, muscle twitching, slurred speech

May include headache, salivation, nausea, vomiting, abdominal pain, diarrhea

Blood cholinesterase levels not reliable indication

Severe symptoms may include coma, seizures, hypotonicity, hypertension, cardio/respiratory depression

**TREATMENT**

Ensure clear airway

Administer atropine

Decontaminate concurrently

Consider pralidoxime in mixed poisonings

Consider GI decontamination

**CONTRAINDICATED**

Adrenergic amines without specific indication (*e.g.*, hypotension)

CHAPTER 6

# N-Methyl Carbamate Insecticides

## Toxicology

The **N-methyl carbamate** esters cause reversible carbamylation of acetylcholinesterase (AChE) enzyme, allowing accumulation of acetylcholine, the neuromediator substance, at parasympathetic neuroeffector junctions (muscarinic effects), at skeletal muscle myoneural junctions and autonomic ganglia (nicotinic effects) and in the brain (CNS effects). The carbamyl-acetylcholinesterase combination dissociates more readily than the phosphoryl-acetylcholinesterase complex produced by organophosphate compounds. This lability has several important consequences: (1) it tends to limit the duration of N-methyl carbamate poisonings, (2) it accounts for the greater difference between symptom-producing and lethal doses than exists in the case of most organophosphate compounds and (3) it frequently invalidates the measurement of blood cholinesterase activity as a diagnostic index of poisoning (see below).

Carbamates are absorbed by inhalation, ingestion and through the skin, although the last tends to be the less-toxic route. For example, carbofuran has a rat oral $LD_{50}$ of 5 mg/kg, compared to a rat dermal $LD_{50}$ of 120 mg/kg, which makes the oral route approximately 24 times more toxic when ingested.[1] The $LD_{50}$ is only one measure of pesticide toxicity. The dose must also be considered since a compound with a high $LD_{50}$ can produce life-threatening symptoms if a large enough dose is ingested. N-methyl carbamates are hydrolyzed enzymatically by the liver, and the degradation products are excreted by the kidneys and the liver.

At cholinergic nerve junctions with smooth muscle and gland cells, high acetylcholine concentration causes muscle contraction and secretion, respectively. At skeletal muscle junctions, excess acetylcholine may be excitatory (cause muscle twitching), but may also weaken or paralyze the cell by depolarizing the end-plate. In the brain, elevated acetylcholine concentrations may cause sensory and behavioral disturbances, incoordination, seizures and depressed motor function including lethargy and coma.[2,3,4] The N-methyl carbamates are lipophilic and penetrate the central nervous system as evidenced by distribution throughout all tissues including the brain on post mortum.[5,6] Respiratory depression combined with pulmonary edema is the usual cause of death from poisoning by N-methyl carbamate compounds.

## Signs and Symptoms of Poisoning

As with organophosphate poisoning, the signs and symptoms are based on excessive cholinergic stimulation. Carbamate poisonings tend to be of shorter duration than organophosphate poisonings because of the reversibility of the AChE binding and the more rapid metabolism of carbamates.[7] However, as mentioned in the next section of this chapter, blood cholinesterase levels may be misleading because of *in vitro* reactivation of a carbamylated enzyme.[8,9] This falsely normal or near-normal level can make the diagnosis more difficult in the acute presentation in the absence of an exposure history.

Malaise, muscle weakness, dizziness and sweating are commonly reported early symptoms. Miosis with blurred vision, incoordination, muscle twitching and slurred speech are reported. Headache, salivation, nausea, vomiting, abdominal pain and diarrhea are often prominent. Transient hyperbilirubinemia may occur.[10] Acute pancreatitis has also been reported in some of the cases of **aldicarb** and **methomyl** poisoning. Some cases of pancreatitis have required surgical drainage of a pancreatitic pseudocyst.[3,11,12,13]

The most severe manifestations of carbamate poisoning occur in the respiratory and central nervous (CNS) systems. CNS findings include coma, seizures and hypotonicity, and nicotinic effects including hypertension and cardiorespiratory depression. The respiratory depression also results from skeletal muscle impairment in which the chest wall cannot expand for adequate respiration. Dyspnea, bronchospasm and bronchorhea with eventual pulmonary edema are other serious signs.[3,14] Data indicate that children and adults differ in their clinical presentation. Children are more likely than adults to present with the CNS symptoms above. While children can develop the classic muscarinic signs, the absence of them does not exclude the possibility of carbamate poisoning in the presence of CNS depression.[2,4,15,16]

## Confirmation of Poisoning

If there are strong clinical indications of acute N-methyl carbamate poisoning, and/or a history of carbamate exposure, treat the patient immediately. Do not wait for laboratory confirmation.

Blood for plasma pseudocholinesterase and RBC AchE should be obtained. Unless a substantial amount of N-methyl carbamate has been absorbed and a blood sample is taken within an hour or two, it is unlikely that blood cholinesterase activities will be found depressed. Even under the above circumstances, a rapid test for enzyme activity must be used to detect an effect, because enzyme reactivation occurs *in vitro* as well as *in vivo*.

Absorption of some N-methyl carbamates can be confirmed by analysis of urine for unique metabolites; alpha-naphthol from **carbaryl**, isopropoxyphenol from **propoxur**, carbofuran phenol from **carbofuran**, and aldicarb sulfone, sulfoxide and nitrile from **aldicarb**. These complex analyses, when available, can be useful in identifying the responsible agent and following the course of carbamate disposition.

## Treatment of N-methyl Carbamate Insecticide Toxicosis

CAUTION: ***Persons attending the victim should avoid direct contact with heavily contaminated clothing and vomitus. Wear rubber gloves while washing pesticide from skin and hair. Vinyl gloves provide no protection.***

1. Ensure that a clear airway exists. Intubate the patient and aspirate the secretions with a large bore suction device if necessary. Administer oxygen by mechanically assisted pulmonary ventilation if respiration is depressed. Improve tissue oxygenation as much as possible before administering atropine, so as to minimize the risk of ventricular fibrillation. In severe poisonings, it may be necessary to support pulmonary ventilation mechanically for several days.

2. Administer atropine sulfate intravenously or intramuscularly if intravenous injection is not possible. Atropine can be administered in small volume doses through an endotracheal tube if initial IV access is difficult to obtain. Carba-

### COMMERCIAL PRODUCTS

***Highly Toxic***[1]

aldicarb (Temik)

aminocarb[2] (Matacil)

bendiocarb[2] (Ficam, Dycarb, Multamat, Niomil, Tattoo, Turcam)

carbofuran (Furadan)

cloethocarb[2] (Lance)

formetanate (Carzol)

isolan[2] (Primin)

methiocarb (Mesurol)

methomyl (Lannate, Nudrin)

oxamyl (Vydate L, DPX 1410)

***Moderately Toxic***[1]

bufencarb[2] (Metalkamate, Bux)

carbaryl (Sevin, Dicarbam)

dimetan[2] (Dimethan)

dioxacarb[2] (Elecron, Famid)

isoprocarb[2] (Etrofolan, MIPC)

pirimicarb[2] (Pirimor, Abol, Aficida, Aphox, Fernos, Rapid)

promecarb[2] (Carbamult)

propoxur (Aprocarb, Baygon, various flea/tick products)

trimethacarb[2] (Landrin, Broot)

---

*[1] "Highly toxic" N-methyl carbamates have listed oral $LD_{50}$ values (rat) less than or equal to 50 mg/kg body weight; "moderately toxic" agents have $LD_{50}$ values in excess of 50 mg/kg and less than 500 mg/kg.*

*[2] Products no longer registered in the United States.*

mates may reverse with smaller dosages of atropine than those required to reverse organophosphates, though the required dosage is still considerably larger than that required to atropinize a non-poisoned patient.[17,18] A common dosing pitfall is giving too little atropine initially to achieve timely atropinization. Severely poisoned individuals may exhibit remarkable tolerance to atropine and require large doses.[14] (See dosage below.)

The objective of atropine antidotal therapy is to antagonize the effects of excessive concentrations of acetylcholine at end-organs having muscarinic receptors. Atropine does not reactivate AChE or accelerate excretion or breakdown of carbamate. Multiple doses of atropine may be necessary, as recrudescence of poisoning can occur if tissue concentrations of toxicant remain high when the antidotal effect wears off. Atropine is effective against muscarinic manifestations, but is ineffective against nicotinic actions, specifically muscle weakness and twitching, and respiratory depression. Despite these limitations, atropine is often a lifesaving agent in N-methyl carbamate poisonings.

Reassess the clinical situation after an adequate loading dose has been given. If symptoms persist, but the history is consistent with carbamate poisoning, then continue atropine therapy. However, if the clinical picture is unclear, clinicians should reassess and consider alternative causes of poisoning, such as pyrethroid insecticide poisoning, which may present a similar clinical picture.

In moderately severe poisoning (hypersecretion and other end-organ manifestations without central nervous system depression) the following dosage schedules have proven effective:

## Dosage of Atropine

### Adults and Children Over 12 Years

- **Initial Dose:** ***1-3 mg IV. Repeat in 3-5 minutes if no change in clinical symptoms. Dose may be doubled with each administration until the patient is atropinized. Once adequate atropinization has been achieved, the patient can be maintained on an atropine continuous infusion at about 10%-20% of the loading dose and titrated to effect.[18,19,20,21] Clear breath sounds and absent pulmonary secretions are the primary end point. Other signs of atropinization including flushing, dry mouth and dilated pupils; tachycardia (pulse of 140 per minute) may occur. Early in therapy, monitor for improving blood pressure and heart rate (above 80 beats/ minute), normal pupil size and drying of the skin and axillae.[20,21] Autoinjectors containing 2.0 mg atropine for IM injection are also available.***

WARNING: *Poisonings in which liquid carbamate pesticide concentrates have been ingested may be complicated by hydrocarbon aspiration. Pulmonary edema and poor oxygenation in these cases will not respond to atropine and should be treated as a case of acute respiratory distress syndrome.*

*continued next page*

*Dosage of Atropine, continued*

### Children Under 12 Years

- **Initial Dose:** ***0.02 mg/kg body weight IV. As with adults, double the dose every 5 minutes until pulmonary secretions are controlled. Consider continuous infusion at 10%-20% of the required loading dose and titrate. Signs of atropinization, including: flushing, dry mouth, dilated pupils and heart rates vary depending on age of child, with young toddlers having a rate approaching 200. Crackles in the lung bases nearly always indicate inadequate atropinization, and pulmonary improvement may not parallel other signs. Continuation of, or return of, cholinergic signs indicate the need for more atropine.***

Reversal of muscarinic manifestations, rather than a specific dosage, is the object of atropine therapy.

*NOTE*: ***Persons not poisoned or only slightly poisoned by N-methyl carbamates may develop signs of atropine toxicity from large doses, such as fever, muscle fibrillations and delirium. If these signs appear and become the predominant clinical effects, atropine administration should be discontinued, at least temporarily, while the severity of poisoning is reevaluated.***

3. Save a urine sample for metabolite analysis if there is need to identify the agent responsible for the poisoning.

4. Consider pralidoxime in cases of mixed carbamate/organophosphate poisoning and cases of an unknown pesticide with muscarinic symptoms on presentation (see **Chapter 5, *Organophosphate Insecticides,*** subsection *Treatment,* item 5, page 49.[22,23] Pralidoxime has been used in some cases of carbamate poisoning, although other cases have resolved from supportive care alone.[24,25] Pralidoxime is probably of little value in N-methyl carbamate poisonings and is not indicated in isolated carbamate poisonings. Atropine alone usually is effective.

5. Decontaminate concurrently with whatever resuscitative and antidotal measures are needed to preserve life. Contamination of the eyes should be removed by flushing with copious amounts of clean water. For asymptomatic individuals who are alert and physically able, skin decontamination should occur as previously outlined in **Chapter 3, *General Principles*.** Specifically, skin and hair should be washed with soap and water. Attending personnel must take precautions including rubber gloves to avoid contamination. Contaminated clothing should be promptly removed, bagged and laundered before returning, and items such as shoes, boots and headgear should be discarded.

6. Consider gastrointestinal decontamination if N-methyl carbamate has been ingested in a quantity sufficient to cause probable poisoning. If the patient has presented with a recent ingestion and still asymptomatic, adsorption of poison with activated charcoal may be beneficial. If the patient has already vomited or

is symptomatic, which is highly likely in significant poisonings, attention should be placed on oxygen, airway management and atropine. In significant ingestions, diarrhea and/or vomiting are so constant that charcoal adsorption and catharsis are not included.

7. Observe patient closely for at least 24-48 hours to ensure that symptoms (sweating, visual disturbances, vomiting, diarrhea, chest and abdominal distress, and sometimes pulmonary edema) do not recur as atropinization is withdrawn. The observation period should be longer in the case of mixed pesticide ingestion, because of the prolonged and delayed symptoms associated with organophosphate poisoning. As the dosage of atropine is reduced over time, check the lung bases frequently for crackles. Atropinization must be reestablished promptly if crackles are heard or if there is a return of miosis, sweating or other signs of poisoning.

8. Monitor pulmonary ventilation carefully, particularly in poisonings by large doses of N-methyl carbamates, even after recovery from muscarinic symptomatology, to forestall respiratory failure.

9. Monitor cardiac status in severely poisoned patients by continuous ECG recording.

10. Give adrenergic amines (n-morphine, succinlycholine, theophylline, phenothiazines and reserpine) only if there is a specific indication, such as marked hypotension. Otherwise, they are probably contraindicated in N-methyl carbamate poisoning cases.

11. Treat cases in which liquid concentrates of some carbamates formulated in a petroleum product base have been ingested as acute respiratory distress syndrome. Hydrocarbon aspiration may complicate these poisonings. Pulmonary edema and poor oxygenation in these cases will not respond to atropine.

12. Do not administer atropine prophylactically to workers exposed to N-methyl carbamate pesticides. Prophylactic dosage may mask early symptoms and signs of carbamate poisoning and thus allow the worker to continue exposure and possible progression to more severe poisoning. Atropine itself may enhance the health hazards of the agricultural work setting, impairing heat loss (due to reduced sweating) and impairing the ability to operate mechanical equipment due to blurred vision caused by mydriasis.

## References

1. Micromedex I. Registry of Toxic Effects of Chemical Substances: NIOSH; 1991.

2. Lifshitz M, Shahak E, Bolotin A, Sofer S. Carbamate poisoning in early childhood and in adults. *J Toxicol Clin Toxicol.* 1997;35(1):25-27.

3. Ragouc-Sengler C, Tracqui A, Chavonnet A, et al. Aldicarb poisoning. *Hum Exp Toxicol.* 2000;19:657-662.

4. Zwiener RJ, Ginsburg CM. Organophosphate and carbamate poisoning in infants and children. *Pediatrics.* Jan 1988;81(1):121-126.

5. Tsatsakis AM, Bertsias GK, Mammas IN, Stiakakis I, Georgopoulos DB. Acute fatal poisoning by methomyl caused by inhalation and transdermal absorption. *Bull Environ Contam Toxicol.* Apr 2001;66(4):415-420.

6. Yamazaki M, Terada M, Kuroki H, Honda K, Matoba R, Mitsukuni Y. Pesticide poisoning initially suspected as a natural death. *J Forensic Sci.* Jan 2001;46(1):165-170.

7. Echobichon DJ. Toxic effect of pesticides. In: Klaassen CD, ed. *Casarett & Doull's Toxicology: The Basic Science of Poisons.* 5th ed. New York: McGraw-Hill; 1996:659.

8. Jokanovic M, Maksimovic M. Abnormal cholinesterase activity: understanding and interpretation. *Eur J Clin Chem Clin Biochem.* Jan 1997;35(1):11-16.

9. Rotenberg M, Almog S. Evaluation of the decarbamylation process of cholinesterase during assay of enzyme activity. *Clin Chim Acta.* Sep 15 1995;240(2):107-116.

10. Saadeh AM. Metabolic complications of organophosphate and carbamate poisoning. *Trop Doct.* Jul 2001;31(3):149-152.

11. Brahmi N, Blel Y, Kouraichi N, Abidi N, Thabet H, Amamou M. Acute pancreatitis subsequent to voluntary methomyl and dichlorvos intoxication. *Pancreas.* Nov 2005;31(4):424-427.

12. Rizos E, Liberopoulos E, Kosta P, Efremidis S, Elisaf M. Carbofuran-induced acute pancreatitis. *JOP.* Jan 2004;5(1):44-47.

13. Weizman Z, Sofer S. Acute pancreatitis in children with anticholinesterase insecticide intoxication. *Pediatrics.* Aug 1992;90(2 Pt 1):204-206.

14. Nelson LS, Perrone J, DeRoos F, Stork C, Hoffman RS. Aldicarb poisoning by an illicit rodenticide imported into the United States: Tres Pasitos. *J Toxicol Clin Toxicol.* 2001;39(5):447-452.

15. Lifshitz M, Shahak E, Sofer S. Carbamate and organophosphate poisoning in young children. *Pediatr Emerg Care.* Apr 1999;15(2):102-103.

16. Verhulst L, Waggie Z, Hatherill M, Reynolds L, Argent A. Presentation and outcome of severe anticholinesterase insecticide poisoning. *Arch Dis Child.* May 2002;86(5):352-355.

17. Goswamy R, Chaudhuri A, Mahashur AA. Study of respiratory failure in organophosphate and carbamate poisoning. *Heart Lung.* Nov-Dec 1994;23(6):466-472.

18. Perera PM, Shahmy S, Gawarammana I, Dawson AH. Comparison of two commonly practiced atropinization regimens in acute organophosphorus and carbamate poisoning, doubling doses vs. ad hoc: a prospective observational study. *Hum Exp Toxicol.* Jun 2008;27(6):513-518.

19. Eddleston M, Buckley NA, Checketts H, et al. Speed of initial atropinisation in significant organophosphorus pesticide poisoning--a systematic comparison of recommended regimens. *J Toxicol Clin Toxicol.* 2004;42(6):865-875.

20. Eddleston M, Dawson A, Karalliedde L, et al. Early management after self-poisoning with an organophosphorus or carbamate pesticide - a treatment protocol for junior doctors. *Crit Care.* Dec 2004;8(6):R391-397.

21. Roberts DM, Aaron CK. Management of acute organophosphorus pesticide poisoning. *BMJ.* Mar 24 2007;334(7594):629-634.

22. Kurtz PH. Pralidoxime in the treatment of carbamate intoxication. *Am J Emerg Med.* Jan 1990;8(1):68-70.

23. Lifshitz M, Sofer S, Shahak E, Rotenberg M, Almog S, Tamiri T. Carbamate poisoning and oxime treatment in children: a clinical and laboratory study. *Pediatrics.* Apr 1994;93(4):652-655.

24. Park CH, Kim KI, Park SK, Lee CH. Carbamate poisoning: high resolution CT and pathologic findings. *J Comput Assist Tomogr.* Jan-Feb 2000;24(1):52-54.

25. CDC. Poisonings Associated with Illegal Use of Aldicarb as a Rodenticide — New York City, 1994–1997. *MMWR Morb Mortal Wkly Rep.* 1997;46(41):961-963.

CHAPTER 7

# Organochlorines

The U.S. Environmental Protection Agency has sharply curtailed the availability of most organochlorines. While use continues in many other regions of the world, in the United States only dicofol and endosulfan are still registered as pesticides. Lindane is still marketed as a second-line agent for treatment of lice and scabies under the trade names Kwell and Thionex, although it is no longer recommended by the American Academy of Pediatrics and has been banned in California. This is the result of multiple cases of acute neurological toxicity either from ingestion or in persons treated for scabies or lice.[1,2,3,4,5,6,7] In recent years, the most frequently reported serious or fatal events were from endosulfan.[8,9,10,11,12,13,14]

## Toxicology

**Organochlorines** are absorbed from the gut, by the lungs and across the skin in varying degrees. **Hexachlorocyclohexane, lindane, the cyclodienes (aldrin, dieldrin, endrin, chlordane, heptachlor)** and **endosulfan** are efficiently absorbed across the skin, while dermal absorption efficiencies of **DDT, dicofol, methoxychlor, toxaphene** and **mirex** are substantially less.[15] Lindane has an estimated 9.3% dermal absorption rate[16] and is absorbed even more efficiently across abraded skin.[4,17] This becomes especially important when taking into account its use on children with severe dermatitis associated with scabies. Fat and fat solvents enhance gastrointestinal, and probably dermal, absorption of organochlorines. Many formulations of organochlorines are in hydrocarbon solvents that probably promote absorption. While most of the solid organochlorines are not highly volatile, pesticide-laden aerosols or dust particles trapped in respiratory mucous and subsequently swallowed may be vehicles leading to significant gastrointestinal absorption.

Following exposure to some organochlorines (notably DDT), a significant part of the absorbed dose is stored in fat tissue as the parent compound. Most organochlorines are in some degree dechlorinated, oxidized and then conjugated. The chief route of excretion is biliary, although nearly all organochlorines yield measurable urinary metabolites. Unfortunately, many of the unmetabolized pesticides are efficiently reabsorbed by the intestine (enterohepatic circulation), substantially retarding fecal excretion. Metabolic dispositions of DDT and DDE (a DDT degradation product), the beta isomer of hexachlorocyclohexane, dieldrin, heptachlor epoxide and mirex tend to be slow, leading to storage in body fat. Storable lipophilic compounds are likely to be excreted in maternal milk.[6,18,19] In contrast, rapid metabolic disposition of lindane, methoxychlor, dienochlor, endrin, chlorobenzilate, dicofol, toxaphene, perthane and endosulfan reduce the likelihood that these organochlorines will be detected as residues in body fat, blood or milk.

The chief acute toxic action of the organochlorine pesticides is on the central nervous system, where these compounds induce a hyperexcitable state in the brain leading to convulsions or other less severe signs of neurologic toxicity such as myoclonic jerking, paresthesias, tremor, ataxia and hyperreflexia.[20] Convulsions caused by cyclodienes may recur over periods of several days and are also characteristic of acute organochlorine poisoning. Agents such as DDT and methoxychlor tend

### HIGHLIGHTS

Only dicofol, endosulfan, lindane still registered for use in U.S.

Absorbed from the gut, by the lungs and across skin

Fat and fat solvents enhance absorption

Most are dechlorinated, oxidized, then conjugated

Chief toxic action is on CNS

### SIGNS & SYMPTOMS

Sensory disturbances: hyperesthesia & paresthesias of face & extremities

Possible headache, dizziness, nausea, vomiting, tremor, confusion

Cyclodienes & toxaphene poisoning may result in seizures (including delayed ones) without other symptoms

Severe poisonings: convulsions, respiratory depression, coma

### TREATMENT

Manage convulsions

Administer oxygen

Decontaminate skin

Consider GI decontamination

Monitor cardiac, pulmonary status

### CONTRAINDICATED

Epinephrine, other adrenergic amines, atropine in most cases

Animal or vegetable oils by mouth

## COMMERCIAL PRODUCTS

aldrin*

BHC* (HCH, hexachlor, hexachloran)

chlordane* (multiple trade names)

chlorobenzilate*

DDT* (multiple trade names)

dicofol (multiple trade names)

dieldrin*

dienochlor (Pentac)*

endosulfan (Thionex)

endrin*

heptachlor*

hexacholorobenzene*

lindane (gamma BHC or HCH)*

methoxychlor (Marlate)*

mirex*

toxaphene*

**All U.S. registrations are suspended.*

to cause the less severe effects, while the cyclodienes, mirex and lindane are associated with the more severe seizures and fatalities.[15] Convulsions may cause death by interfering with pulmonary gas exchange and by generating severe metabolic acidosis.

High tissue concentrations of organochlorines increase myocardial irritability, predisposing to cardiac arrhythmia. When tissue organochlorine concentrations drop below threshold levels, recovery from the poisoning occurs. Organochlorines are not cholinesterase inhibitors.

High tissue levels of some organochlorines (notably DDT, DDE and cyclodienes) have been shown to induce hepatic microsomal drug-metabolizing enzymes.[21] This tends to accelerate excretion of the pesticides themselves but may also stimulate biotransformation of endogenous steroid hormones and exogenous therapeutic drugs, occasionally necessitating reevaluation of required dosages of therapeutic drugs in persons intensively exposed to organochlorines. Human absorption of organochlorine sufficient to cause enzyme induction is likely to occur only as a result of prolonged, intensive exposure.

Ingestion of hexachlorobenzene-treated wheat has been associated with human dermal toxicity diagnosed as porphyria cutanea tarda. The skin forms blisters and becomes very sensitive to sunlight. Subsequent poor healing results in scarring and contracture formation.[22] Unlike other organochlorine compounds, no cases of convulsions caused by the fungicide hexachlorobenzene have been reported in the medical literature. Lindane and chlordane have been infrequently associated with hematological disorders, including aplastic anemia and megaloblastic anemia.[23,24,25]

Evidence has emerged that the organochlorines interact with endocrine receptors, particularly estrogen and androgen receptors. *In vitro* studies and animal experimentation suggest that organochlorines may alter the function of the endocrine system by these interactions.[26,27] In addition, the potential for carcinogenicity has resulted in regulatory action to limit use or remove registration for multiple organochlorines. An extensive literature has accumulated relevant to neurodevelopmental and neurologic effects of chronic low-level exposure to organochlorines.[28,29,30,31,32,33,34] These chronic health implications on the endocrine system and nervous system, and carcinogenic potential are discussed in **Chapter 21, *Chronic Effects*.**

## Signs and Symptoms of Poisoning

Early manifestations of poisoning by some organochlorine pesticides, particularly DDT, are often sensory disturbances: hyperesthesia and paresthesias of the face and extremities. Headache, dizziness, nausea, vomiting, incoordination, tremor and mental confusion are also reported. More severe poisoning results in myoclonic jerking movements, often followed by generalized tonic-clonic convulsions. Coma and respiratory depression may follow the seizures.

Poisoning by the cyclodienes and toxaphene is more likely to begin with the sudden onset of convulsions, often not preceded by the premonitory manifestations mentioned above. Seizures caused by cyclodienes may appear as long as 48 hours after exposure and then may recur periodically over several days following the initial episode. Since lindane and toxaphene are more rapidly biotransformed in the body and excreted, they are less likely than dieldrin, aldrin and chlordane to cause delayed or recurrent seizures.

There have been reports of mixed poisonings, where anticholinesterase agents such as organophosphates and anticholinesterase carbamates have been mixed with organochlorines. In such cases the cholinergic symptoms may be prominent on presentation, but aggressive treatment of the cholinergic findings leave the subject with the symptoms of the organochlorine poisoning, which need additional treatment.[35,36]

Medical providers should be alert to the possibility of such mixed poisonings in the diagnosis and management of pesticide poisonings.

## Confirmation of Poisoning

Organochlorine pesticides and/or their metabolites can sometimes be identified in blood by gas-liquid chromatographic examination of samples taken within a few days of significant pesticide absorption. Such tests are performed by a limited number of government, university and private laboratories, which can usually be contacted through poison control centers or health departments. Some organochlorine pesticides or their metabolic products (notably DDT, dieldrin, mirex, heptachlor epoxide and chlordecone) persist in tissues and blood for weeks or months after absorption, but others are likely to be excreted in a few days, limiting the likelihood of detection. Blood levels tend to correlate more with acute toxicity, while levels found in adipose tissue and breast milk usually reflect more long-term and historic exposure.[37]

Chromatographic methods make possible detection of most organochlorines at concentrations much lower than those associated with symptoms of toxicity. Therefore, a positive finding in a blood sample does not, of itself, justify a diagnosis of acute poisoning. Current general population tissue concentration levels for many of the organochlorines are available from the Centers for Disease Control and Prevention's Biomonitoring Program and may be helpful in interpreting findings.[38]

Lindane tissue concentration reports appear in the literature more frequently than other compounds. The time of the blood sampling in relation to exposure time must be taken into account when interpreting blood levels. In one study, lindane levels were measured at 10.3 ng/mL in healthy volunteers 3 days after application to the skin.[39] In a study with childhood dermal absorption in children with scabies and a non-affected control group, lindane peaked at 28 ng/mL 6 hours after application in the affected group and at 24 ng/mL in the control group. At 48 hours, levels were 6 ng/mL and 5 ng/mL, respectively. Findings from this study also provide evidence for increased absorption across abraded skin.[17] A child with severely abraded skin was treated for scabies and developed seizures. Three days after exposure, his lindane level was 54 ng/mL.[4] Most reports of acute toxicity from lindane involve blood levels of 130 ng/mL or greater, with the most severe and fatal cases involving levels exceeding 500 ng/mL.[2]

DDT, DDE and a few other organochlorines are still found at very low levels in blood samples from the general U.S. population, presumably due to past and/or current low-level contamination of food by these environmentally persistent pesticides.

Overall, blood organochlorine levels have the most readily available information for understanding the acute clinical implications of exposures. Measurements of urinary metabolites of some organochlorine pesticides can be useful in monitoring occupational exposures; however, the analytical methods are complex and are not likely to detect amounts of metabolites generated by minimal exposures.

## Treatment of Organochlorine Toxicosis

1. Observe persons with suspected very high exposure to organochlorine pesticides by any route for sensory disturbances, incoordination, speech slurring, mental aberrations and involuntary motor activity that would warn of imminent convulsions.

2. If convulsions occur, place the victim in the left lateral decubitus position with the head down. Move away furniture or other solid objects that may be a source of injury. If jaw movements are violent, place padded tongue blades between the teeth to protect the tongue. Whenever possible, remove dentures and other removable dental work. Aspirate oral and pharyngeal secretion and, when possible, insert an oropharyngeal airway to maintain an open passage unobstructed by the tongue. Minimize noise and any manipulation of the patient that may trigger seizure activity.

3. Administer oxygen by mask. Maintain pulmonary gas exchange by mechanically assisted ventilation whenever respiration is depressed.

4. Control convulsions. Seizures in patients caused by organochlorine toxicity are likely to be prolonged and difficult to control. Status epilepticus is common. For this reason, patients with seizures that do not respond immediately to anticonvulsants should be transferred as soon as possible to a trauma center and will generally require intensive care admission until seizures are controlled and neurologic status is improved. Initial therapy with benzodiazepines should be instituted.

### Dosage of Diazepam

- ***Adults: 5-10 mg IV and repeat every 5-10 minutes to maximum of 30 mg.***
- ***Children: 0.2 to 0.5 mg/kg every 5 minutes to maximum of 10 mg in children over 5 years, and a maximum of 5 mg in children under 5 years.***

Although lorazepam is widely accepted as a treatment of choice for status epilepticus, there are no reports of its use for organochlorine intoxication. Some cases have required aggressive seizure management including the addition of phenobarbital and the induction of pentobarbital coma.

5. Decontaminate skin thoroughly, as outlined in **Chapter 3, *General Principles*.**

6. Consider gastric decontamination procedures as outlined in **Chapter 3** if organochlorine has been ingested in a quantity sufficient to cause poisoning and the patient presents within an hour. If the patient presents more than an hour after ingestion, activated charcoal may still be beneficial. If the victim is convulsing, it is almost always necessary first to control seizures before attempting gastric decontamination. Activated charcoal administration has been advocated in such poisonings, but there is little human or experimental evidence to support this modality.

7. Particularly in poisonings by large doses of organochlorine, monitor pulmonary ventilation carefully to forestall respiratory failure. Assist pulmonary ventilation mechanically with oxygen whenever respiration is depressed. Since these compounds are often formulated in a hydrocarbon vehicle, hydrocarbon aspiration may occur with ingestion of these agents. The hydrocarbon aspiration should be managed in accordance with accepted medical practice as a case of acute respiratory distress syndrome and will usually require intensive care management.

8. Monitor cardiac status of severely poisoned patients by continuous ECG recording to detect arrhythmia.

9. Do not give epinephrine, other adrenergic amines or atropine unless absolutely necessary. The enhanced myocardial irritability induced by chlorinated hydrocarbons predisposes to ventricular fibrillation.

10. Do not give animal or vegetable oils or fats by mouth. They enhance gastrointestinal absorption of the lipophilic organochlorines.

11. Control seizures and myoclonic movements that sometimes persist for several days following acute poisoning by the more slowly excreted organochlorines. Phenobarbital orally is likely to be effective. Dosage should be based on manifestations in the individual case and on information contained in the package insert.

12. Use cholestryamine resin to accelerate the biliary-fecal excretion of the more slowly eliminated organochlorine compounds.[40]

### Dosage of Cholestryamine Resin

- ***Adults: 4-gram doses, 4 times a day, before meals and at bedtime.***
- ***Children: 240 mg/kg/24 hours, divided, every 8 hours.***

The dose may be mixed with a pulpy fruit or liquid. It should never be given in its dry form and must always be administered with water, other liquids or a pulpy fruit. Prolonged treatment (several weeks or months) may be necessary.

13. During convalescence, enhance carbohydrate, protein and vitamin intake by diet or parenteral therapy.

## References

1. Unintentional topical lindane ingestions--United States, 1998-2003. *MMWR Morb Mortal Wkly Rep.* Jun 3 2005;54(21):533-535.
2. Aks SE, Krantz A, Hryhrczuk DO, Wagner S, Mock J. Acute accidental lindane ingestion in toddlers. *Ann Emerg Med.* Nov 1995;26(5):647-651.
3. Fischer TF. Lindane toxicity in a 24-year-old woman. *Ann Emerg Med.* Nov 1994;24(5):972-974.
4. Friedman SJ. Lindane neurotoxic reaction in nonbullous congenital ichthyosiform erythroderma. *Arch Dermatol.* Aug 1987;123(8):1056-1058.
5. Solomon BA, Haut SR, Carr EM, Shalita AR. Neurotoxic reaction to lindane in an HIV-seropositive patient. An old medication's new problem. *J Fam Pract.* Mar 1995;40(3):291-296.
6. Solomon LM, Fahrner L, West DP. Gamma benzene hexachloride toxicity: a review. *Arch Dermatol.* Mar 1977;113(3):353-357.
7. Tenenbein M. Seizures after lindane therapy. *J Am Geriatr Soc.* Apr 1991;39(4):394-395.
8. Brandt VA, Moon S, Ehlers J, Methner MM, Struttmann T. Exposure to endosulfan in farmers: two case studies. *Am J Ind Med.* Jun 2001;39(6):643-649.
9. Eyer F, Felgenhauer N, Jetzinger E, Pfab R, Zilker TR. Acute endosulfan poisoning with cerebral edema and cardiac failure. *J Toxicol Clin Toxicol.* 2004;42(6):927-932.
10. Kucuker H, Sahin O, Yavuz Y, Yurumez Y. Fatal Acute Endosulfan Toxicity: A Case Report. *Basic Clin Pharmacol Toxicol.* 2008;104:49-51.
11. Oktay C, Goksu E, Bozdemir N, Soyuncu S. Unintentional toxicity due to endosulfan: a case report of two patients and characteristics of endosulfan toxicity. *Vet Hum Toxicol.* Dec 2003;45(6):318-320.
12. Parbhu B, Rodgers G, Sullivan JE. Death in a toddler following endosulfan ingestion. *Clin Toxicol (Phila).* Nov 2009;47(9):899-901.
13. Roberts DM, Dissanayake W, Rezvi Sheriff MH, Eddleston M. Refractory status epilepticus following self-poisoning with the organochlorine pesticide endosulfan. *J Clin Neurosci.* Sep 2004;11(7):760-762.
14. Yavuz Y, Yurumez Y, Kucuker H, Ela Y, Yuksel S. Two cases of acute endosulfan toxicity. *Clin Toxicol (Phila).* Jun-Aug 2007;45(5):530-532.
15. Echobichon DJ. Toxic effects of pesticides. In: Klaassen CD, ed. *Casarett & Doull's Toxicology: The Basic Science of Poisons.* 5th ed. New York: McGraw-Hill; 1996:649-655.
16. Feldmann RJ, Maibach HI. Percutaneous penetration of some pesticides and herbicides in man. *Toxicol Appl Pharmacol.* Apr 1974;28(1):126-132.
17. Ginsburg CM, Lowry W, Reisch JS. Absorption of lindane (gamma benzene hexachloride) in infants and children. *J Pediatr.* Dec 1977;91(6):998-1000.
18. Rogan WJ. Pollutants in breast milk. *Arch Pediatr Adolesc Med.* Sep 1996;150(9):981-990.
19. Stevens MF, Ebell GF, Psaila-Savona P. Organochlorine pesticides in Western Australian nursing mothers. *Med J Aust.* Feb 15 1993;158(4):238-241.
20. Joy RM. The effects of neurotoxicants on kindling and kindled seizures. *Fundam Appl Toxicol.* Feb 1985;5(1):41-65.
21. Hunter J, Maxwell JD, Stewart DA, Williams R, Robinson J, Richardson A. Increased hepatic microsomal enzyme activity from occupational exposure to certain organochlorine pesticides. *Nature.* Jun 16 1972;237(5355):399-401.
22. Booth NH, McDowell JR. Toxicity of hexachlorobenzene and associated residues in edible animal tissues. *J Am Vet Med Assoc.* Mar 15 1975;166(6):591-595.

23. Furie B, Trubowitz S. Insecticides and blood dyscrasias. Chlordane exposure and self-limited refractory megaloblastic anemia. *JAMA.* Apr 19 1976;235(16):1720-1722.

24. Rauch AE, Kowalsky SF, Lesar TS, Sauerbier GA, Burkart PT, Scharfman WB. Lindane (Kwell)-induced aplastic anemia. *Arch Intern Med.* Nov 1990;150(11):2393-2395.

25. Rugman FP, Cosstick R. Aplastic anaemia associated with organochlorine pesticide: case reports and review of evidence. *J Clin Pathol.* Feb 1990;43(2):98-101.

26. Fry DM. Reproductive effects in birds exposed to pesticides and industrial chemicals. *Environ Health Perspect.* Oct 1995;103 Suppl 7:165-171.

27. Vonier PM, Crain DA, McLachlan JA, Guillette LJ, Jr., Arnold SF. Interaction of environmental chemicals with the estrogen and progesterone receptors from the oviduct of the American alligator. *Environ Health Perspect.* Dec 1996;104(12):1318-1322.

28. Dick FD. Parkinson's disease and pesticide exposures. *Br Med Bull.* 2006;79-80:219-231.

29. Jurewicz J, Hanke W. Prenatal and childhood exposure to pesticides and neurobehavioral development: Review of epidemiological studies. *Int J Occup Med Environ Health.* 2008;21(2):121-132.

30. Kamel F, Engel LS, Gladen BC, Hoppin JA, Alavanja MC, Sandler DP. Neurologic symptoms in licensed pesticide applicators in the Agricultural Health Study. *Hum Exp Toxicol.* Mar 2007;26(3):243-250.

31. Kanthasamy AG, Kitazawa M, Kanthasamy A, Anantharam V. Dieldrin-induced neurotoxicity: relevance to Parkinson's disease pathogenesis. *Neurotoxicology.* Aug 2005;26(4):701-719.

32. Ribas-Fito N, Torrent M, Carrizo D, Julvez J, Grimalt JO, Sunyer J. Exposure to hexachlorobenzene during pregnancy and children's social behavior at 4 years of age. *Environ Health Perspect.* Mar 2007;115(3):447-450.

33. Rosas LG, Eskenazi B. Pesticides and child neurodevelopment. *Curr Opin Pediatr.* Apr 2008;20(2):191-197.

34. Sagiv SK, Nugent JK, Brazelton TB, et al. Prenatal organochlorine exposure and measures of behavior in infancy using the Neonatal Behavioral Assessment Scale (NBAS). *Environ Health Perspect.* May 2008;116(5):666-673.

35. Cable GG, Doherty S. Acute carbamate and organochlorine toxicity causing convulsions in an agricultural pilot: a case report. *Aviat Space Environ Med.* Jan 1999;70(1):68-72.

36. Thunga G, Sam KG, Khera K, Xavier V, Verma M. Profile of acute mixed organophosphorus poisoning. *Am J Emerg Med.* Jun 2009;27(5):628 e621-623.

37. Frank R, Braun HE. Organochlorine residues in bird species collected dead in Ontario 1972-1988. *Bull Environ Contam Toxicol.* Jun 1990;44(6):932-939.

38. National Report on Human Exposure to Environmental Chemicals. Centers for Disease Control and Prevention. http://www.cdc.gov/exposurereport/. Accessed on 1/2/11.

39. Hosler J, Tschanz C, Hignite CE, Azarnoff DL. Topical application of lindane cream (Kwell) and antipyrine metabolism. *J Invest Dermatol.* Jan 1980;74(1):51-53.

40. Cohn WJ, Boylan JJ, Blanke RV, Fariss MW, Howell JR, Guzelian PS. Treatment of chlordecone (Kepone) toxicity with cholestyramine. Results of a controlled clinical trial. *N Engl J Med.* Feb 2 1978;298(5):243-248.

## HIGHLIGHTS

Derived from living systems

*Bacillus thuringiensis* is the most important live agent

Generally of low-order toxicity

Poison control center advice can help avoid potentially harmful treatment

## SIGNS & SYMPTOMS

Highly variable based on specific agents

Several cause GI irritation

Nicotine may have serious CNS effects

Nicotine and sabadilla may have cardiovascular effects

## TREATMENT

Specific to the agent

Skin, eye, GI decontamination may be indicated

Nicotine and sabadilla may require aggressive support

***Avermectin***
## COMMERCIAL PRODUCTS

avermectin B1 (Abamectin)

emamectin benzoate (salt of avermectin)

ivermectin

CHAPTER 8

# Biologicals and Insecticides of Biological Origin

This chapter concerns several widely used insecticidal products of natural origin, and also certain agents usually identified as biological control agents. This latter group includes many living control agents, though only the bacterial agent *Bacillus thuringiensis* will be discussed in detail, as it is one of the most widely used. Other agents, such as parasitic wasps and insects, are so host specific they pose little or no risk to man.

Many of the pesticides in this chapter, with the notable exception of nicotine, are relatively less toxic to mammals than to insects. Consequently, there may be no findings of toxicity following ingestion of these compounds. While clinicians should always consider calling their regional poison control center (1-800-222-1222) for advice on any poisoning, it may be of particular value in the case of some of these biological pesticides, where no treatment is warranted and poison control center advice can help avoid potentially harmful treatments.

Agents are presented in alphabetical order.

## AVERMECTIN

### Source and Products

Avermectin and related products are synthetically derived from the toxin of the soil bacterium *Streptomyces avermitilis*. They are used for control of mites, fire ants (ant bait stations) and other insects. **Ivermectin** is used as an antihelminth and a miticide.

### Toxicology and Signs and Symptoms of Poisoning

**Avermectins** work by stimulating the gamma amino butyric acid (GABA) receptor, thereby inhibiting nerve conduction to nerves and muscles. The result is insect paralysis and death within a few days. Mammalian GABA receptors reportedly have a much lower affinity for avermectins than insect GABA receptors.[1] Reports of acute toxicity are uncommon in the medical literature. Clinical manifestations appear to most prominently involve the nervous, GI and respiratory systems. Patients may initially present with nausea, vomiting, salivation, diarrhea and dizziness. More severe manifestations may include aspiration pneumonia, respiratory failure, hypotension and coma.[1,2,3] Rhabdomyolysis has also been reported.[2] One case study of 19 patients demonstrated a dose/response relationship, with the most severe toxicity occurring in patients who ingested in the range of 67 mg/kg to 227 mg/kg. One exception was a patient who ingested 15 mg/kg and had severe toxic symptoms. Most patients who ingested less than 40 mg/kg exhibited either mild or no toxicity.[2]

### Treatment

1. Provide supportive treatment as there is no antidotal therapy.

2. Remove skin contamination with soap and water. Remove eye contamination by flushing the eyes with clean water or saline.

3. If ingested, consider gastrointestinal decontamination as outlined in **Chapter 3, *General Principles.***

## AZADIRACHTIN

### Source and Products

This compound is a biologically obtained insecticide derived from the neem tree *(Azadirachta indica).* It is an insect growth regulator that interferes with the molting hormone ecdysone.

### Toxicology and Signs and Symptoms of Poisoning

**Azadirachtin** causes severe dermal and gastrointestinal irritation. Central nervous system stimulation and depression have been seen. This agent is primarily used and manufactured in India, so little use or exposures are expected in the United States.

### Treatment

1. Wash skin thoroughly with soap and water if skin has been exposed.

2. Do not use gastric emptying or catharsis because of severe gastrointestinal irritation. Do not use activated charcoal when there is severe GI irritation because of potential need for gastrointestinal endoscopy.

## *BACILLUS THURINGIENSIS*

### Source and Products

Several strains of *Bacillus thuringiensis* are pathogenic to some insects. The bacterial organisms are cultured and then harvested in spore form for use as insecticide. Production methods vary widely. Proteinaceous and nucleotide-like toxins generated by the vegetative forms (which infect insects) are responsible for the insecticidal effect. The spores are formulated as wettable powders, flowable concentrates and granules for application to field crops and for control of mosquitoes and black flies.

### Toxicology and Signs and Symptoms of Poisoning

The varieties of ***Bacillus thuringiensis*** used commercially survive when injected into mice, and at least one of the purified insecticidal toxins is toxic to mice. Infections of humans have been extremely rare. A single case report of ingestion by volunteers of *Bacillus thuringiensis var. galleriae* resulted in fever and gastrointestinal symptoms. However, this specific agent is not registered as a pesticide. *B. thuringiensis* products are exempt from tolerances on raw agricultural commodities in the United States. Neither irritative nor sensitizing effects have been reported in workers preparing and applying commercial products. A single case of corneal ulcer caused by a splash of *B. thuringiensis* suspension into the eye has been reported.[4]

*Bacillus thuringiensis*
**COMMERCIAL PRODUCTS**

Several varieties are available:

*kurstaki*

*israelensis*

*aizawai*

*morrisoni*

*tenebrionis*

## Treatment

1. Remove skin contamination with soap and water. Remove eye contamination by copious flushing of the eyes with clean water or saline. If irritation persists, or if there is any indication of infection, refer patient for further treatment.

2. Observe a patient who has ingested a *B. thuringiensis* product for manifestations of bacterial gastroenteritis: abdominal cramps, vomiting and diarrhea. The illness is likely to be self limited if it occurs at all. The patient should be treated symptomatically and fluid support provided as appropriate.

# EUGENOL

## Source and Products

This compound is derived from clove oil, which is found in the dried flower bud of *Eugenia caryophyllata*. It is used as an insect attractant. It is also used in numerous dental products, which accounts for some of the reports of toxicity.[5]

## Toxicology

**Eugenol** is similar in its clinical effects to phenol in terms of its caustic properties. Although it works as an anesthetic, in large doses, it can cause burns to epithelial surfaces.[6] Sloughing of mucous membranes occurred as an allergic reaction to a small dose applied topically in the mouth.[5] Gastric mucosal lesions have been reported in animals, but no lesions were seen on endoscopy after clove oil ingestion.[7] Large doses may result in coma, metabolic acidosis, seizures, liver dysfunction and disseminated intravascular coagulation.[8,9,10] Large ingestions can be particularly toxic to children. The mechanism of liver toxicity appears to be similar to that of acetaminophen poisoning, in which eugenol is metabolized by the cytochrome-p450 system to produce a toxicologically active quinone metabolite and a resultant glutathione depletion.[11,12]

## Treatment

1. Provide supportive treatment as necessary as there is no antidote.

2. Consider gastrointestinal decontamination as outlined in **Chapter 3,** ***General Principles,*** for ingestions**.** If mucosal burns are present, consider endoscopy to look for other ulcerations.

There is one report of the use of n-acetylcysteine, using the same dose prescribed for acetaminophen ingestion. It is of note that the patient's hepatic transaminase levels began to decrease sharply after initiating NAC therapy.[9] Without further study, it is difficult to recommend this as routine treatment.

# GIBBERELLIC ACID (GIBBERELLIN, $GA_3$)

## Source and Products

Gibberellic acid is not a pesticide, but it is commonly used in agricultural production as a growth-promoting agent. It is a metabolic product of a cultured fungus, formulated in tablets, granules and liquid concentrates for application to soil beneath growing plants and trees.

## Toxicology

Experimental animals tolerate large oral doses of **gibberellic acid** without apparent adverse effect. No human poisonings have been reported. Sensitization has not been reported, and irritant effects are not remarkable.

## Treatment

1. Provide supportive treatment for any toxic effects in humans, as there is no known antidote.
2. Wash contamination from skin with soap and water. Flush contamination from eyes with clean water or saline. If irritation occurs, refer patient for further medical treatment.
3. For significant ingestion, consider gastrointestinal decontamination as outlined in **Chapter 3, *General Principles***, although that may not be necessary. Poison control centers may be helpful to guide whether any therapy is indicated based on the ingestion.

## NICOTINE

## Source and Products

Nicotine is an alkaloid contained in the leaves of many species of plants, but is usually obtained commercially from the tobacco plant *(Nicotiana tabacum)*. A 95% solution of the free alkaloid in organic solvent has been marketed in the past as a greenhouse fumigant. Another product used for the same purpose is a 40% aqueous solution of nicotine sulfate. Significant volatilization of nicotine occurs from both products. Commercial nicotine insecticides have long been known as Black Leaf 40. This formulation was discontinued in 1992, although old preparations of nicotine insecticides may still be found on occasion.[13] The last remaining registered nicotine product will be discontinued as of 2013 by request of the registrant.[14] Today, most nicotine poisonings are the result of ingestion of tobacco products and ingestion and/or incorrect use of nicotine replacement products such as nicotine gum and transdermal patches.[15,16] However, ingestions from old pesticide products may still occur.[17]

## Toxicology

**Nicotine** alkaloid is efficiently absorbed by the gut, lung and skin. Extensive biotransformation occurs in the liver, with 70%-75% occurring as a first-pass effect.[18] Both the liver and kidney participate in the formation and excretion of multiple end-products, which are excreted within a few hours. Estimates of the half-life of nicotine range from about 1 hour in smokers to as much as 2 hours in non-smokers.[19,20]

Toxic action is complex. At low doses, autonomic ganglia are stimulated. At higher doses, blockade of autonomic ganglia and skeletal muscle neuromuscular junctions and direct effects on the central nervous system occur. Paralysis and vascular collapse are prominent features of acute poisoning, but death is often due to respiratory paralysis, which may ensue promptly after the first symptoms of poisoning.[14,17] Nicotine is not an inhibitor of cholinesterase enzyme.

*Nicotine*
### HIGHLIGHTS

Efficiently absorbed by gut, lung, skin

CNS impacts

Respiratory impacts

### SIGNS & SYMPTOMS

Salivation, sweating, dizziness, nausea, vomiting, diarrhea

Possible burning in mouth/ throat, agitation, confusion, headache, abdominal pain

Cardiovascular symptoms at high dosage/exposure

### TREATMENT

Decontaminate eyes, skin

Maintain airway

Limit GI absorption

IV atropine if indicated

Control seizures

## Signs and Symptoms of Poisoning

Early and prominent symptoms of poisoning include salivation, sweating, dizziness, nausea, vomiting and diarrhea. Burning sensations in the mouth and throat, agitation, confusion, headache and abdominal pain are reported. Cardiovascular symptoms are prominent with high dosages of exposure. In severe poisoning, cardiovascular collapse is manifested by bradycardia or other arrhythmias and hypotensive shock.[13,17,21,22,23] Patients may have dyspnea, then respiratory failure and unconsciousness.[13,21,22,23] In some cases, hypertension and tachycardia may precede hypotension and bradycardia.[21,22] Seizures may also occur.[13,17,22] In one case of ingestion of a large dose of nicotine alkaloid pesticide, the patient developed asystole within 2 minutes. He later developed seizures and refractory hypotension.[13] A child developed seizures, respiratory depression and hypoxic encephalopathy after ingesting a nicotine-containing pesticide.[17]

If symptoms of poisoning appear during exposure to an airborne nicotine insecticide, the person should be removed from the contaminated environment immediately and any skin areas that may be contaminated should be washed. The victim should then be transported to the nearest treatment facility. Although mild poisoning may resolve without treatment, it is often difficult to predict the ultimate severity of poisoning at the onset.

## Confirmation of Poisoning

Urine, plasma and salivary content of the metabolite cotinine can be used to confirm absorption of nicotine. However, these studies generally need to be sent to a reference lab and are not clinically useful in acute toxicity. Treatment should be based on clinical presentation and findings. If necessary, lab confirmation can be done at a later date.

## Treatment of Nicotine Toxicosis

1. If liquid or aerosol spray has come in contact with skin, wash the area thoroughly with soap and water. If eyes have been contaminated, flush them thoroughly with clean water or saline. If irritation persists, refer patient for specialized medical treatment.

2. If there is any indication of loss of respiratory drive, maintain pulmonary ventilation by mechanical means. Toxic effects of nicotine other than respiratory depression are usually survivable. Maintaining adequate gas exchange is therefore of paramount importance.

3. If a nicotine-containing product has been ingested recently, take immediate steps to limit gastrointestinal absorption. If the patient is fully alert, immediately administer activated charcoal orally as outlined in the **Chapter 3**, ***General Principles***. This is probably the best initial step in management. Since most patients who ingest nicotine have significant vomiting, activated charcoal is not always necessary. Do not administer cathartics or syrup of ipecac.

4. Manage patients with severe poisoning in the intensive care environment, preferably with toxicology consultation if available. Monitor cardiac status by electrocardiography and measure blood pressure frequently. Cardiopulmonary resuscitation may be necessary. Vascular collapse may require administration of vasopressors. Infusions of electrolyte solutions, plasma and/or blood may also be required to combat shock.

5. Treat excessive parasympathetic stimulation, such as severe hypersecretion (especially salivation and diarrhea) or bradycardia, with intravenous atropine sulfate. There is no specific antidote for nicotine poisoning.

**Dosage of Atropine Sulfate**

- ***Adults and children over 12 years:*** **0.5-1.0 mg slow IV, repeated every 5 minutes if necessary.**
- ***Children under 12 years:*** **0.02 mg/kg body weight, slow IV, repeated every 5 minutes if necessary. (Minimum dose of 0.1 mg.)**

6. Control seizures as outlined in **Chapter 3,** ***General Principles.***

## ROTENONE

### Source and Products

Although this natural substance is present in a number of plants, the source of most rotenone used in the United States is the dried derris root imported from Central and South America. It is formulated as dusts, powders and sprays (less than 5% active ingredient) for use in gardens and on food crops. Many products contain piperonyl butoxide as a synergist, and other pesticides are included in some commercial products. Rotenone degrades rapidly in the environment. Emulsions of rotenone are applied to lakes and ponds to kill fish.

### Toxicology and Manifestations of Poisoning

Although **rotenone** is toxic to the nervous systems of insects, fish and birds, commercial rotenone products have presented little hazard to man over many decades. Neither fatalities nor systemic poisonings in humans have been reported in relation to ordinary use. However, there is one report of a fatality in a child who ingested a product called Gallocide, which contains rotenone and etheral oils, including clove oil (eugenol). She developed a gradual loss of consciousness over 2 hours and died of respiratory arrest.[24]

There have been some reports of toxic symptoms following occupational exposure. Eye irritation is the most common. In addition, numbness of oral mucous membranes has been reported in workers who got dust from the powdered derris root in their mouths. Dermatitis, respiratory tract irritation, headaches and peripheral neuropathy have also been reported.[25]

When rotenone has been injected into animals, tremors, vomiting, incoordination, seizures and respiratory arrest have been observed. These effects have not been reported in occupationally exposed humans.

### Treatment of Rotenone Toxicosis

1. Provide supportive treatment, as there is no specific antidote.

2. Remove skin contamination by washing with soap and water. Remove eye contamination by flushing the eye thoroughly with clean water or saline. Wash out any dust in the mouth. If irritation persists, refer for further medical treatment.

3. If a large amount of a rotenone-containing product has been swallowed and retained, consider gastric decontamination as outlined in **Chapter 3, *General Principles.***

## SABADILLA (VERATRUM ALKALOID)

### Source and Products

Sabadilla consists of the powdered ripe seeds of a South American lily. Its only remaining registered use in the United States is for agricultural application to citrus fruits, avocados and mangos.[26] Insecticidal alkaloids are those of the *Veratrum* plant. The concentration of alkaloids in commercial sabadilla is usually less than 0.5%. Little or no sabadilla is used in the United States today, but it is probably used in other countries. Although poisoning by medicinal *Veratrum* preparations may have occurred in the remote past, systemic poisoning by sabadilla preparations used as insecticides has been very rare. Much of the toxic encounters with *Veratrum* alkaloid occur from the inadvertent ingestion of the *Veratrum* plant or a related plant from the genus *Zigadenus*.[27,28]

### Toxicology

**Sabadilla** dust is very irritating to the upper respiratory tract, causing sneezing, and is also irritating to the skin. *Veratrum* alkaloids are apparently absorbed across the skin and gut, and probably by the lung as well. *Veratrum* alkaloids have a digitalis-like action on the heart muscles (impaired conduction and arrhythmia).[27,29,30]

### Signs and Symptoms of Poisoning

The prominent symptoms of *Veratrum* alkaloid poisoning are severe nausea and vomiting, increased salivation and mental status changes. Cardiovascular effects may be severe, including hypotension and bradycardia. Other arrhythmias or A-V block may occur in large ingestions. These symptoms often resolve within 24 hours.[27,29,30]

### Treatment of Sabadilla Toxicosis

1. Wash contaminated skin thoroughly with soap and water. Flush eyes, if affected, with copious amounts of clean water or saline. If skin or eye irritation persists, refer patient for further medical treatment.

2. Consider gastric decontamination as outlined in **Chapter 3, *General Principles*.**

3. If there is a suspicion that significant amounts of sabadilla alkaloids have been absorbed, monitor cardiac activity for arrhythmia and conduction defects with an ECG. Place patient with severe toxicity in intensive care. Treat bradycardia with atropine.[29,30]

*Spinosyns*
**COMMERCIAL PRODUCTS**

Spinosad (a mixture of Spinosyn A and Spinosyn D)

**Dosage of Atropine Sulfate**

- ***Adults and children over 12 years: 0.5-1.0 mg slow IV, repeated every 5 minutes, if necessary.***
- ***Children under 12 years: 0.02 mg/kg body weight, slow IV, repeated every 5 minutes, if necessary. (Minimum dose of 0.1 mg.)***

## SPINOSYNS

### Source and Products

Spinosad is a biologically based synthetic pesticide that is used to control a variety of insects including fleas, mites, fire ants, caterpillars, fruit flies and leaf beetle larvae. It has recently been approved to treat head lice in humans.[31] The spinosyns are derived from the rare soil-dwelling actinomycete bacterium called *Saccharopolyspora spinosa.*

### Toxicology and Manifestations of Poisoning

**Spinosad** must be ingested by the target pest to control it. It causes rapid excitation of the insect's nervous system and is relatively fast acting. Spinosyns interfere with nicotinic function and also disrupt GABA function in central nervous system neurons; however, they do not bind to the receptor sites.[32] Spinosad has low mammalian oral toxicity ($LD_{50}$ rat is >3,000 mg/kg).[33] Similar to fipronil, spinosyns have a much higher affinity for insects than for mammals. There have not been reports of human toxicity in the medical literature.

### Treatment

1. Provide supportive treatment should toxic effects occur in humans, as there is no known antidote.
2. Wash contamination from skin with soap and water. Flush contamination from eyes with clean water or saline. If irritation occurs, refer for further medical treatment.
3. For significant ingestion, consider gastrointestinal decontamination as outlined in **Chapter 3, *General Principles***, although that may not be necessary. Poison control centers may be helpful to guide whether any therapy is indicated based on the ingestion.

## STREPTOMYCIN

### Source and Products

Streptomycin sulfate and nitrate are used as pesticides for the control of a variety of commercially important bacterial plant pathogens. Streptomycin is an antibiotic derived from the growth of *Streptomyces griseus*.

## Toxicology

**Streptomycin** shares a toxic profile with the aminoglycoside antibiotics commonly used to treat human diseases. Its major modes of toxicity are nephrotoxicity and ototoxicity. Fortunately, it is poorly absorbed from the gastrointestinal tract, so systemic toxicity is unlikely with ingestion. It may cause some minor nausea and GI upset.

## Treatment

If a large amount has been ingested and 1 hour or less has passed, consider gastric decontamination as outlined in **Chapter 3, *General Principles.***

## References

1. Yen TH, Lin JL. Acute poisoning with emamectin benzoate. *J Toxicol Clin Toxicol.* 2004;42(5):657-661.
2. Chung K, Yang CC, Wu ML, Deng JF, Tsai WJ. Agricultural avermectins: an uncommon but potentially fatal cause of pesticide poisoning. *Ann Emerg Med.* Jul 1999;34(1):51-57.
3. Soyuncu S, Oktay C, Berk Y, Eken C. Abamectin intoxication with coma and hypotension. *Clin Toxicol (Phila).* 2007;45(3):299-300.
4. Samples JR, Buettner H. Corneal ulcer caused by a biologic insecticide *(Bacillus thuringiensis). Am J Ophthalmol.* Feb 1983;95(2):258-260.
5. Barkin ME, Boyd JP, Cohen S. Acute allergic reaction to eugenol. *Oral Surg Oral Med Oral Pathol.* Apr 1984;57(4):441-442.
6. Isaacs G. Permanent local anaesthesia and anhidrosis after clove oil spillage. *Lancet.* Apr 16 1983;1(8329):882.
7. Lane BW, Ellenhorn MJ, Hulbert TV, McCarron M. Clove oil ingestion in an infant. *Hum Exp Toxicol.* Jul 1991;10(4):291-294.
8. Brown SA, Biggerstaff J, Savidge GF. Disseminated intravascular coagulation and hepatocellular necrosis due to clove oil. *Blood Coagul Fibrinolysis.* Oct 1992;3(5):665-668.
9. Eisen JS, Koren G, Juurlink DN, Ng VL. N-acetylcysteine for the treatment of clove oil-induced fulminant hepatic failure. *J Toxicol Clin Toxicol.* 2004;42(1):89-92.
10. Hartnoll G, Moore D, Douek D. Near fatal ingestion of oil of cloves. *Arch Dis Child.* Sep 1993;69(3):392-393.
11. Mizutani T, Satoh K, Nomura H. Hepatotoxicity of eugenol and related compounds in mice depleted of glutathione: structural requirements for toxic potency. *Res Commun Chem Pathol Pharmacol.* Jul 1991;73(1):87-95.
12. Thompson D, Constantin-Teodosiu D, Egestad B, Mickos H, Moldeus P. Formation of glutathione conjugates during oxidation of eugenol by microsomal fractions of rat liver and lung. *Biochem Pharmacol.* May 15 1990;39(10):1587-1595.
13. Lavoie FW, Harris TM. Fatal nicotine ingestion. *J Emerg Med.* May-Jun 1991;9(3):133-136.
14. United States Environmental Protection Agency. Reregistration Eligibility Decision (RED) for nicotine. 2008. http://www.epa.gov/pesticides/reregistration/REDs/nicotine_red.pdf. Accessed December 30, 2012.
15. Pereira CB, Strupp M, Eggert T, Straube A, Brandt T. Nicotine-induced nystagmus: three-dimensional analysis and dependence on head position. *Neurology.* Nov 28 2000;55(10):1563-1566.
16. Wain AA, Martin J. Can transdermal nicotine patch cause acute intoxication in a child? A case report and review of literature. *Ulster Med J.* May 2004;73(1):65-66.

17. Rogers AJ, Denk LD, Wax PM. Catastrophic brain injury after nicotine insecticide ingestion. *J Emerg Med.* Feb 2004;26(2):169-172.

18. Svensson CK. Clinical pharmacokinetics of nicotine. *Clin Pharmacokinet.* Jan 1987;12(1):30-40.

19. Feyerabend C, Ings RM, Russel MA. Nicotine pharmacokinetics and its application to intake from smoking. *Br J Clin Pharmacol.* Feb 1985;19(2):239-247.

20. Kyerematen GA, Damiano MD, Dvorchik BH, Vesell ES. Smoking-induced changes in nicotine disposition: application of a new HPLC assay for nicotine and its metabolites. *Clin Pharmacol Ther.* Dec 1982;32(6):769-780.

21. Malizia E, Andreucci G, Alfani F, Smeriglio M, Nicholai P. Acute intoxication with nicotine alkaloids and cannabinoids in children from ingestion of cigarettes. *Hum Toxicol.* Apr 1983;2(2):315-316.

22. Sanchez P, Ducasse JL, Lapeyre-Mestre M, et al. Nicotine poisoning as a cause of cardiac arrest? *J Toxicol Clin Toxicol.* 1996;34(4):475-476.

23. Woolf A, Burkhart K, Caraccio T, Litovitz T. Self-poisoning among adults using multiple transdermal nicotine patches. *J Toxicol Clin Toxicol.* 1996;34(6):691-698.

24. De Wilde AR, Heyndrickx A, Carton D. A case of fatal rotenone poisoning in a child. *J Forensic Sci.* Oct 1986;31(4):1492-1498.

25. United States Environmental Protection Agency. Reregistration Eligibility Decision (RED) for Rotenone. 2007; http://www.epa.gov/oppsrrd1/REDs/rotenone_red.pdf.

26. United States Environmental Protection Agency. Reregistration Eligibility Decision (RED) Exposure and Risk Assessment on Lower Risk Pesticide Chemicals Sabadilla Alkaloids. 2004; http://www.epa.gov/oppsrrd1/REDs/sabadilla_red.pdf.

27. Dunnigan D, Adelman RD, Beyda DH. A young child with altered mental status. *Clin Pediatr (Phila).* Jan-Feb 2002;41(1):43-45.

28. Heilpern KL. Zigadenus poisoning. *Ann Emerg Med.* Feb 1995;25(2):259-262.

29. Jaffe AM, Gephardt D, Courtemanche L. Poisoning due to ingestion of *Veratrum viride* (false hellebore). *J Emerg Med.* Mar-Apr 1990;8(2):161-167.

30. Quatrehomme G, Bertrand F, Chauvet C, Ollier A. Intoxication from *Veratrum album. Hum Exp Toxicol.* Mar 1993;12(2):111-115.

31. Kirst HA. The spinosyn family of insecticides: realizing the potential of natural products research. *J Antibiot (Tokyo).* Mar;63(3):101-111.

32. Sparks TC, Crouse G.D., Durst, G. Natural products as insecticides: the biology, biochemistry and quantitative structure-- activity relationships of spinosyns and spinodoids. *Pest Manag Sci.* 2001;57:896-905.

33. Stebbins KE, Bond DM, Novilla MN, Reasor MJ. Spinosad insecticide: subchronic and chronic toxicity and lack of carcinogenicity in CD-1 mice. *Toxicol Sci.* Feb 2002;65(2):276-287.

## HIGHLIGHTS

Multiple agents with widely varying toxicity

Agents of concern include borates, fluorides, pyrethroids

Neonicotinoids are a newer class that merits attention due to widespread use and toxicity

## SIGNS & SYMPTOMS

Variable and highly related to the specific agent

Boric acid, fluorides, n-phenylpyrazones and neonicotinoids should be suspected in cases with CNS symptoms

## TREATMENT

Specific to agent

Skin/eye decontamination

Consider GI decontamination based on quantity and time interval factors

Severe CNS symptoms may require intensive care management



CHAPTER 9

# Other Insecticides and Acaricides

This chapter concerns insecticides and acaricides having toxicologic characteristics distinct from the insecticides discussed in previous chapters. It discusses benzyl benzoate, borates, chlordimeform, chlorobenzilate, cyhexatin, fluorides, fipronil (an n-phenylpyrazone insecticide), haloaromatic substituted urea compounds, methoprene, neonicotinoids, propargite and sulfur.

## BENZYL BENZOATE

Incorporated into lotions and ointments, this agent has been used for many years in veterinary and human medicine against mites and lice.

### Toxicology

Apart from occasional cases of skin irritation, adverse effects have been few. The efficiency of skin absorption is not known. Absorbed **benzyl benzoate** is rapidly biotransformed to hippuric acid that is excreted in the urine. Oral toxicity in animals is low, with $LD_{50}$ values in the 2-3 grams/kg range in rats and cats. When given in large doses to laboratory animals, benzyl benzoate causes excitement, incoordination, paralysis of the limbs, convulsions, respiratory paralysis and death.[1] Very few human exposures have been reported to the National Poison Data System.

### Treatment

1. If significant irritant effect appears, discontinue use of product and cleanse skin with soap and water. Treat eye contamination by irrigating exposed eyes with copious amounts of clean water or saline for at least 15 minutes. Remove contact lenses, if present, prior to irrigation. If irritation persists after irrigation, obtain specialized medical treatment in a healthcare facility.

2. If a potentially toxic amount has been swallowed and retained and the patient is seen soon after exposure, consider gastrointestinal decontamination. If seizures occur, control may require treatment with benzodiazepines.

## BORIC ACID AND BORATES

Boric acid and borate products can be formulated as tablets and powder to kill larvae in livestock confinement areas and cockroaches in residences. Rarely, solutions are sprayed as a nonselective herbicide.

### Toxicology

When determining toxicity of **boric acid** from ingestion, it is important to distinguish between acute and chronic exposure. Chronic ingestion is more likely to cause significant toxicity than acute exposure.[2,3] **Borates** are well absorbed by the gut and by abraded or burned skin, but not by intact skin.[4] The kidney efficiently excretes them. The residence half-life in humans averages 13 hours, in a range of 4-28 hours.[2]

## Signs and Symptoms of Poisoning

Generally, boric acid is of lower toxicity when compared to other insecticides that are widely used in the United States. A series of 784 patients has been described with no fatalities and minimal toxicity. Only 12% of these patients had symptoms of toxicity, mostly to the gastrointestinal tract.[2] However, fatal poisonings have been reported.[3,5,6] A large number of poisonings in newborns occurred in the 1950s and 1960s and often resulted in death.[7,8] Historically, many poisonings have resulted from injudicious uses in human medicine aimed at suppressing bacterial growth, such as compresses for burns, powders for diaper rash, and irrigation solutions.[4,9]

Boric acid powders and pellets scattered on the floors of homes can present a hazard to children. Their frequent use for roach control increases access for ingestion. Consequently, cases of suicidal or accidental ingestion continue to be reported in the medical literature.[5,6,10,11,12] One toddler died following a massive ingestion of boric acid powder that had been stored in a bathroom cabinet.[5] An 82-year-old man accidentally ingested 30 mL of boric acid instead of the magnesium sulfate he was supposed to take for a colonoscopy prep.[10] Three cases of apparent suicide attempts in adults have been reported.[11,12,13] Borax dust is moderately irritating to skin. Inhaled dust caused irritation of the respiratory tract among workers in a borax plant. Symptoms included nasal irritation, mucous membrane dryness, cough, shortness of breath and chest tightness.[14,15]

The gastrointestinal tract, renal system, skin, vascular system and brain are the principal organs and tissues affected. Nausea, persistent vomiting, abdominal pain and diarrhea reflect a toxic gastroenteritis.[2,3,9] In severe poisonings, a beefy red skin rash, most often affecting palms, soles, buttocks and scrotum, has been described. It has been characterized as a "boiled lobster appearance." The intense erythema is followed by extensive exfoliation.[3,8,11,16] This may be difficult to distinguish from staphylococcal scalded skin syndrome.[16] Reversible alopecia has been reported following exposure to boric acid and related compounds.[17,18,19]

Headache, agitation, weakness, lethargy, restlessness and tremors may occur, but are less frequent than gastrointestinal effects.[2,10] Seven infants who were exposed to a mixture of borax and honey on their pacifiers developed seizures.[20] Unconsciousness and respiratory depression signify life-threatening brain injury. Cyanosis, weak pulse, hypotension and cold clammy skin indicate shock, which is sometimes the cause of death in borate poisoning.[3,6,9] Hypotension and at times hypertension may occur even in milder cases where victims fully recover.[10,11]

Acute renal failure (oliguria or anuria) may be a consequence of shock, of direct toxic action on renal tubule cells, or possibly of both. It occurs in severe borate poisoning.[3,6,8,16] Metabolic acidosis may be a consequence of the acid itself, seizure activity or metabolic abnormalities.[3] Fever is sometimes present in the absence of infection.

## Confirmation of Poisoning

Borate can be measured in serum by colorimetric methods, as well by high temperature atomic spectrometric methods. Studies of serum levels of boric acid and boron in non-poisoned individuals ranged from 0.0 to 0.2 mg/dL in adults and from 0.0 to 0.125 mg/dL in children.[9,21,22] Levels reported in toxic incidents have varied widely and it is felt that serum levels are of little use in guiding therapy.[2,9,21]



***Boric Acid/Borates***

### COMMERCIAL PRODUCTS

Boric acid, sodium tetraborate decahydrate, sodium polyborates (discontinued 1992)

### HIGHLIGHTS

Chronic ingestion more likely to cause significant toxicity than acute

Absorbed by gut and abraded/burned (not intact) skin

### SIGNS & SYMPTOMS

Nausea, vomiting, abdominal pain, diarrhea

Severe poisonings: erythema ("boiled lobster") and exfoliation

CNS symptoms may be present

### TREATMENT

Skin/eye decontamination

Consider GI contamination

Large or protracted ingestion may require IV fluids and cardiac monitoring

## Treatment

1. Decontaminate the skin with soap and water as outlined in **Chapter 3, *General Principles***. Treat eye contamination by irrigating the exposed eye(s) with copious amounts of clean water or saline for at least 15 minutes. Remove contact lenses, if present, prior to irrigation. If irritation persists after irrigation, send patient for specialized medical treatment in a healthcare facility.

2. In acute poisonings, if a large amount has been ingested and the patient is seen within 1 hour of exposure, gastrointestinal decontamination may be considered as outlined in **Chapter 3**. It is important to keep in mind that vomiting and diarrhea are common, and severe poisoning may be associated with seizures.

3. If massive ingestion of borate (several grams) has occurred or if borate ingestion has extended over several days, administer IV fluids such as D5NS or Lactated Ringers to sustain urinary excretion of borate. Monitor fluid balance and serum electrolytes (including acid base status) regularly. Monitor cardiac status by ECG. Test the urine for protein and cells to detect renal injury, and monitor serum concentration of borate if possible. Metabolic acidosis may be treated with sodium bicarbonate. If shock develops, treat as appropriate. Administer oxygen continuously. If oliguria (less than 25-30 mL urine formed per hour) occurs, intravenous fluids must be slowed or stopped to avoid overloading the circulation. Such patients should usually be referred to a center capable of providing intensive care for critically ill patients.

4. Consider hemodialysis in severe poisonings, if patient fails to respond to conventional therapy. Dialysis has been demonstrated to enhance the clearance of boric acid even in the presence of normal renal function.[2,7,10,12] There is no consensus on its use. Forced diuresis has also been successfully used in early stages of poisoning.[13]

   Peritoneal dialysis was performed historically in borate poisoning and thought to be as effective as, and safer than, exchange transfusion in removing borate.[8,23] Exchange transfusion has been reported to be effective in chronic exposures. No large study of efficacy has been done. Exchange transfusion and peritoneal dialysis are rarely used today in acute poison management.[2]

5. Control seizures as recommended for other agents and as outlined in **Chapter 3**.

## CHLORDIMEFORM

Formulations are emulsifiable concentrates and water-soluble powders. Chlordimeform demonstrates good dermal absorption and can be inhaled. It is an ovicide and acaricide. All registrations in the United States are currently canceled.

## Toxicology

In a reported episode of occupational exposure to **chlordimeform**, several workers developed hematuria. Hemorrhagic cystitis, probably due to chloraniline biodegradation products, was the source of the blood in the urine. Symptoms reported by the affected workers included gross hematuria, dysuria, urinary frequency and urgency, penile discharge, abdominal and back pain, a generalized "hot" sensation, sleepiness, skin rash and desquamation, a sweet taste and anorexia. Symptoms persisted for 2-8 weeks after exposure was terminated.[24] In a single case, methemoglobinemia was reported.[25]

Chlordimeform is not a cholinesterase inhibitor.

## Confirmation of Poisoning

Although methods do exist for measurement of urinary excretion products, these tests are not generally available in the clinical setting.

## Treatment

1. Decontaminate skin thoroughly with soap and water, as outlined in **Chapter 3, *General Principles***. Decontaminate eyes by irrigating exposed eyes with copious amounts of clean water or saline for at least 15 minutes. Remove contact lenses if present prior to irrigation. If irritation persists after irrigation, send patient for specialized medical treatment in a healthcare facility.

2. If chlordimeform has been ingested no more than an hour prior to treatment consider gastrointestinal decontamination as outlined in **Chapter 3**. Patients are at risk for fluid loss and subsequent electrolyte disturbances; young children are especially susceptible. Monitor fluid balance, electrolytes and acid base status closely.

3. Patients exposed should have serial urinalyses for protein and red cells to detect injury to the urinary tract. Resolution of hematuria ordinarily can be expected in 2-8 weeks. Relief from other symptoms usually can be expected earlier.

# CHLOROBENZILATE

Chlorobenzilate is a chlorinated hydrocarbon acaricide, usually formulated as an emulsion or wettable powder for application in orchards. All U.S. registrations have been canceled.

## Toxicology

**Chlorobenzilate** is moderately irritating to the skin and eyes.

Although structurally similar to DDT, chlorobenzilate is much more rapidly excreted following absorption, chiefly in the urine as the benzophenone and benzoic acid derivatives. Based on observation of dosed animals, extreme absorbed doses may cause diarrhea, tachypnea, tremors, ataxia and muscle weakness.[26]

Limited human acute poisoning data are available. A case of toxic encephalopathy in a male following unprotected pesticide application in a field for 14 days at 10 hours per day has been reported. His symptoms included muscle pain, weakness, fever and mental status changes, progressing to a tonic-clonic seizure. He recovered without apparent sequelae within 6 days. Treatment included respiratory support and seizure management.[27]

Chlorobenzilate is not a cholinesterase inhibitor.

*Fluorides*

**COMMERCIAL PRODUCTS**

Cryolite

Kryocide

**HIGHLIGHTS**

Most cases of poisoning today are sources other than insecticides

Highly toxic sodium fluoride and sodium fluosilicate products no longer registered for use

**SIGNS & SYMPTOMS**

Hypocalcemia with possible tetany

Cardiac arrhythmia, shock

Possible CNS impacts

**TREATMENT**

Skin, eye, possible GI decontamination

May require intensive care treatment

Treat hypocalcemia with calcium gluconate or calcium chloride

### Treatment of Chlorobenzilate Poisoning

1. Decontaminate the skin with soap and water as outlined in the **Chapter 3, *General Principles***. Treat eye contamination by irrigating exposed eyes with copious amounts of clean water or saline for at least 15 minutes. Remove contact lenses, if present, prior to irrigation. If irritation persists after irrigation, send patient for specialized medical treatment in a healthcare facility.

2. If a large amount of chlorobenzilate was ingested within a few hours prior to treatment, consider gastrointestinal decontamination as outlined in **Chapter 3**.

3. Treat seizures as outlined in **Chapter 3**.

## CYHEXATIN

All U.S. registrations of this chemical have been canceled.

### Toxicology

**Tricyclohexyl tin hydroxide** is formulated as a 50% wettable powder for control of mites on ornamentals, hops, nut trees and some fruit trees. It is moderately irritating, particularly to the eyes. While information on the systemic toxicity of this specific tin compound is lacking, it should probably be assumed that cyhexatin can be absorbed to some extent across the skin, and that substantial absorbed doses would cause nervous system injury (see organotin compounds on page 154 in **Chapter 16, *Fungicides***).

### Treatment

1. Promptly decontaminate skin by washing with soap and water and decontaminate eyes by irrigating with clean water or saline for at least 15 minutes. Remove contact lenses, if present, prior to irrigation.

2. Manage poisonings by ingestion on the assumption that cyhexatin is toxic, even though rodent $LD_{50}$ values are fairly high and no human poisonings have been reported in the medical literature. Management should be as with other organotin compounds (see page 154 in **Chapter 16, *Fungicides***).

## FLUORIDES

Sodium fluoride is a crystalline mineral once widely used in the United States for control of larvae and crawling insects in homes, barns, warehouses and other storage areas. It is highly toxic to all plant and animal life. No commercial products are available at this time.

**Sodium fluosilicate** (sodium silico fluoride) has been used to control ectoparasites on livestock, as well as crawling insects in homes and work buildings. It is approximately as toxic as sodium fluoride. Commercial products containing sodium fluosilicate are no longer registered for use.

**Sodium fluoaluminate** (sodium aluminofluoride, Cryolite) is a stable mineral containing fluoride. It is used as an insecticide on some vegetables and fruits. Cryolite has very low water solubility, does not yield fluoride ion on decomposition and presents very little toxic hazard to mammals, including man.

Most cases of fluoride poisoning now are related to hydrofluoric acid, sulfur fluoride or excess fluorosis from sources other than insecticides, such as well water and toothpaste. Hydrofluoric acid is an important industrial toxicant but is not used as a pesticide. The clinical symptoms from hydrofluoric acid poisoning are essentially the same as described below for fluoride pesticides and are related to the fluoride ion's effects on potassium, calcium and magnesium.[28] Fluoroacetate is discussed in **Chapter 18, *Rodenticides***. Sulfuryl fluoride is discussed in **Chapter 17, *Fumigants***.

## Toxicology

**Sodium fluoride** and **sodium fluosilicate** used as insecticides present a serious hazard to humans because of high inherent toxicity and the possibility that children crawling on floors of treated dwellings will ingest the material. They are both used in the water fluoridation process, which is more likely to be a source of exposure than the insecticide. In a series of 87 pediatric cases of fluoride poisoning reported to a poison center, only one child had ingested an insecticide. Of note, that child was also the only fatality in the case series.[29]

Fluorides are readily and quickly absorbed from the GI tract with near complete bioavailability.[30,31] Plasma fluoride levels peak at around 30 minutes following ingestion. Fluoride is also distributed to the bone and saliva.[31,32] Excretion is chiefly in the urine. Within the first 24 hours of intoxication, renal clearance of fluoride from the blood is rapid. However, patients continue to excrete large amounts of fluoride for several days. The fluoride ion binds calcium and magnesium, leading to life-threatening cardiac toxicity in severe cases. Children are at relatively greater risk because of their smaller body mass compared to adults in relation to the amount ingested.[33]

## Signs and Symptoms of Poisoning

The toxic effects of fluoride in mammals are multiple and may be life threatening. The primary effects from fluoride result from an inhibition of critical intracellular enzymes and the direct effect on ionized calcium in extra-cellular fluid. The absorbed fluoride ion reduces extracellular fluid concentrations of calcium and magnesium. Hypocalcemia commonly occurs, sometimes severe enough to result in tetany or cardiac toxicity leading death.[29,34,35,36,37]

While sodium fluoride supplementation is available in the form of liquid drops, there is a rather narrow therapeutic range; chronic, mild fluorosis is present with an intake of 0.1 mg/kg/day. Most evidence of minor skeletal fluorosis will disappear as the fluoride supplementation is stopped, except for the teeth mottling.[38] Acutely toxic dosages usually start at about 3-10 mg/kg, with GI symptoms being the first to develop.[29,39] Ingested fluoride is transformed in the stomach to hydrofluoric acid, which has a corrosive effect on the epithelial lining of the gastrointestinal tract. Thirst, abdominal pain, vomiting and diarrhea are usual symptoms. Hemorrhagic gastroenteritis, ulceration, erosions and edema are commons signs.[40]

Cardiac arrhythmia and shock are often prominent features of severe poisoning. Hypotension and severe arrhythmia including ventricular fibrillation may also occur.[37,41] These probably result from combinations of effects of fluid and electrolyte disturbances including hypocalcemia,[29,34,35,36,37] hyperkalemia[41] and direct actions of fluoride on heart and vascular tissues. Fluoride may directly affect the central nervous system resulting in headache, muscle weakness, stupor, convulsions and coma.[33,34,37] Respiratory failure and ventricular arrhythmias are common causes of death.[33,37]

## Confirmation of Poisoning

A population drinking water with a concentration of 1 mg per liter will have a plasma inorganic fluoride concentration between 0.01-0.03 mg per liter[34] and rarely above 0.10 milligram per liter. In fatal cases of poisoning, plasma levels of 3.5 mg per liter and higher have been recorded, although survival has been reported in patients with levels as high as 14 mg per liter.[34,37] While not specific for fluoride poisoning, a low serum calcium level can be helpful in making the diagnosis.[29]

## Treatment

1. Decontaminate the skin with soap and water as outlined in **Chapter 3, *General Principles***. Treat eye contamination by irrigating exposed eyes with copious amounts of clean water or saline for at least 15 minutes. Remove contact lenses, if present, prior to irrigation. If irritation persists after irrigation, send patient for specialized medical treatment in a health carefacility.

2. If sodium fluoride or sodium fluosilicate has been ingested, consider gastric decontamination as outlined in **Chapter 3**. It should be noted that activated charcoal will not bind the fluoride ion.

3. Severe complications such as hypotension, shock, cardia arrhythymia or cyanosis should be treated in an intensive care setting. Monitor serum electrolytes (sodium, potassium, ionized calcium, magnesium, fluoride and bicarbonate) and correct as needed. Calcium and magnesium replacement are of primary consideration.[29,36]

   If the victim is fully alert and the amount ingested is less than 8 mg/kg of fluoride, consider giving the victim milk.[29] Milk provides calcium ions that will bind to fluoride, thereby reducing absorption. Magnesium-based antacids have also been used to neutralize the acid and facilitate the production of poorly absorbed salts.[37] There are no data on the optimum amounts to be administered.

4. If hypocalcemia is demonstrated, or if it appears likely that a significant amount of fluoride has been absorbed, aggressive calcium repletion may be required. Give 10 mL of 10% calcium gluconate intravenously slowly and repeat as necessary to keep the calcium in the normal or supranormal range:

### Dosage of Calcium Gluconate

Supplied as 100 mg/mL (10% solution)

- ***Adults and children over 12 years: 10 mL of 10% solution, given slowly, intravenously. Repeat as necessary.***
- ***Children under 12 years: 200-500 mg/kg/24 hr divided Q6 hr. Repeat dosage as needed.***

*Haloaromatic Substituted Urea Insecticides*

**COMMERCIAL PRODUCTS**

diflubenzuron
(brand names include, but are not limited to, Dimilin, Micromite, Vigilante)

teflubenzuron
(brand names include, but are not limited to, Nomolt, Dart, Diaract)

Severe cases may require use of 10% calcium chloride:

**Dosage of Calcium Chloride**

- ***Adults and children over 12 years: 5 to 10 mL (500 to 1,000 mg) intravenously over 1 to 5 minutes; may repeat after 10 minutes.***
- ***Children under 12 years: 0.2 to 0.3 mL/kg (20 to 30 mg/kg) per dose, up to a maximum single dose of 5 mL (500 mg) intravenously over 5 to 10 minutes, repeated up to four times or until serum calcium increases.***

These patients should be managed in the intensive care setting.

5. If hypomagnesaemia is present, administer magnesium sulfate.
6. Consider hemodialysis, as it may be beneficial in patients with significant toxicity.[37]
7. Refer patients with evidence of burns in their oral cavity for surgical evaluation and endoscopy, since these compounds can cause severe burns to the esophagus and stomach.
8. If a very large amount of sodium fluoaluminate (Cryolite) has been ingested, although it is much less toxic than other fluorides, measure serum calcium to ensure that hypocalcemia has not occurred. If it has, intravenous calcium may be required (see 4 above).

## HALOAROMATIC SUBSTITUTED UREA INSECTICIDES

Haloaromatic substituted urea compounds control insects by impairing chitin deposition in the larval exoskeleton. They are formulated in wettable powders, oil dispersible concentrate and granules for use in agriculture and forestry and in settings where fly populations tend to be large, such as feedlots. **Diflubenzuron** is the most commonly used product in this class, and most human data are based on this active ingredient.

### Toxicology

There is limited absorption of **haloaromatic substituted urea compounds** across the skin and intestinal lining of mammals, after which enzymatic hydrolysis and excretion rapidly eliminate the pesticide from tissues. Irritant effects are not reported and systemic toxicity is low. Based on animal studies, methemoglobinemia is a risk from the metabolite of diflubenzuron (4-chloroaniline).[65,66] There has been a report of occupational exposure to 4-chloroaniline that resulted in methemoglobinemia, although it is not clear that the source of 4-chloroaniline was diflubenzuron.[67]

### Treatment

1. Decontaminate the skin with soap and water as outlined in **Chapter 3, *General Principles.*** Treat eye contamination by irrigating exposed eyes with copious

amounts of clean water or saline for at least 15 minutes. Remove contact lenses, if present, prior to irrigation. If irritation persists after irrigation, obtain specialized medical treatment in a healthcare facility. Sensitization reactions may require steroid therapy.

2. If large amounts of propargite have been ingested and the patient is seen within an hour, consider gastrointestinal decontamination as discussed in **Chapter 3, *General Principles*.**

3. If methemoglobinemia is severe (>30%), or the patient is cyanotic, administer methylene blue.

**Dosage of Methylene Blue**

- ***Adults and children: 1-2 mg/kg of 1% methylene blue, slow IV, in symptomatic patients. Additional doses may be required, given as a slow IV push over a few minutes, every 4 hours as needed. (It is formulated as a 1% solution with 1 mL containing 10 mg of methylene blue.)***

## METHOPRENE

Methoprene is a long-chain hydrocarbon ester active as an insect growth regulator. It is effective against several insect species. Formulations include slow-release briquettes, sprays, foggers, soluble concentrate, suspension concentrate and baits.

### Toxicology

**Methoprene** is neither an irritant nor a sensitizer in humans or laboratory animals. Systemic toxicity in laboratory animals is very low. No human poisonings or adverse reactions in exposed workers have been reported.

### Treatment

1. Wash contaminated skin with soap and water. Treat eye exposures by irrigating exposed eyes with copious amounts of clean water or saline for at least 15 minutes. Remove contact lenses, if present, prior to irrigation. If irritation persists after irrigation, send patient to a healthcare facility for further medical attention.

2. If a very large amount of methoprene has been ingested, consider GI decontamination as outlined in **Chapter 3, *General Principles*.**

## N-PHENYLPYRAZONE INSECTICIDES

Fipronil is a broad-spectrum n-phenylpyrazole insecticide first registered by the U.S. Environmental Protection Agency in 1996. It is used for pests on agricultural crops and for lawn treatments. It is also commonly used for ant and cockroach control in the form of bait stations and as a topical application to domestic animals for flea and tick control.

## Toxicology

**Fipronil's** mechanism of action is by inhibition of GABA-gated chloride channels. This inhibits passage of chloride ions, thus producing hyperexcitability. This effect is similar to the mechanism of action for the organochlorine insecticides, the difference being that fipronil acts only on the $GABA_A$ channels, while organochlorines inhibit both the $GABA_A$ and $GABA_C$ channels.[42,43,44]

Fipronil is well absorbed by the GI tract in its parent form. It is rapidly metabolized to a sulfone compound. This metabolite is toxicologically active like the parent compound. It also binds to the same GABA receptors as fipronil, but at a much higher affinity.[45]

Animal studies demonstrate that fipronil has a selectively higher toxicity for insects than mammals, mostly attributed to a much more selective affinity for insect $GABA_A$ channels than vertebrate $GABA_A$ channels.[43,45]

## Signs and Symptoms of Poisoning

Despite the higher selective affinity for insects, there have been some reports of acute human toxicity. Patients may present with nausea and vomiting within several hours of ingestion. These appear to be self limiting, and no long-term gastrointestinal effects have been reported.[46] Consistent with the fact that the central nervous system is the primary target of fipronil, neurological symptoms have been the most commonly observed health effects.[46,47,48] Neurologic symptoms have been confirmed in cases of human poisoning following ingestion. Patients may present with altered mental status.[47] In severe cases, unconsciousness and generalized tonic-clonic seizures may also occur.[47,48,49] Most episodes of seizures or altered mental status appeared to be self limiting and have resolved within hours.[46,47]

One study analyzed pesticide surveillance data from 2001-2007, where 103 acute illnesses were identified with fipronil exposures in 11 states. The annual number of reported cases was shown to increase over time. The findings showed that the great majority of cases demonstrated mild clinical effects or short duration, thus confirming some of the previous observations. The reported effects in this study included conjunctivitis, headache, dizziness, nausea, vomiting, abdominal pain, oropharyngeal pain, cough, sweating, sensory impairment, weakness, drowsiness, agitation and seizure.[48] Of note, pet-care products were related to more than one-third of cases and accounted for the majority of childhood cases (64%).

There are no data available for signs and symptoms of chronic or subacute poisoning or exposure. However, the study of pesticide surveillance data also suggests that with occupational exposure, there is greater likelihood of repeated exposure to higher concentrations, thereby resulting in more severe effects.[48]

## Confirmation of Poisoning

The parent compound can be measured in plasma and in urine, although the test is not widely available in most hospitals. Levels reported with acute symptomatic human poisoning have been recorded as 1,600 µg/L–3,740 µg/L.[46] The levels peaked by approximately 3-4 hours following ingestion. Reported levels of the sulfone metabolite were not available.

## Treatment

1. Provide supportive care, as there is no specific antidote.

***N-Phenylpyrazones***

### COMMERCIAL PRODUCTS

Fipronil
(brand names include, but are not limited to Maxforce, Over'nOut!, Frontline, Frontline Topspot, Combat, Chipco Choice)

### HIGHLIGHTS

Inhibits $GABA_A$ channels

Well absorbed by GI tract

### SIGNS & SYMPTOMS

Nausea, vomiting

CNS impacts

Unconsciousness, seizures

### TREATMENT

Supportive care

GI decontamination

Control seizures with benzodiazepines

Control extreme agitation with lorazepam or propofol

*Neonicotinoids*
**HIGHLIGHTS**

Introduced in U.S. market in 1990s

Large (11-15%) and growing market share

Developed by modifying nicotine

Displace ACh from nAChRs

**SIGNS & SYMPTOMS**

Resembles acute nicotine poisoning

Usually ingestion or inhalation

Disorientation, confusion, agitation, headache, drowsiness, dizziness, weakness, tremor, unconsciousness

GI symptoms (vomiting, sore throat, nausea, diarrhea, abdominal pain) may be from formulation solvent

Respiratory toxicity can also occur

**TREATMENT**

Supportive treatment

Consider GI decontamination

Control extreme agitation with lorazepam or propofol

Consider IC setting for patients with mental status changes or severe poisoning

2. Send patients with significant mental status changes to an intensive care setting. At least initially, they are better managed there.

3. Use GI decontamination within the guidelines outlined in **Chapter 3, *General Principles***. There are insufficient data on the efficacy of activated charcoal.

4. Control seizures as early as possible with benzodiazepines.[46]

5. Control extreme agitation with lorazepam or propofol.

## NEONICOTINOID INSECTICIDES

Neonicotinoids are a relatively new class of insecticides, developed in the mid 1980s and introduced in the U.S. market in the early 1990s. They are quickly growing in widespread use and were recently noted to have 11%-15% U.S. market share of insecticide use.[50] They are well absorbed into plants and consequently are used in agriculture for piercing-sucking insects such as aphids and other crop-damaging insects. They are also used for flea control on domestic pets. They act fairly selectively on insects, with comparably less acute toxicity to mammals. As noted below, however, they are not free from human toxicity. Imidacloprid is the most widely used insecticide in this class, while most others have limited use in the commercial U.S. market. Reported clinical toxicity in humans is rare. However, increasing use of this insecticide and its potential toxicity among humans warrants a heightened awareness about these compounds and their toxicity.

In one report of two fatal intoxications with **imidacloprid**, the diagnosis was made post mortem by liquid chromatography/mass spectrometric quantification of insecticide. No clinical descriptions of symptoms were available, as both patients were found dead.

### Toxicology

Similar to synthetic pyrethroids being derived from naturally occurring pyrethrins, **neonicotinoids** were developed by modifying nicotine. Modifications include the **nitromethylene, nitroimine** or **cyanoimine** groups, which provide better activity and stability than nicotine. They are not very effective as contact insecticides but rather derive their effectiveness by being absorbed into the plant and migrating to the growing plant tip. They then affect insects that attempt to pierce the plant.

The toxicology of the neonicotinoids and special chemistry of the selective affinity of these insecticides is discussed in great detail in two recent reviews.[50,51] They act on nicotinic acetylcholine receptors (nAChRs) by displacing acetylcholine (ACh) from the receptor. Compared to other insecticides, most notably the organophosphate class, the neonicotinoids exhibit a relatively more selective affinity towards insect nAChRs than mammalian nAChRs.[50]

The acute oral $LD_{50}$ in rats of the neonicotinoids varies from 182 mg/kg (acetamiprid) to 2,400 mg/kg (dinotefuran). At an $LD_{50}$ of >5,000 mg/kg, clothianidin appears to be an outlier of this group.[50] While all neonicotinoids appear to selectively target insect nAChRs, imidacloprid and others that specifically contain the nitroimine group – thiamethoxam, clothianidin and dinotefuran – have a significantly higher affinity for the insect target site.[51] Of this subgroup, imidacloprid has the lowest rat $LD_{50}$ and by far the highest market share. Thiamethoxam on the other hand, while having a high $LD_{50}$, has a much lower NOAEL than imidacloprid and is considered a likely human carcinogen.[50]

Mammalian toxicity is thought to be centrally mediated. Toxic effects are similar to that of nicotine. Vertebrate alpha-4-Beta-2 nAChRs are the primary target. Prolonged or chronic exposures will up-regulate the receptors without changing receptor affinity. Perhaps most notably, the neonicotinoids also have some responses outside the target nAChRs. Following binding to the nAChR, the protein kinase cascade may be activated, which could decrease neurologic functions. Some also have an analgesic effect similar to that of nicotine.[50]

*In vitro* studies of human intestinal cells find that imidacloprid is well absorbed. These pesticides are relatively highly soluble in water, and most are excreted unchanged by the kidney. Most do undergo significant metabolism in insects, and the same occurs in mammals. However, the process in mammals is slow and likely an insignificant part of their elimination process in humans. One notable metabolic process of imidacloprid is reduction by the P450 system in humans to a nitroso derivative. In animal studies conducted in mice this metabolic byproduct enters the brain.[52] It is not known whether this byproduct or the active ingredient may be responsible for toxic effects.[50]

## Signs and Symptoms of Poisoning

Human data are currently limited to several reports of clinical poisoning, at least some of which have led to death, as confirmed by autopsy.[53,54,55,56] Toxic effects bear some resemblance to those of acute nicotine poisoning except for GI corrosive injuries, which may be related to solvent effects.[52] Human poisoning appears most likely following ingestion or inhalation. Most clinical effects are based on excessive nicotinic stimulation. Patients have presented with disorientation, confusion and agitation – severe enough to require sedation – headache, drowsiness, dizziness, weakness, tremor and, in some situations, loss of consciousness.[53,54,55] No seizures have been reported, and chronic residual neuropsychiatric effects have not been studied.

In a series of 68 patients, gastrointestinal effects following oral ingestion of an imidacloprid formulation were the most commonly reported and included vomiting, sore throat, nausea, diarrhea and abdominal pain.[57] Following ingestion, ulceration was noted in the posterior pharynx, esophagus and stomach. It was not clear if the effects were due to the toxicity of the active ingredient or the accompanying solvent. There is evidence that a solvent found in some formulations, N-methyl pyrrolide (NMP), has a severe irritant effect.[56] This emphasizes the importance of identifying and understanding the effects of inert ingredients in any pesticide exposure. Unfortunately, identification of such ingredients is usually difficult as they are not disclosed on the label and it is necessary to contact the formulator directly to determine which inert ingredients are in the formulation.

Toxicity to the respiratory system can also occur. Signs and symptoms include labored breathing, chest tightness, dyspnea, hypoxia and aspiration pneumonia. In severe cases, respiratory failure has ensued, requiring mechanical ventilation.[57,58,59]

Rhabdomyolysis may occur in severe poisoning; with creatine phosphokinase levels being reported as high as 1,200 U/L. Renal function and serum electrolytes were normal in this case. Patients will usually present with tachycardia due to nicotinic receptor over-stimulation of the autonomic nervous system.[53]

Cardiovascular effects include tachycardia, bradycardia, hypertension, hypotension and palpitations.[55] One case of fatal arrhythmia has been reported in which the patient presented within hours of ingesting 200 mL of imidacloprid. She initially had sinus tachycardia that rapidly progressed to ventricular tachycardia and subsequently ventricular fibrillation. At the time of presentation, this patient was noted to have a normal cardiac enzyme panel. Primary coronary artery disease could not be completely ruled out because of several coronary risk factors.[55]

***Neonicotinoids***
**COMMERCIAL PRODUCTS**

acetamiprid

clothianidin

dinotefuran

imidacloprid
(brand names include, but are not limited to Merit, Admire, Provado, Gaucho, Imicide, Premise, Advantage),

thiacloprid

thiamethoxam

## Confirmation of Poisoning

Imidacloprid can be detected by liquid chromatography/mass spectroscopy, which was used to identify the cause of death in two patients found dead with no obvious initial cause.[60] However, the test is not widely available, and there are insufficient data on toxic levels to predict severity of toxicity.

## Treatment

1. Provide supportive treatment, as there is no specific antidote for neonicotinoid poisoning. Patients with significant mental status changes should ideally be managed in the intensive care setting, at least initially.
2. Use GI decontamination within the guidelines previously outlined in **Chapter 3, *General Principles*.**
3. Control extreme agitation with lorazepam or propofol.
4. Consider cardiac monitoring, especially in patients with risk factors for coronary artery disease.
5. In a severe poisoning, send patient to an intensive care setting for respiratory support.

## PROPARGITE

Formulations are wettable powders and emulsifiable concentrates. Propargite is an acaricide with residual action.

## Toxicology

**Propargite** exhibits very little systemic toxicity in animals. No systemic poisonings have been reported in humans. However, many workers having dermal contact with this acaricide, especially during the summer months, have experienced skin irritation and some have had documented positive skin testing.[61,62] Eye irritation has also occurred.[61] For this reason, stringent measures should be taken to prevent inhalation or any skin or eye contamination by propargite. Epidemiological studies have related this pesticide to an increased risk for cancer.[63,64] This is discussed in **Chapter 21, *Chronic Effects*.**

## Confirmation of Poisoning

There is no readily available method for detecting absorption of propargite.

## Treatment

1. Decontaminate the skin with soap and water as outlined in **Chapter 3, *General Principles*.** Treat eye contamination by irrigating exposed eyes with copious amounts of clean water or saline for at least 15 minutes. Remove contact lenses, if present, prior to irrigation. If irritation persists after irrigation, specialized medical treatment in a healthcare facility should be obtained. Sensitization reactions may require steroid therapy.
2. If large amounts of propargite have been ingested and the patient is seen within an hour, consider gastrointestinal decontamination as discussed in **Chapter 3**.

## SULFUR

Elemental sulfur is an acaricide and fungicide widely used on orchard, ornamental, vineyard, vegetable, grain and other crops. It is prepared as dust in various particle sizes and applied as such, or formulated with various minerals to improve flowability or applied as an aqueous emulsion or wettable powder.

### Toxicology

**Elemental sulfur** is moderately irritating to the skin and is associated with occupationally related irritant dermatitis.[68] Airborne dust is irritating to the eyes and the respiratory tract. In hot, sunny environments, there may be some oxidation of foliage-deposited sulfur to gaseous sulfur oxides, which are very irritating to the eyes and respiratory tract. Ingested sulfur powder induces catharsis and has been used medicinally (usually with molasses) for that purpose. Some hydrogen sulfide is formed in the large intestine and this may present a degree of toxic hazard; the characteristic smell of rotten eggs may aid in the diagnosis. An adult has survived ingestion of 200 grams.[69]

Ingested colloidal sulfur is efficiently absorbed by the gut and is promptly excreted in the urine as inorganic sulfate.

### Treatment

1. Remove skin contamination by washing with soap and water as outlined in **Chapter 3, *General Principles*.** Treat contamination of the eyes by irrigating exposed eyes with clean saline or water for at least 15 minutes. If present, remove contact lenses prior to irrigation. If eye irritation persists after irrigation, obtain specialized treatment in a healthcare facility.

2. Unless an extraordinary amount of sulfur (several grams) has been ingested shortly prior to treatment, there is probably no need for gastrointestinal decontamination. Absorbability of sulfur on activated charcoal has not been tested.

3. Administer oral or intravenous glucose and/or electrolyte solutions as appropriate if diarrhea is severe. The most serious consequence of sulfur ingestion is likely to be that of catharsis, resulting in dehydration and electrolyte depletion, particularly in children.

*Sulfur*

#### COMMERCIAL PRODUCTS

*Many commercial products are produced by many manufacturers. It is one of the agents approved by USDA for use by organic growers.*

#### HIGHLIGHTS

Widely used organic acaricide/fungicide

#### SIGNS & SYMPTOMS

Skin/eye/respiratory irritant

#### TREATMENT

Decontaminate skin and eyes

Oral or IV glucose/ electrolytes if diarrhea is severe

### References

1. Graham BE, Kuizenga MH. Toxicity studies on benzyl benzoate and related benzyl compounds. *J Pharmacol Exp Ther.* Aug 1945;84:358-362.
2. Litovitz TL, Klein-Schwartz W, Oderda GM, Schmitz BF. Clinical manifestations of toxicity in a series of 784 boric acid ingestions. *Am J Emerg Med.* May 1988;6(3):209-213.
3. Restuccio A, Mortensen ME, Kelley MT. Fatal ingestion of boric acid in an adult. *Am J Emerg Med.* Nov 1992;10(6):545-547.
4. Ducey J, Williams DB. Transcutaneous absorption of boric acid. *J Pediatr.* Dec 1953;43(6):644-651.
5. Hamilton RA, Wolf BC. Accidental boric acid poisoning following the ingestion of household pesticide. *J Forensic Sci.* May 2007;52(3):706-708.
6. Ishii Y, Fujizuka N, Takahashi T, et al. A fatal case of acute boric acid poisoning. *J Toxicol Clin Toxicol.* 1993;31(2):345-352.

7. Goldbloom RB, Goldbloom A. Boric acid poisoning; report of four cases and a review of 109 cases from the world literature. *J Pediatr.* Dec 1953;43(6):631-643.

8. Wong LC, Heimbach MD, Truscott DR, Duncan BD. Boric Acid Poisoning: Report of 11 Cases. *Can Med Assoc J.* Apr 25 1964;90:1018-1023.

9. Linden CH, Hall AH, Kulig KW, Rumack BH. Acute ingestions of boric acid. *J Toxicol Clin Toxicol.* 1986;24(4):269-279.

10. Corradi F, Brusasco C, Palermo S, Belvederi G. A case report of massive acute boric acid poisoning. *Eur J Emerg Med.* Feb 2010;17(1):48-51.

11. Lung D, Clancy C. "Boiled lobster" rash of acute boric acid toxicity. *Clin Toxicol (Phila).* May 2009;47(5):432.

12. Naderi AS, Palmer BF. Successful treatment of a rare case of boric acid overdose with hemodialysis. *Am J Kidney Dis.* Dec 2006;48(6):e95-97.

13. Teshima D, Taniyama T, Oishi R. Usefulness of forced diuresis for acute boric acid poisoning in an adult. *J Clin Pharm Ther.* Oct 2001;26(5):387-390.

14. Garabrant DH, Bernstein L, Peters JM, Smith TJ, Wright WE. Respiratory effects of borax dust. *Br J Ind Med.* Dec 1985;42(12):831-837.

15. Hu X, Wegman DH, Eisen EA, Woskie SR, Smith RG. Dose related acute irritant symptom responses to occupational exposure to sodium borate dusts. *Br J Ind Med.* Oct 1992;49(10):706-713.

16. Schillinger BM, Berstein M, Goldberg LA, Shalita AR. Boric acid poisoning. *J Am Acad Dermatol.* Nov 1982;7(5):667-673.

17. Beckett WS, Oskvig R, Gaynor ME, Goldgeier MH. Association of reversible alopecia with occupational topical exposure to common borax-containing solutions. *J Am Acad Dermatol.* Apr 2001;44(4):599-602.

18. Shillinert BM, Berstein M, Goldberg LA, Shalita AR. Boric acid poisoning. *J Am Acad Dermatol.* 1992;7:667-673.

19. Stein KM, Odom RB, Justice GR, Martin GC. Toxic alopecia from ingestion of boric acid. *Arch Dermatol.* Jul 1973;108(1):95-97.

20. O'Sullivan K, Taylor M. Chronic boric acid poisoning in infants. *Arch Dis Child.* Sep 1983;58(9):737-739.

21. Fisher RS, Freimuth HC. Blood boron levels in human infants. *J Invest Dermatol.* Feb 1958;30(2):85-86.

22. Imbus HR, Cholak J, Miller LH, Sterling T. Boron, cadmium, chromium, and nickel in blood and urine. A survey of American working men. *Arch Environ Health.* Feb 1963;6:286-295.

23. Segar WE. Peritoneal dialysis in the treatment of boric acid poisoning. *N Engl J Med.* Apr 21 1960;262:798-800.

24. Folland DS, Kimbrough RD, Cline RE, Swiggart RC, Schaffner W. Acute hemorrhagic cystitis. Industrial exposure to the pesticide chlordimeform. *JAMA.* Mar 13 1978;239(11):1052-1055.

25. Arima T, Morooka H, Tanigawa T, Imai M, Tsunashima T, Kita S. Methemoglobinemia induced by chlorphenamidine. *Acta Med Okayama.* Feb 1976;30(1):57-60.

26. Horn HJ, Weir RJ. Inhalation toxicology of chlorine trifluoride. I. Acute and subacute toxicity. *AMA Arch Ind Health.* Nov 1955;12(5):515-521.

27. Ravindran M. Toxic encephalopathy from chlorobenzilate poisoning: report of a case. *Clin Electroencephalogr.* Oct 1978;9(4):170-172.

28. Martinez MA, Ballesteros S, Piga FJ, Sanchez de la Torre C, Cubero CA. The tissue distribution of fluoride in a fatal case of self-poisoning. *J Anal Toxicol.* Oct 2007;31(8):526-533.

29. Augenstein WL, Spoerke DG, Kulig KW, et al. Fluoride ingestion in children: a review of 87 cases. *Pediatrics.* Nov 1991;88(5):907-912.

30. Drummond BK, Curzon ME, Strong M. Estimation of fluoride absorption from swallowed fluoride toothpastes. *Caries Res.* 1990;24(3):211-215.

31. Trautner K, Einwag J. Human plasma fluoride levels following intake of dentifrices containing aminefluoride or monofluorophosphate. *Arch Oral Biol.* 1988;33(8):543-546.

32. McIvor ME. Acute fluoride toxicity. Pathophysiology and management. *Drug Saf.* Mar-Apr 1990;5(2):79-85.

33. Heifetz SB, Horowitz HS. Amounts of fluoride in self-administered dental products: safety considerations for children. *Pediatrics.* Jun 1986;77(6):876-882.

34. Gessner BD, Beller M, Middaugh JP, Whitford GM. Acute fluoride poisoning from a public water system. *N Engl J Med.* Jan 13 1994;330(2):95-99.

35. Harchelroad F, Goetz C. Systemic fluoride intoxication with leukocytosis and pyrexia. *Vet Hum Toxicol.* 1993;35(4):351.

36. Swanson L, Filandrinos DT, Shevlin JM, Willett JR. Death from accidental ingestion of an ammonium and sodium bifluoride glass etching compound. *Vet Hum Toxicol.* 1993;35(4):351.

37. Yolken R, Konecny P, McCarthy P. Acute fluoride poisoning. *Pediatrics.* Jul 1976;58(1):90-93.

38. Grandjean P, Thomsen G. Reversibility of skeletal fluorosis. *Br J Ind Med.* Nov 1983;40(4):456-461.

39. Spoerke DG, Bennett DL, Gullekson DJ. Toxicity related to acute low dose sodium fluoride ingestions. *J Fam Pract.* Jan 1980;10(1):139-140.

40. Spak CJ, Sjostedt S, Eleborg L, Veress B, Perbeck L, Ekstrand J. Tissue response of gastric mucosa after ingestion of fluoride. *BMJ.* Jun 24 1989;298(6689):1686-1687.

41. Baltazar RF, Mower MM, Reider R, Funk M, Salomon J. Acute fluoride poisoning leading to fatal hyperkalemia. *Chest.* Oct 1980;78(4):660-663.

42. Bloomquist JR. Ion channels as targets for insecticides. *Annu Rev Entomol.* 1996;41:163-190.

43. Ratra GS, Casida JE. GABA receptor subunit composition relative to insecticide potency and selectivity. *Toxicol Lett.* Jul 6 2001;122(3):215-222.

44. Ratra GS, Kamita SG, Casida JE. Role of human GABA(A) receptor beta3 subunit in insecticide toxicity. *Toxicol Appl Pharmacol.* May 1 2001;172(3):233-240.

45. Hainzl D, Cole LM, Casida JE. Mechanisms for selective toxicity of fipronil insecticide and its sulfone metabolite and desulfinyl photoproduct. *Chem Res Toxicol.* Dec 1998;11(12):1529-1535.

46. Mohamed F, Senarathna L, Percy A, et al. Acute human self-poisoning with the N-phenylpyrazole insecticide fipronil--a GABAA-gated chloride channel blocker. *J Toxicol Clin Toxicol.* 2004;42(7):955-963.

47. Fung HT, Chan KK, Ching WM, Kam CW. A case of accidental ingestion of ant bait containing fipronil. *J Toxicol Clin Toxicol.* 2003;41(3):245-248.

48. Lee SJ, Mulay P, Diebolt-Brown B, et al. Acute illnesses associated with exposure to fipronil--surveillance data from 11 states in the United States, 2001-2007. *Clin Toxicol (Phila).* Aug 2010;48(7):737-744.

49. Chodorowski Z, Anand JS. Accidental dermal and inhalation exposure with fipronil--a case report. *J Toxicol Clin Toxicol.* 2004;42(2):189-190.

**50.** Tomizawa M, Casida JE. Neonicotinoid insecticide toxicology: mechanisms of selective action. *Annu Rev Pharmacol Toxicol.* 2005;45:247-268.

**51.** Matsuda K, Buckingham SD, Kleier D, Rauh JJ, Grauso M, Sattelle DB. Neonicotinoids: insecticides acting on insect nicotinic acetylcholine receptors. *Trends Pharmacol Sci.* Nov 2001;22(11):573-580.

**52.** Chao SL, Casida JE. Interaction of imidacloprid metabolites and analogs with the nicotinic acetylcholine receptor of mouse brain in relation to toxicity. *Pest Biochem Physio.* 1997;58:77-88.

**53.** Agarwal R, Srinivas R. Severe neuropsychiatric manifestations and rhabdomyolysis in a patient with imidacloprid poisoning. *Am J Emerg Med.* Sep 2007;25(7):844-845.

**54.** David D, George IA, Peter JV. Toxicology of the newer neonicotinoid insecticides: imidacloprid poisoning in a human. *Clin Toxicol (Phila).* Jun-Aug 2007;45(5):485-486.

**55.** Huang NC, Lin SL, Chou CH, Hung YM, Chung HM, Huang ST. Fatal ventricular fibrillation in a patient with acute imidacloprid poisoning. *Am J Emerg Med.* Nov 2006;24(7):883-885.

**56.** Wu IW, Lin JL, Cheng ET. Acute poisoning with the neonicotinoid insecticide imidacloprid in N-methyl pyrrolidone. *J Toxicol Clin Toxicol.* 2001;39(6):617-621.

**57.** Mohamed F, Gawarammana I, Robertson TA, et al. Acute human self-poisoning with imidacloprid compound: a neonicotinoid insecticide. *PLoS One.* 2009;4(4):e5127.

**58.** Panigrahi AK, Subrahmanyam DK, Mukku KK. Imidacloprid poisoning: a case report. *Am J Emerg Med.* Feb 2009;27(2):256 e255-256.

**59.** Phua DH, Lin CC, Wu ML, Deng JF, Yang CC. Neonicotinoid insecticides: an emerging cause of acute pesticide poisoning. *Clin Toxicol (Phila).* Apr 2009;47(4):336-341.

**60.** Proenca P, Teixeira H, Castanheira F, et al. Two fatal intoxication cases with imidacloprid: LC/MS analysis. *Forensic Sci Int.* Oct 4 2005;153(1):75-80.

**61.** Saunders LD, Ames RG, Knaak JB, Jackson RJ. Outbreak of Omite-CR-induced dermatitis among orange pickers in Tulare County, California. *J Occup Med.* May 1987;29(5):409-413.

**62.** Verma G, Sharma NL, Shanker V, Mahajan VK, Tegta GR. Pesticide contact dermatitis in fruit and vegetable farmers of Himachal Pradesh (India). *Contact Dermatitis.* Nov 2007;57(5):316-320.

**63.** Mills PK, Yang RC. Agricultural exposures and gastric cancer risk in Hispanic farmworkers in California. *Environ Res.* Jun 2007;104(2):282-289.

**64.** Reynolds P, Von Behren J, Gunier RB, Goldberg DE, Hertz A, Harnly ME. Childhood cancer and agricultural pesticide use: an ecologic study in California. *Environ Health Perspect.* Mar 2002;110(3):319-324.

**65.** Agency USEP. Pesticide Tolerance for Diflubenzuron. Washington, D.C.1996.

**66.** Ehlhardt WJ, Woodland JM, Worzalla JF, et al. Comparison of metabolism and toxicity to the structure of the anticancer agent sulofenur and related sulfonylureas. *Chem Res Toxicol.* Sep-Oct 1992;5(5):667-673.

**67.** Pizon AF, Schwartz AR, Shum LM, et al. Toxicology laboratory analysis and human exposure to p-chloroaniline. *Clin Toxicol (Phila).* Feb 2009;47(2):132-136.

**68.** O'Malley MA. Skin reactions to pesticides. *Occup Med.* Apr-Jun 1997;12(2):327-345.

**69.** Schwartz SM, Carroll HM, Scharschmidt LA. Sublimed (inorganic) sulfur ingestion. A cause of life-threatening metabolic acidosis with a high anion gap. *Arch Intern Med.* Jul 1986;146(7):1437-1438.

*Section III*

# HERBICIDES

## HIGHLIGHTS

Hundreds of products

Names can confuse; check label for composition

Sometimes mixed with fertilizers

Irritate skin, eyes, respiratory & GI systems

Severe metabolic acidosis from ingesting large amounts

## SIGNS & SYMPTOMS

Vomiting, diarrhea

Headache, confusion, bizarre or aggressive behavior

Peculiar odor on breath

Body temperature may be elevated

Muscle weakness, peripheral neuropathy, loss of reflexes

## TREATMENT

Decontaminate skin, hair, eyes

Consider GI decontamination

IV fluids

Consider urine alkalinization

CHAPTER 10

# Chlorophenoxy Herbicides

Several hundred commercial products contain chlorophenoxy herbicides in various forms, concentrations and combinations. In some cases, the same name is used for products with different ingredients. The exact composition must therefore be determined from the product label. Chlorophenoxy compounds are sometimes mixed into commercial fertilizers to control growth of broadleaf weeds. Sodium, potassium and alkylamine salts are commonly formulated as aqueous solutions, while the less water-soluble esters are applied as emulsions. Low molecular weight esters are more volatile than the acids, salts or long-chain esters.

## Toxicology

Some of the **chlorophenoxy** acids, salts and esters are moderately irritating to skin, eyes and respiratory and gastrointestinal linings. In a few individuals, local cutaneous depigmentation has apparently resulted from protracted dermal contact with chlorophenoxy compounds.[1]

The chlorophenoxy compounds are well absorbed from the gastrointestinal tract.[2] They are less well absorbed from the lung. Cutaneous absorption appears to be minimal.[3] They are not significantly stored in fat. Excretion occurs almost entirely by way of urine. Apart from some conjugation of the acids, there is limited biotransformation in the body.[2,3] The compounds are highly protein bound.[3] Under normal conditions, the average half-life of **2,4-D** in humans is between 13 and 39 hours,[2,4,6] that of **2,4,5-T** about 24 hours[7] and that of **MCPP** about 17 hours.[8] However, half-life varies markedly with urinary pH, with excretion being greatly enhanced in an alkaline urine,[4,6,9] and with a half-life as prolonged as 70-90 hours with acidic urine.[9] Half-life is also longer with large doses and prolonged exposure.

A unique finding in a recent study is that chlorophenoxy herbicides, particularly 2,4-DP, 2,4-D and MCPP, inhibit the human taste receptor for sweets. Interestingly, this was not found in animal studies. While not necessarily a toxic effect, this finding could potentially be of use in diagnosing a poisoning from one of these herbicides.[10]

Ingestion of large amounts of chlorophenoxy acids has resulted in severe metabolic acidosis in humans. Such cases have been associated with electrocardiographic changes, myotonia, muscle weakness, myoglobinuria and elevated serum creatine phosphokinase, all reflecting injury to striated muscle. The medical literature contains a few reports of peripheral neuropathy, some following dermal exposures to 2,4-D[11,12,13] and another following ingestion.[14] Chlorophenoxy acids are weak uncouplers of oxidative phosphorylation; therefore, extraordinary doses may produce hyperthermia from increased production of body heat.[6]

In the manufacture of some of these herbicides, other more toxic substances can be formed at excessive temperatures. These include **chlorinated dibenzo dioxin (CDD)** and **chlorinated dibenzo furan (CDF)**. The 2,3,7,8-tetra CDD form is extraordinarily toxic to multiple mammalian tissues; it is formed only in the synthesis of 2,4,5-T. However, 2,3,7,8 tetra CDD has been found as a contaminant in samples of 2,4-D, 2,4-DB and MCPA.[15] These byproducts are discussed in **Chapter 21, *Chronic Effects***. Chloracne (a chronic, disfiguring skin condition) has been seen in workers engaged in the manufacture of 2,4,5-T and certain other chlorinated organic compounds, although it is thought to be related to the resulting 2,3,7,8-tetra CDD exposure as opposed to acute 2,4-D or 2,4,5-T toxicity. Although chloracne along with other dermal effects has been reported in an herbicide applicator,[16] it has not been reported in other occupational exposures except the manufacture of these agents.

## Signs and Symptoms of Poisoning

Human poisoning from chlorophenoxy compounds was reviewed in detail in 2000.[17] In a large case series resulting from intentional self-poisoning from MCPA, most patients (85%) had minimal signs of poisoning, with mild gastrointestinal symptoms being the most commonly reported.[18] Other non-specific, mild findings from the California pesticide illness surveillance system include nausea, abdominal pain, headache, generalized weakness and dizziness.[19]

Manifestations of systemic toxicity of chlorophenoxy compounds are known mainly from clinical experience with cases of deliberate suicidal ingestion of large quantities. While most clinical reports involve exposure to 2,4-D and mecoprop, it is reasonable to assume that all chlorophenoxy herbicides will share a similar clinical picture. Most reports of fatal outcomes involve renal failure, acidosis, electrolyte imbalance and a resultant multiple organ failure.[5,9,20] The agents most often involved in these incidents have been 2,4-D and mecoprop.

Patients will present within a few hours of ingestion with vomiting, diarrhea, headache, confusion and bizarre or aggressive behavior. In a large case series resulting from intentional self-poisoning from MCPA, most patients (85%) had minimal signs of poisoning, with mild gastrointestinal symptoms being the most commonly reported.[18] Mental status changes occur, with progression to coma in severe cases.[4,6,9,18] Moderate cerebral edema has also been reported following intentional ingestion.[21] A peculiar odor is often noticed on the breath. Body temperature may be moderately elevated, but this is rarely a life-threatening feature of the poisoning. The respiratory drive is not depressed. Conversely, hyperventilation is sometimes evident, probably secondary to the metabolic acidosis that occurs. Convulsions occur very rarely. With effective urinary excretion of the toxicant, consciousness usually returns in 48-96 hours.[4,6,9]

Muscle weakness and peripheral neuropathy have been reported after occupational exposure.[9] The presentations are variable. Myotonia and muscle weakness may persist for months after acute poisoning.[6] Additional findings include loss of reflexes and fasciculation.[4 6,8,20] Electromyography and nerve conduction studies in some recovering patients have demonstrated a mild proximal neuropathy and myopathy.

As mentioned above, there are significant metabolic changes from the chlorophenoxy compounds. Metabolic acidosis is manifest as a low arterial pH and bicarbonate content. The urine is usually acidic. Skeletal muscle injury, if it occurs, is reflected in elevated creatine phosphokinase and, sometimes, myoglobinuria. Moderate elevations of blood urea nitrogen and serum creatinine are commonly found as the toxicant is excreted. Cases of renal failure are reported, often with an accompanying hyperkalemia or hypocalcemia, and were thought to result in the cardiovascular instability that led to death.[5,20] Tachycardia is commonly observed and hypotension has also been reported.[4,5,9] T-wave flattening has also been observed.[6] Mild leukocytosis and biochemical changes indicative of liver cell injury have been reported.

### COMMERCIAL PRODUCTS

2,4-D or 2,4-dichlorophenoxyacetic acid

2,4-DP or 2,4-dichlorophenoxypropionic acid (Dichlorprop)

2,4-DB, or 2,4-dichlorophenoxybutyric acid

2,4,5-T, or 2,4,5-trichlorophenoxy acid

4-chloro-2-methyl-phenoxyacetic acid (MCPA)

MCPB

MCPP (Mecoprop)

2-methyl-3, 6 dichlorobenzoic acid (Dicamba)

## Confirmation of Poisoning

Gas-liquid chromatographic methods are available for detecting chlorophenoxy compounds in blood and urine. These analyses are useful in confirming and assessing the magnitude of chlorophenoxy absorption. Poisoning episodes characterized by unconsciousness have shown initial blood chlorophenoxy concentrations ranging from 80 to more than 1,000 mg per liter.[4] Urine samples should be collected as soon as possible after exposure because the herbicides may be almost completely excreted in 24-72 hours in most cases. Urine samples can also confirm overexposure. In a study of asymptomatic herbicide applicators, their urinary excretion of chlorophenoxy compounds rarely exceeded 1-2 mg/L.[22] The half-life may be much longer in cases of intoxication depending on the extent of absorption and urine pH. Analyses can be performed at competent laboratories, usually known to local poison control centers. If the clinical scenario indicates that excessive exposure to chlorophenoxy compounds has occurred, initiate appropriate treatment measures immediately, not waiting for chemical confirmation of toxicant absorption.

## Treatment of Chlorophenoxy Toxicosis

1. Decontaminate skin and hair by bathing with soap and water and shampooing. Individuals with chronic skin disease or known sensitivity to these herbicides should either avoid using them or take strict precautions to avoid contact (respirator, gloves, etc.).

2. Flush contaminating chemicals from eyes with copious amounts of clean water for 10-15 minutes. If irritation persists, an ophthalmologic examination should be performed.

3. If any symptoms of illness occur during or following inhalation of spray, remove victim from contact with the material for at least 2-3 days. Allow subsequent contact with chlorophenoxy compounds only if effective respiratory protection is practiced.

4. Consider gastric decontamination procedures as outlined in **Chapter 3, *General Principles.*** If substantial amounts of chlorophenoxy compounds have been ingested, spontaneous emesis may occur.

5. Administer intravenous fluids to accelerate excretion of the chlorophenoxy compound and to limit concentration of the toxicant in the kidney. A urine flow of 4-6 mL/minute is desirable. Intravenous saline/dextrose has sufficed to rescue comatose patients who drank 2,4-D and mecoprop several hours before hospital admission.

   CAUTION: ***Monitor urine protein and cells, BUN, serum creatinine, serum electrolytes and fluid intake/output carefully to ensure that renal function remains unimpaired and that fluid overload does not occur.***

6. Alkalinize the urine to maintain a pH between 7.6 and 8.8. Urinary alkalinization has been used successfully in management of suicidal ingestions of chlorophenoxy compounds, especially when initiated early.[4,6,9] Although the term "forced alkaline diuresis" has been used previously to describe this treatment, the preferred terminology is now "urinary alkalinization" to emphasize the impor-

tance of urine pH manipulation for clearing the weak acid.[23] Alkalinizing the urine by including sodium bicarbonate (44-88 mEq per liter) in the intravenous solution accelerates excretion of 2,4-D and mecoprop excretion substantially, because the weak acid is in an ionized state in the renal tubule and thus cannot diffuse back across the tubule into the blood. Renal clearance is minimal at an acidic pH of 5.1 (0.14 mL/min) compared to clearance at a pH of 8.3 (63 mL/min).[6,23]

Controversy and lack of controlled clinical studies exist surrounding the most effective way to induce clearance of 2,4-D and mecoprop. The AACT and EAPCCT position paper recommends that urine alkalinization and high urine flow (forced diuresis) be considered.[23] A Cochrane Database of Systemic Reviews notes the lack of evidence, based on the lack of randomized, controlled trials for this treatment. The author concluded that it is "not unreasonable to attempt urinary alkalinization" given the prolonged toxicity and potential for death, and that, "well conducted randomized, controlled trials are required."[24] No patients in a large case series reported by the same author as the Cochrane Review article were treated with urinary alkalinization, although it should be noted that 85% showed signs of minimal toxicity.[18]

7. Include potassium chloride as needed to offset increased potassium losses, with 20-40 mEq of potassium chloride to each liter of intravenous solution. High urine flow, approximately 200 mL/h, improves clearance, although an even higher flow rate may be required for maximal 2,4-D clearance.[6,23] Renal failure has occurred in patients with severe intoxication despite urinary alkalinization. In one case of renal failure, the urinary alkalinization was begun 26 hours after ingestion,[9] and in another it was initiated on day 2 of the hospitalization.[20] Therefore, it is crucial to carefully monitor renal function, as well as serum electrolytes, especially potassium and calcium.

8. Consider hemodialysis in severe cases, particularly where excess fluid administration is not advised.[17] Hemodialysis has been used in four patients who survived intoxication.[25] It is not recommended as first-line therapy.

9. Include electromyography and nerve conduction studies in the follow-up clinical examination to detect any neuropathic changes and neuromuscular junction defects.

## References

1. Garry VF, Tarone RE, Kirsch IR, et al. Biomarker correlations of urinary 2,4-D levels in foresters: genomic instability and endocrine disruption. *Environ Health Perspect.* May 2001;109(5):495-500.

2. Kohli JD, Khanna RN, Gupta BN, Dhar MM, Tandon JS, Sircar KP. Absorption and excretion of 2,4-dichlorophenoxyacetic acid in man. *Xenobiotica.* 1974;4(2):97-100.

3. Arnold EK, Beasley VR. The pharmacokinetics of chlorinated phenoxy acid herbicides: a literature review. *Vet Hum Toxicol.* Apr 1989;31(2):121-125.

4. Friesen EG, Jones GR, Vaughan D. Clinical presentation and management of acute 2,4-D oral ingestion. *Drug Saf.* Mar-Apr 1990;5(2):155-159.

5. Keller T, Skopp G, Wu M, Aderjan R. Fatal overdose of 2,4-dichlorophenoxyacetic acid (2,4-D). *Forensic Sci Int.* Mar 1994;65(1):13-18.

6. Prescott LF, Park J, Darrien I. Treatment of severe 2,4-D and mecoprop intoxication with alkaline diuresis. *Br J Clin Pharmacol.* Jan 1979;7(1):111-116.

7. Gehring PJ, Kramer CG, Schwetz BA, Rose JQ, Rowe VK. The fate of 2,4,5-trichlorophenoxyacetic acid (2,4,5-T) following oral administration to man. *Toxicol Appl Pharmacol.* Nov 1973;26(3):352-361.

8. Meulenbelt J, Zwaveling JH, van Zoonen P, Notermans NC. Acute MCPP intoxication: report of two cases. *Hum Toxicol.* May 1988;7(3):289-292.

9. Flanagan RJ, Meredith TJ, Ruprah M, Onyon LJ, Liddle A. Alkaline diuresis for acute poisoning with chlorophenoxy herbicides and ioxynil. *Lancet.* Feb 24 1990;335(8687):454-458.

10. Maillet EL, Margolskee RF, Mosinger B. Phenoxy herbicides and fibrates potently inhibit the human chemosensory receptor subunit T1R3. *J Med Chem.* Nov 12 2009;52(21):6931-6935.

11. Berkley MC, Magee KR. Neuropathy following exposure to a dimethylamine salt of 2, 4-D. *Arch Intern Med.* Mar 1963;111:351-352.

12. Berwick P. 2,4-dichlorophenoxyacetic acid poisoning in man. Some interesting clinical and laboratory findings. *JAMA.* Nov 9 1970;214(6):1114-1117.

13. Goldstein NP, Jones PH, Brown JR. Peripheral neuropathy after exposure to an ester of dichlorophenoxyacetic acid. *J Am Med Assoc.* Nov 7 1959;171:1306-1309.

14. O'Reilly JF. Prolonged coma and delayed peripheral neuropathy after ingestion of phenoxyacetic acid weedkillers. *Postgrad Med Journal.* 1984;60:76-77.

15. Holt E, Weber R, Stevenson G, Gaus C. Polychlorinated Dibenzo-p-Dioxins and Dibenzofurans (PCDD/Fs) Impurities in Pesticides: A Neglected Source of Contemporary Relevance. *Environ Sci Technol.* Jul 15 2010;44(14):5409-5415.

16. Poskitt LB, Duffill MB, Rademaker M. Chloracne, palmoplantar keratoderma and localized scleroderma in a weed sprayer. *Clin Exp Dermatol.* May 1994;19(3):264-267.

17. Bradberry SM, Watt BE, Proudfoot AT, Vale JA. Mechanisms of toxicity, clinical features, and management of acute chlorophenoxy herbicide poisoning: a review. *J Toxicol Clin Toxicol.* 2000;38(2):111-122.

18. Roberts DM, Seneviratne R, Mohammed F, et al. Intentional self-poisoning with the chlorophenoxy herbicide 4-chloro-2-methylphenoxyacetic acid (MCPA). *Ann Emerg Med.* Sep 2005;46(3):275-284.

19. Regulation CDoP. *California Pesticide Illness Query (CalPIQ)* August 9, 2010 2009.

20. Kancir CB, Andersen C, Olesen AS. Marked hypocalcemia in a fatal poisoning with chlorinated phenoxy acid derivatives. *J Toxicol Clin Toxicol.* 1988;26(3-4):257-264.

21. Brahmi N, Mokhtar HB, Thabet H, Bouselmi K, Amamou M. 2,4-D (chlorophenoxy) herbicide poisoning. *Vet Hum Toxicol.* Dec 2003;45(6):321-322.

22. Kolmodin-Hedman B, Hoglund S, Akerblom M. Studies on phenoxy acid herbicides. I. Field study. Occupational exposure to phenoxy acid herbicides (MCPA, dichlorprop, mecoprop and 2,4-D) in agriculture. *Arch Toxicol.* Dec 1983;54(4):257-265.

23. Proudfoot AT, Krenzelok EP, Vale JA. Position paper on urine alkalinization. *J Toxicol Clin Toxicol.* 2004;42(1):1-26.

24. Roberts DM, Heilmair R, Buckley NA, et al. Clinical outcomes and kinetics of propanil following acute self-poisoning: a prospective case series. *BMC Clin Pharmacol.* 2009;9:3.

25. Durakovic Z, Durakovic A, Durakovic S, Ivanovic D. Poisoning with 2,4-dichlorophenoxyacetic acid treated by hemodialysis. *Arch Toxicol.* 1992;66(7):518-521.

CHAPTER 11

# Pentachlorophenol and Dinitrophenolic Pesticides

## PENTACHLOROPHENOL

Pentachlorophenol (PCP) is presently registered in the United States only as a restricted use pesticide for use as a "heavy duty" wood preservative. It is registered only for use in pressure treatment of utility poles. Heavy duty wood preservatives are defined as those that are applied by pressure treatment rather than by brushing or other surface applications. PCP is a general biocide that has been used as an herbicide, algaecide, defoliant, wood preservative, germicide, fungicide and molluscicide.[1] As a function of the manufacturing process, PCP is contaminated with chlorinated dibenzodioxans (CDDs), chlorinated dibenzofurans (CDFs) and hexachlorobenzene (HCB). These contaminants are toxic and persistent, but their levels in PCP preparations are usually low enough to limit the concern to chronic rather than acute effects. Technical PCP also contains lower chlorinated phenols (4%-12%). Incomplete combustion of PCP-treated wood may lead to further formation of these contaminant compounds.

Pentachlorophenol volatilizes from treated wood. It has a significant phenolic odor, which becomes quite strong when the material is heated. Though not registered for indoor use, heavily treated interior surfaces may be a source of exposure sufficient to cause irritation of eyes, nose and throat.

### Toxicology

**Pentachlorophenol (PCP)** is readily absorbed across the skin, the lungs and the lining of the gastrointestinal tract. USEPA data submitted in support of reregistration of PCP report a dermal $LD_{50}$ >3,980 mg/kg, suggesting very low dermal toxicity. In animals, the dermal $LD_{50}$ has been reported as the same order of magnitude as the oral.[2] With acute exposure it is rapidly excreted, mainly in the urine as unchanged PCP and as PCP glucuronide. In chronic exposures as well as a volunteer study, the elimination half-life has been reported to be very prolonged, up to 20 days. The long half-life was attributed to the low urinary clearance because of high protein binding.[3] It is widely distributed to other tissues in the body, including kidney, liver, heart and adrenal glands.

The primary acute toxicological mechanism appears to be increased cellular oxidative metabolism resulting from the uncoupling of oxidative phosphorylation.[1,4] Heat production is increased and leads to clinical hyperthermia with profuse sweating and electrolyte disturbances. This clinical state may mimic the signs and symptoms of hyperthyroidism. Large doses are toxic to the liver, kidneys and nervous system. Due to depletion of ATP, severe rhabdomyolysis may occur. Numerous additional mechanisms may contribute to chronic toxicity

Based on laboratory experimentation in animals, PCP has been reported to have fetotoxic and embryotoxic properties and to bind to various hormone receptors.[5,6] Epidemiologic evidence suggests exposed women may be at risk for miscarriages, and maternal or paternal exposure can increase risk for reduced birth weight and infant malformations.[7,8]

*Pentachlorophenol*
**HIGHLIGHTS**

Limited use in pressure-treated utility poles

Volatilizes from treated wood

Skin, lung, GI absorption

Low urinary clearance

Distributes to kidney, liver, heart, adrenals

Prenatal implications

**SIGNS & SYMPTOMS**

Mucosal membrane irritation

Fatigue, headache, lack of concentration

Contact dermatitis, chloracne

Wide variety of non-specific symptoms

Tachycardia, increased respiratory rate typical in serious poisonings

*Pentachlorophenol & Dinotrophenolic Pesticides*
**TREATMENT**

Control hyperthermia

Support oxygen, fluids

Stabilize electrolytes

Decontaminate skin, eyes

Consider ICU management

*Pentachlorophenol*
**COMMERCIAL PRODUCTS**

chlorophen
PCP
penchlorol
penta
pentacon
penwar
sinituho
The sodium salt is sodium pentachlorophenate

Albuminuria, glycosuria, aminoaciduria and elevated BUN reflect renal injury. Liver enlargement, anemia and leucopenia have been reported in some intensively exposed workers. Elevated serum alkaline phosphatase, AST and LDH enzymes indicate significant insult to the liver, including both cellular damage and some degree of biliary obstruction.

## Signs and Symptoms of Poisoning

The most common effects of airborne PCP include mucosal membrane irritation of the eyes, nose and throat, producing conjunctivitis, rhinitis and pharyngitis.[9,11] Additional common features include fatigue, lack of concentration and headache.[10,11,12] In adequate concentration, PCP is irritating to skin. Effects include irritation, contact dermatitis or, more rarely, diffuse urticaria or chloracne.[11,13,14] Contact dermatitis is common among workers having contact with PCP. In a study of employees involved in the manufacture of PCP, chloracne was found in 7% of the workers, and the risk was significantly higher among employees with documented skin contact compared to employees without skin contact.[14] Urticaria has also been reported as an uncommon response in exposed persons. Individual cases of exfoliative dermatitis of the hands and diffuse urticaria and angioedema of the hands have been reported in intensively exposed workers. Several infant deaths occurred in a nursery where a PCP-containing diaper rinse had been used.[15] Severe poisoning and death have occurred as a result of intensive PCP exposure.[10,16,17]

Acute poisoning occurs with systemic absorption that can occur by any route of sufficient dosage, although most occupational poisonings occur through dermal contact.[16,17] Most of the signs and symptoms of PCP are non-specific and, therefore, the diagnosis can be difficult. Symptoms include abdominal pain, anorexia, intense thirst, dizziness, restlessness and altered mental status. Workers exposed over long periods may experience weight loss. Serious poisoning may be manifested by hyperthermia, muscle spasm, tremor, respiratory distress, chest tightness and altered mental status, including lethargy and coma.[1,10,16,17] Tachycardia and increased respiratory rate are usually apparent. Most adult fatalities have occurred in persons working in hot environments where hyperthermia is poorly tolerated. In severe poisonings that have resulted in death, severe hyperthermia with temperatures up to 108°F has been reported.[16] Multiorgan system failure (seizures and coma, hepatic necrosis, renal failure, cardiovascular collapse and rhabdomyolysis) are often contributing factors in fatal outcomes.[15,16]

PCP has been classified as B2 (probable human carcinogen). Cases of aplastic anemia and leukemia have been reported that were associated temporally with PCP exposure. Causal relationships in these cases were not established.[18] For more information, see the cancer section in **Chapter 21, *Chronic Effects*.**

Peripheral neuropathies have also been reported in some cases of long-term occupational exposure; however, a causal relationship has not been supported by longitudinal studies.[19] Studies of health effects in a community where a wood treatment plant is located have suggested an association with long-term adverse health effects. Residents in the community had a higher prevalence of cancer, respiratory disease and neurological disorders than those in the control group. It is unclear from the study, however, whether PCP or creosote, another wood preservative (see **Chapter 19, *Miscellaneous Pesticides, Solvents and Adjuvants***), was the primary pesticide of concern.[20]

## Confirmation of Poisoning

CAUTION: ***If poisoning is suspected on the basis of exposure, symptoms and signs, do not postpone treatment until diagnosis is confirmed.***

PCP can be measured in plasma, urine and adipose tissue by gas-liquid chromatography. **Plasma levels can be much higher than urine levels (ratio of blood to urine is 1.0 to 2.5), so care must be taken to interpret results.**[19,21] There is no clear-cut determination of what constitutes an abnormally high level of PCP, and there is great variability among different references. Most information on the extent of serum levels in relation to toxicity is based on individual cases or small series of patients. Reports exist of asymptomatic infants with serum levels as high as 26 parts per million (ppm);[15,21] however, most other reports of non-occupational exposure in the general public have levels in the parts per billion range.[1,22,23,24] Food is probably the main source of this nanogram-level dosage.[1] Serum levels among occupationally exposed persons often exceed 1 ppm.[1] A report of a lethal case describes a plasma level of 16 ppm,[17] but most cases generally involve serum levels in the range of 100 ppm or higher.[16] It is reasonable to assume that levels greater than 1 ppm are consistent with an unusual exposure and that levels approaching 100 ppm are cause for great concern.

## DINOTROPHENOLIC PESTICIDES

Dinitrophenolic pesticides have many uses in agriculture worldwide: herbicides (weed killing and defoliation), acaricides, nematocides, ovicides and fungicides. Relatively insoluble in water, most technical products are dissolved in organic solvents and are formulated for spray application as emulsions. There are some wettable powder formulations. Only dinocap is currently registered in the United States.

### Toxicology

Nitroaromatic compounds are highly toxic to humans and animals with $LD_{50}$s in the range of 25 to 50 mg/kg.[25] Most **dinitrophenols** are well absorbed from the gastrointestinal tract, across the skin and by the lung when fine droplets are inhaled.[26]

Dinitrophenols undergo some biotransformation in humans, chiefly reduction (one nitro group to an amino group) and conjugation at the phenolic site. Although dinitrophenols and metabolites appear consistently in the urine of poisoned individuals, hepatic excretion is probably the main route of disposition. Elimination is slow, with a documented half-life in humans between 5-14 days.[25] Blood and tissue concentrations tend to increase progressively if an individual is substantially exposed on successive days.

The basic mechanism of toxicity is stimulation of oxidative metabolism in cell mitochondria, by the uncoupling of oxidative phosphorylation. This leads to hyperthermia, tachycardia, headache, malaise and dehydration and, in time, depletes carbohydrate and fat stores. The major systems prone to toxicity are the hepatic, renal and nervous systems. The dinitrophenols are more active as uncouplers than chlorophenols such as pentachlorophenol. Hyperthermia and direct toxicity to the central nervous system cause restlessness and headache and, in severe cases, seizures, coma and cerebral edema. The higher the ambient temperature, such as in an agriculture environment, the more difficult it is to dissipate the heat.[25,26] Liver parenchyma and renal tubules show degenerative changes. Albuminuria, pyuria, hematuria and azotemia are signs of renal injury.

***Dinotrophenolic Pesticides***

**HIGHLIGHTS**

Only dinocap currently registered in U.S.

Absorbed from GI, skin, lung

Hepatic excretion with 5 to 14-day half-life

**SIGNS & SYMPTOMS**

*Non-specific and may include*

Sweating

Thirst

Fever

Headache

Confusion

Malaise

Restlessness

Serious poisoning

Hyperthermia

Tachycardia

Tachypnea

Renal failure

Bright yellow staining of skin, hair

**TREATMENT**

Same as Pentachlorophenol

**CONTRAINDICATED**

Antipyretic therapy with salicylates

Atropine

***Dinotrophenolic Pesticides***
**COMMERCIAL PRODUCTS**

*While there are many dinitrophenolic pesticides, dinocap is the only one that is still actively registered in the United States.*

*Other products (no longer registered in the United States) included:*

3Dinitrophenol (Chemox PE)

dinitrocresol (DNOC, DNC, Chemsect DNOC, Elgetol 30, Nitrador, Selinon, Sinox, Trifocide)

dinobuton (Acrex, Dessin, Dinofen, Drawinol, Talan)

dinopenton

dinoprop (Crotothane, Karathane)

dinosam (DNAP, Chemox General),

dinoseb (DNBP, dinitro, Basanite, Caldon, Chemox General, Chemox PE, Chemsect DNBP, Dinitro, Dinitro-3, Dinitro General Dynamyte, Elgetol 318, Gebutox, Hel-Fire, Kiloseb, Nitropone C, Premerge, Snox General, Subitex, Unicrop DNBP, Vertac, Dinitro Weed Killer 5, Vertac General Weed Killer, Vertac Selective Weed Killer)

dinoseb acetate (Aretit)

*continued next page*

Cataracts occur in laboratory animals given dinitrophenols and have occurred in humans, both as a result of ill-advised medicinal use and as a consequence of chronic occupational exposure.[27] Cataract formation is sometimes accompanied by glaucoma.

## Signs and Symptoms of Poisoning

Most patients present within a few hours of exposure with generalized non-specific signs and symptoms including profuse sweating, thirst, fever, headache, confusion, malaise and restlessness. The skin may appear warm and flushed as hyperthermia develops, along with tachycardia and tachypnea, all of which indicate a serious degree of poisoning. Apprehension, anxiety, manic behavior, seizures and coma reflect cerebral injury, with the latter two signifying an immediately life-threatening intoxication. Respiratory distress and cyanosis are consequences of the stimulated metabolism and tissue anoxia. Renal failure may occur early in cases of severe exposure. Liver damage is first manifested by jaundice, and cell death can occur within 48 hours and is dose dependent.[28] Death may occur within 24 to 48 hours after exposure in cases of severe poisoning.[26] In cases of survival of severe poisoning, complete resolution of symptoms may be slow due to the toxicant's long half-life.[26,29]

A characteristic bright yellow staining of skin and hair is often present with topical exposure and can be an important diagnostic clue to the clinician.[25,26,29] Yellow staining of the sclerae and urine indicate absorption of potentially toxic amounts. Weight loss occurs in persons continually exposed to relatively low doses of dinitrophenols.[25,27]

## Confirmation of Poisoning

If poisoning is probable, do not await confirmation before commencing treatment, but save urine and blood specimens on ice at a temperature below 20°C in the event confirmation is necessary later. Unmetabolized dinitrophenols can be identified spectrophotometrically, or by gas-liquid chromatography, in the serum at concentrations well below those that have been associated with acute poisonings. The data on exposure and systemic levels of compounds in this group are limited and most reports specify the compound dinitro-ortho-cresol. In general, blood levels of 10 µg/dL or greater are usually seen when systemic toxicity is evident.[25,30] One fatal case occurred with a level of 75 µg/dL.[30] Blood analysis is useful in confirming the cause of poisoning. Monitor levels routinely during acute intoxication to better establish a decay curve and determine when therapy can be safely discontinued.

## Treatment of Poisoning

Treatment of pentacholorophenol and dinitrophenol and its derivatives is the same, though there are some differences in toxicity as noted above.

1. Provide support treatment, including oxygen, fluid replacement and, most important, control of hyperthermia. There is no specific antidote for PCP or dinitrophenol toxicity.

2. Since these patients require aggressive control of hyperthermia, administer sponge baths and use fans to increase evaporation.[31] Cooling blankets and ice packs to body surfaces may also be used. In fully conscious patients, administer cold, sugar-containing liquids by mouth as tolerated. Antipyretic therapy with salicylates is **strongly contraindicated**, as salicylates also uncouple oxidative phosphorylation. Other antipyretics are thought to be of no use because of the

peripherally mediated mechanism of hyperthermia in poisoning of this nature. Note that profuse sweating is common in this poisoning, indicating that central acting antipyretics would have no effect. Neither the safety nor the effectiveness of the other antipyretics has been tested.

3. Administer oxygen continuously by mask to minimize tissue anoxia. Unless there are manifestations of cerebral or pulmonary edema or of inadequate renal function, administer intravenous fluids to restore hydration and support physiologic mechanisms for heat loss and toxicant disposition. Monitor serum electrolytes, adjusting IV infusions to stabilize electrolyte concentrations. Follow urine contents of albumin and cells, and keep an accurate hourly record of intake/output to forestall fluid overload if renal function declines.

   CAUTION: ***In the presence of cerebral edema and/or impaired renal function, intravenous fluids must be administered very cautiously to avoid increased intracranial pressure and pulmonary edema. Central monitoring of venous and pulmonary wedge pressures may be indicated. This is particularly important when cardiac dysfunction or heart failure is observed. Such critically ill patients should be treated in an intensive care unit.***

4. Decontaminate the skin with soap and water, as outlined in **Chapter 3, *General Principles.***

5. Treat eye contamination by irrigating the exposed eyes with copious amounts of clean water or saline for at least 15 minutes. Remove contact lenses, if present, prior to irrigation. Send patient for further medical attention if irritation or other injury persists.

6. Treat severe systemic poisoning in an intensive care unit setting with appropriate supportive care including respiratory support, intravenous fluids, cardiac monitoring and renal function support as necessary. The toxicant itself and severe electrolyte disturbances may predispose the patient to arrhythmias and myocardial weakness. Atropine is a medication that is absolutely contraindicated, and it is essential not to confuse the clinical signs for dinitrophenol with manifestations for cholinesterase inhibition poisoning.[26]

7. To reduce production of heat in the body, control agitation and involuntary motor activity with sedation. Lorazepam or other benzodiazepines should be effective, although use of these drugs in these poisonings has not been studied. Control seizures as outlined in **Chapter 3**.

8. Although most occupational poisoning is from inhalation, if ingested, consider gastrointestinal decontamination as outlined in **Chapter 3**.

***Dinotrophenolic Commercial Products, cont.***

dinoseb methacrylate (binapacryl, Morocide, Acricid, Ambox, Dapacryl, Endosan, FMC 9044, Hoe 002784, Morrocid, NIA 9044)

dinosulfon

dinoterb acetate

dinoterb salts

dinoterbon

## References


1. Jorens PG, Schepens PJ. Human pentachlorophenol poisoning. *Hum Exp Toxicol.* Nov 1993;12(6):479-495.
2. Pentachlorophenol. National Toxicology Information Program, National Library of Medicine, Bethesda, MD; 2000.
3. Kalman DA, Horstman SW. Persistence of tetrachlorophenol and pentachlorophenol in exposed woodworkers. *J Toxicol Clin Toxicol.* Jun 1983;20(4):343-352.
4. Weinbach EC. The effect of pentachlorophenol on oxidative phosphorylation. *J Biol Chem.* Oct 1954;210(2):545-550.
5. Danzo BJ. Environmental xenobiotics may disrupt normal endocrine function by interfering with the binding of physiological ligands to steroid receptors and binding proteins. *Environ Health Perspect.* Mar 1997;105(3):294-301.
6. Tran DQ, Klotz DM, Ladlie BL, Ide CF, McLachlan JA, Arnold SF. Inhibition of progesterone receptor activity in yeast by synthetic chemicals. *Biochem Biophys Res Commun.* Dec 13 1996;229(2):518-523.
7. DeMaeyer J, Schepens PJ, Jorens PG, Verstaete R. Exposure to pentachlorophenol as a possible cause of miscarriages. *Br J Obstet Gynaecol.* 1995;102:1010-1011.
8. Dimich-Ward H, Hertzman C, Teschke K, et al. Reproductive effects of paternal exposure to chlorophenate wood preservatives in the sawmill industry. *Scand J Work Environ Health.* Aug 1996;22(4):267-273.
9. Klemmer HW, Wong L, Sato MM, Reichert EL, Korsak RJ, Rashad MN. Clinical findings in workers exposed to pentachlorophenol. *Archives of environmental contamination and toxicology.* 1980;9(6):715-725.
10. Proudfoot AT. Pentachlorophenol poisoning. *Toxicol Rev.* 2003;22(1):3-11.
11. Walls CB, Glass WI, Pearce NE. Health effects of occupational pentachlorophenol exposure in timber sawmill employees: a preliminary study. *N Z Med J.* Sep 25 1998;111(1074):362-364.
12. Daniel V, Huber W, Bauer K, et al. Association of elevated blood levels of pentachlorophenol (PCP) with cellular and humoral immunodeficiencies. *Arch Environ Health.* Jan-Feb 2001;56(1):77-83.
13. Kentor PM. Urticaria from contact with pentachlorophenate. *JAMA.* Dec 26 1986;256(24):3350.
14. O'Malley MA, Carpenter AV, Sweeney MH, et al. Chloracne associated with employment in the production of pentachlorophenol. *Am J Ind Med.* 1990;17(4):411-421.
15. Robson AM, Kissane JM, Elvick NH, Pundavela L. Pentachlorophenol poisoning in a nursery for newborn infants. I. Clinical features and treatment. *J Pediatr.* Aug 1969;75(2):309-316.
16. Gray RE, Gilliland RD, Smith EE, Lockard VG, Hume AS. Pentachlorophenol intoxication: report of a fatal case, with comments on the clinical course and pathologic anatomy. *Arch Environ Health.* May-Jun 1985;40(3):161-164.
17. Wood S, Rom WN, White GL, Jr., Logan DC. Pentachlorophenol poisoning. *J Occup Med.* Jul 1983;25(7):527-530.
18. Roberts HJ. Aplastic anemia due to pentachlorophenol. *N Engl J Med.* Dec 31 1981;305(27):1650-1651.
19. Casarett LJ, Bevenue A, Yauger WL, Jr., Whalen SA. Observations on pentachlorophenol in human blood and urine. *Am Ind Hyg Assoc J.* Jul-Aug 1969;30(4):360-366.
20. Dahlgren J, Warshaw R, Thornton J, Anderson-Mahoney CP, Takhar H. Health effects on nearby residents of a wood treatment plant. *Environ Res.* Jun 2003;92(2):92-98.

21. Clayton GD, Clayton FE, eds. *Patty's Industrial Hygiene and Toxicology.* 4th ed. New York: John Wiley & Sons; 1994; No. 2B.

22. Gomez-Catalan J, To-Figueras J, Planas J, Rodamilans M, Corbella J. Pentachlorophenol and hexachlorobenzene in serum and urine of the population of Barcelona. *Hum Toxicol.* Sep 1987;6(5):397-400.

23. Wylie JA, Gabica J, Benson WW, Yoder J. Exposure and contamination of the air and employees of a pentachlorophenol plant, Idaho-1972. *Pest Monit.* 1975;9:150-153.

24. Wagner SL. Pentachlorophenol. *Clinical Toxicology of Agricultural Chemicals.* Corvallis: Oregon State University Press; 1981;131-137.

25. Leftwich RB, Floro JF, Neal RA, Wood AJ. Dinitrophenol poisoning: a diagnosis to consider in undiagnosed fever. *South Med J.* Feb 1982;75(2):182-184.

26. Finkel AJ, ed *Herbicides: Dinitrophenols.* 4 ed. Boston: John Wright PSG, Inc; 1983. Hamilton and Hardy's Industrial Toxicology.

27. Kurt TL, Anderson R, Petty C, Bost R, Reed G, Holland J. Dinitrophenol in weight loss: the poison center and public health safety. *Vet Hum Toxicol.* Dec 1986;28(6):574-575.

28. Palmeira CM, Moreno AJ, Madeira VM. Thiols metabolism is altered by the herbicides paraquat, dinoseb and 2,4-D: a study in isolated hepatocytes. *Toxicol Lett.* Nov 15 1995;81(2-3):115-123.

29. Smith WD. An investigation of suspected dinoseb poisoning after the agricultural use of a herbicide. *Practitioner.* Jun 1981;225(1356):923-926.

30. NIOSH. *Criteria for a Recommended Standard: Occupational Exposure to Dinitro-Ortho-Cresol.* 1978. 78-131.

31. Graham BS, Lichtenstein MJ, Hinson JM, Theil GB. Nonexertional heatstroke. Physiologic management and cooling in 14 patients. *Arch Intern Med.* Jan 1986;146(1):87-90.

*Paraquat & Diquat*
**HIGHLIGHTS**

Toxic doses are life threatening

Impacts GI tract, kidney, lungs, liver, heart, other organs

Pulmonary fibrosis is the usual cause of death in paraquat poisoning

Diquat has greater renal activity

Can be measured in blood and urine by spectrophotometric, gas chromatographic, liquid chromatographic and radioimmunoassay

**SIGNS & SYMPTOMS**

Ingestion (either): burning pain in mouth, throat, chest, upper abdomen; pulmonary edema, pancreatitis, renal & CNS effects

Dermal (paraquat): dry and fissured hands, horizontal ridging or loss of fingernails, ulceration, abrasion

Diquat: CNS toxicity as nervousness, irritability, combativeness, disorientation, diminished reflexes



CHAPTER 12

# Paraquat and Diquat

Paraquat and diquat are identified chemically as dipyridyls.

Paraquat is a synthetic, non-selective contact herbicide, marketed as paraquat, paraquat dichloride salt and bismethylsulfate salt. Liquid technical products range from 20% to 50% concentration, but the formulations used in the field range from 0.07% to 0.14%. It is a restricted use pesticide.

Diquat is usually prepared as the dibromide monohydrate salt, 15%-25% in liquid concentrates, but the formulations in the field are usually 0.23%. Diquat dibromide is a non-selective contact herbicide, desiccant and plant growth regulator for use as a general herbicide for control of broadleaf and grassy weeds in terrestrial non-crop and aquatic areas; as a desiccant in seed crops and potatoes; and for tassel control and spot weed control in sugarcane. Unlike paraquat, it is not registered as a restricted use pesticide.

## PARAQUAT

### Toxicology

When a toxic dose is ingested (see below), **paraquat** has life-threatening effects on the gastrointestinal tract, kidney, liver, heart and other organs. The $LD_{50}$ in humans is approximately 3-5 mg/kg, which translates into as little as 10-15 mL of a 20% solution.[1,2] In spite of the fact that the lung is the primary target organ, toxicity from inhalation is rare.

Although pulmonary toxicity occurs later in paraquat poisoning than other manifestations, it is the most severe and, therefore, mentioned first. Pulmonary effects represent the most lethal and least treatable manifestation of toxicity from this agent. The primary mechanism is through the generation of free radicals with oxidative damage to lung tissue.[1,2] While acute pulmonary edema and early lung damage may occur within a few hours of severe acute exposures,[3,4] the delayed toxic damage of pulmonary fibrosis, the usual cause of death, most commonly occurs 7-14 days after the ingestion.[5] In those patients who ingest a very large amount of concentrated solution (20%), some have died more rapidly from circulatory failure (within 48 hours) prior to the onset of pulmonary fibrosis.[5]

Both types I and II pneumatocytes appear to selectively accumulate paraquat. Biotransformation of the paraquat in these cells results in free-radical production with resulting lipid peroxidation and cell injury.[1,2,3] Hemorrhagic proteinaceous edema fluid and leukocytes infiltrate the alveolar spaces, after which there is rapid proliferation of fibroblasts. There is a progressive decline in arterial oxygen tension and $CO_2$ diffusion capacity. Such a severe impairment of gas exchange causes progressive proliferation of fibrous connective tissue in the alveoli and eventual death from asphyxia and tissue anoxia.[6] One study of survivors suggests that some of the fibrous toxic damage may be reversible, as evidenced by markedly improved pulmonary function tests 3 months after survival.[7]

*Paraquat & Diquat*
**TREATMENT**

Immediate GI decontamination with Bentonite, Fuller's Earth or activated charcoal

Maintain urinary output by administering IV, but monitor fluids in case of renal failure

Decontaminate eyes and skin

**CONTRAINDICATED**

Supplemental oxygen (unless patient develops hypoxemia)

Local skin damage includes a contact dermatitis. Prolonged contact will produce erythema, blistering, abrasion, ulceration and fingernail changes.[8,9] Although absorption across intact skin is slow, abraded or eroded skin allows efficient absorption.

The gastrointestinal (GI) tract is the site of initial or Phase 1 toxicity to the mucosal surfaces following ingestion of the substance. This toxicity is manifested by swelling, edema and painful ulceration of the mouth, pharynx, esophagus, stomach and intestine. With higher levels, other GI toxicity includes centrizonal hepatocellular injury that can cause elevated bilirubin, and hepatocellular enzymes such as AST, ALT, LDH and alkaline phosphatase.

Damage to the proximal renal tubule occurs and is often more reversible than the destruction to lung tissue. However, impaired renal function may play a critical role in determining the outcome of paraquat poisoning. Normal tubule cells actively secrete paraquat into the urine, efficiently clearing it from the blood; but high blood concentrations poison the secretory mechanism and may destroy the cells. Diquat poisoning typically results in greater renal injury than paraquat.[10]

Focal necrosis of the myocardium and skeletal muscle are the main features of toxicity to any type of muscle tissue and typically occurs following the Phase 1 gastrointestinal toxicity.

Ingestion has been reported to cause cerebral edema and brain damage. At necropsy, brain damage was found in the form of moderate neuronal depletion, probably secondary to anoxia, and damage to the central white matter and particularly the brain around the lateral and third ventricles. Examination of the brain by electron microscopy showed edema and destruction of myelin, with abundant myelin breakdown products, and astrocytic fibrous gliosis.[11]

Although much concern has been expressed about effects of smoking paraquat-contaminated marijuana, toxic effects by this mechanism have been either very rare or nonexistent. Most paraquat that contaminates marijuana is pyrolyzed to dipyridyl during smoking, which is a product of leaf (including marijuana) combustion and presents little toxic hazard.

## Signs and Symptoms of Poisoning

Initial clinical signs depend upon the route of exposure. Early symptoms and signs of poisoning by ingested paraquat are burning pain in the mouth, throat, chest and upper abdomen, due to the corrosive effect of paraquat on the mucosal lining. Diarrhea, which is sometimes bloody, can also occur. Giddiness, headache, fever, lethargy and coma are other examples of CNS and systemic findings. Pancreatitis may cause severe abdominal pain. Proteinuria, hematuria, pyuria and azotemia reflect renal injury. Oliguria/anuria indicates acute tubular necrosis. Because the kidneys are almost the exclusive route of paraquat elimination from body tissues, renal failure fosters a buildup of tissue concentration, including the very important concentration in the lung.

Unfortunately, this pathogenic sequence may occur in the first several hours following paraquat ingestion, generating lethal concentrations of paraquat in lung tissue before therapeutic measures to limit absorption and enhance disposition have taken effect. It is probably for this reason that methods for enhancing paraquat disposition several hours following ingestion have had little effect on mortality.[9]

Cough, dyspnea and tachypnea usually appear 2-4 days following paraquat ingestion but may be delayed as long as 14 days. Progressive cyanosis and dyspnea reflect deteriorating gas exchange in the damaged lung. In some cases, the coughing up of frothy sputum (pulmonary edema) is the early and principal manifestation of paraquat lung injury.[9]

Dermal signs are common among agriculture workers with acute skin exposure to paraquat. Particularly in concentrated form, paraquat causes localized injury

*Paraquat*
**COMMERCIAL PRODUCTS**

Bonfire
Firestorm
Gramoxone
Helmquat
Para-Shot
Parazone
Quik-Quat

*Diquat*
**COMMERCIAL PRODUCTS**

Chemsico
Rapid Kill
Razor Burn
Reglone
Touchdown
Weedtrine-D

to tissues with which it comes into contact. Fatal poisonings are reported to have occurred as a result of protracted dermal contamination by paraquat, but this is likely to occur only when the skin's barrier integrity is impaired due to abrasion, erosion or other pathologic processes. In these cases, more efficient systemic absorption can occur. With an intact dermal barrier, paraquat leaves the skin of the hands dry and fissured, and causes horizontal ridging of the fingernails. Chronic exposure may even result in the loss of fingernails. Prolonged contact with skin will create ulceration and abrasion sufficient to allow systemic absorption.[9]

In addition, some agriculture workers can be exposed through prolonged inhalation of spray droplets and develop nosebleeds because of local damage. However, inhalation has not resulted in systemic toxicity because of the low vapor pressure and lower concentration of paraquat field formulations.

Eye contamination with paraquat concentrate or higher concentration diluted solutions results in severe conjunctivitis and sometimes protracted corneal opacification.[12,13]

The hepatic injury from paraquat may be severe enough to cause jaundice, which signifies severe injury. However, hepatotoxicity is rarely a major determinant to clinical outcome. No hepatic signs or symptoms are present other than the abnormal laboratory values mentioned under the toxicology section.

Clinical experience has offered a rough dose-effect scale on which to base prognosis in cases of paraquat ingestion[9]:

1. Less than 20 mg paraquat ion per kg body weight (less than 7.5 mL of 20% [w/v] paraquat concentrate). No symptoms or only gastrointestinal symptoms occur. Recovery is likely.

2. Twenty to 40 mg paraquat ion per kg body weight (7.5-15.0 mL of 20% [w/v] paraquat concentrate). Pulmonary fibroplasia ensues. Death occurs in most cases, but may be delayed 2-3 weeks. Multiple organ damage will occur.

3. More than 40 mg paraquat ion per kg body weight (more than 15.0 mL of 20% [w/v] paraquat concentrate). Multiple organ damage occurs as in Class 2 but is more rapidly progressive. The gastrointestinal effects are often characterized by marked ulceration of the oropharynx. Mortality is essentially 100% in 1-7 days.

## DIQUAT

### Toxicology

**Diquat** poisoning is less common than paraquat poisoning, thus the human reports and animal experimental data for diquat poisoning are less extensive than for paraquat. Systemically absorbed diquat is not selectively concentrated in lung tissue, as is paraquat, and pulmonary injury by diquat is less prominent. In animal studies, diquat causes mild, reversible injury to type I pneumatocytes but does not injure the type II cells. No progressive pulmonary fibrosis has been noted in diquat poisoning.[14,15]

However, diquat has severe toxic effects on the central nervous system that are not typical of paraquat poisoning.[14,15] While laboratory experimentation has suggested that diquat is not directly neurotoxic, there have been relatively consistent pathologic brain changes noted in reported fatal cases of diquat poisoning. These consist of brain stem infarction, particularly involving the pons.[16] It is not clear whether these post-mortem changes represent direct toxicity or secondary effects related to the systemic illness and therapy. (See *Signs and Symptoms* section below for CNS clinical effects.)

## Signs and Symptoms of Poisoning

In many human diquat poisoning cases, clinical signs of neurologic toxicity tend to be the most important. These include nervousness, irritability, restlessness, diminished reflexes, combativeness, disorientation, nonsensical statements and inability to recognize friends or family members. Neurologic effects may progress to coma, accompanied by tonic-clonic seizures, and result in the death of the patient.[14,15] Parkinsonism has also been reported following dermal exposure to diquat.[17]

Except for the CNS signs listed in the preceding paragraph, early symptoms of poisoning by ingested diquat are similar to those from paraquat, reflecting diquat's corrosive effect on tissues. They include burning pain in the mouth, throat, chest and abdomen; intense nausea and vomiting; and diarrhea. If the dosage was small, these symptoms may be delayed 1-2 days. Blood may appear in the vomitus and feces. Intestinal ileus, with pooling of fluid in the gut, has characterized several human poisonings by diquat.[10]

The kidney is the principal excretory pathway for diquat absorbed into the body. Renal damage is, therefore, an important feature of poisonings. Proteinuria, hematuria and pyuria may progress to renal failure and azotemia. Elevations of serum alkaline phosphatase, AST, ALT and LDH reflect liver injury. Jaundice may develop.

If the patient survives several hours or days, circulatory function may fail because of dehydration. Hypotension and tachycardia can occur, with shock resulting in death. Other cardiorespiratory problems may develop such as toxic cardiomyopathy or a secondary infection such as bronchopneumonia.

Diquat is somewhat less damaging to the skin than paraquat, but irritant effects may appear following dermal contamination with the concentrate. There is probably significant absorption of diquat across abraded or ulcerated skin.

The great majority of poisonings by paraquat and diquat (discussed below) have been caused by ingestion with suicidal intent, particularly in Japan[16] and many developing countries. Since 1987, there has been a decline in most countries in the total numbers of suicidal deaths attributed to paraquat and diquat. Nearly all of the relatively few occupationally related poisonings have been survived, but the mortality rate among persons who have swallowed paraquat or diquat remains high.[2,5] Avoidance of this mortality will probably have to rely on preventive strategies or on stopping gastrointestinal absorption very soon after the toxicant has been ingested.

Even though intestinal absorption of dipyridyls is relatively slow, lethal uptake by critical organs and tissues apparently occurs within 18 hours, possibly within 6 hours, following ingestion of toxic quantities of paraquat or diquat. Dipyridyls have large volumes of distribution. Once distribution to tissues has occurred, measures to remove dipyridyls from the blood are very inefficient in reducing the total body burden.

Several strategies are being tested to reduce the frequency of these occurrences. These include the addition of emetics, stenching agents, gelling substances and bittering agents such as sodium denatonium.

## Confirmation of Poisoning

At some treatment facilities, a simple colorimetric test is used to identify paraquat and diquat in the urine and give a rough indication of the magnitude of absorbed dose. To one volume of urine is added 0.5 volume of freshly prepared 1% sodium dithionite (sodium hydrosulfite) in one normal sodium hydroxide (1.0 N NaOH). The color is observed after 1 minute. Development of a blue color indicates the presence of paraquat in excess of 0.5 mg per liter. Both positive and negative controls should be run to ensure that the dithionite has not undergone oxidation in storage.

When urine collected within 24 hours of paraquat ingestion is tested, the dithionite test appears to have some approximate prognostic value: concentrations less than 1 milligram per liter (no color to light blue) generally predict survival, while concentrations in excess of 1 milligram per liter (navy blue to dark blue) often foretell a fatal outcome. Analysis of serum by a sodium dithionite test has been reported to predict outcome in paraquat exposures. In one center a positive test was associated with 100% mortality, while negative or equivocal tests resulted in a 68% survival rate.[18]

Diquat in urine yields a green color with the dithionite test. Although there is less experience with this test in diquat poisonings, the association of bad prognosis with intense color is probably similar.

Paraquat and diquat can be measured in blood and urine by spectrophotometric, gas chromatographic, liquid chromatographic and radioimmunoassay methods. These tests are available in numerous clinical reference laboratories and sometimes by the manufacturing company. Paraquat poisonings in which plasma concentrations do not exceed 2.0, 0.6, 0.3, 0.16 and 0.1 mg per liter at 4, 6, 10, 16 and 24 hours, respectively, after ingestion are likely to survive.[19] A comparison of several methods of measuring plasma paraquat levels revealed comparable results. However, while the positive predictive value for death was quite high, the ability to predict survival was much lower.[20]

## Lung Imaging

It has been reported that high-resolution computerized tomography of the lungs may be of predictive value in acute paraquat poisoning. A calculation is made of areas of ground glass opacities (GGOs) on tomography. In one study no patient survived when the area was greater than 40% and all survived when the area was less than 20%.[21] This study may be useful in evaluating newer therapeutic approaches.

## Treatment of Paraquat and Diquat Toxicosis

1. Flush skin immediately with copious amounts of water to decontaminate. Irrigate the eyes with clean water for a prolonged period to remove material splashed in the eyes. Eye contamination should thereafter be treated by an ophthalmologist. Mild skin reactions usually respond to simple avoidance of further contact, but the irritation may take several weeks to resolve. Severe dermatitis with inflammation, cracking, secondary infection or nail injury should be treated by a dermatologist.

2. If paraquat or diquat has been ingested in any amount, immediately administer an adsorbent. This is the one therapeutic measure most likely to affect the outcome of paraquat or diquat ingestion favorably. Bentonite (7.5% suspension) and Fuller's Earth (15% suspension) are highly effective but sometimes not available.

### Dosage of Bentonite and Fuller's Earth

- ***Adults and children over 12 years: 100-150 gm***
- ***Children under 12 years: 2 gm/kg body weight***

CAUTION: ***Hypercalcemia and fecaliths have sometimes occurred following administration of Fuller's Earth.***

Activated charcoal is nearly as effective, and is widely available. This treatment is discussed in **Chapter 3, *General Principles***.

3. Secure a blood sample as soon as possible for paraquat analysis and urine samples for either paraquat and/or diquat. Serial samples of urine for either agent and plasma for paraquat may be followed for prognostic information.

4. Do not administer supplemental oxygen until the patient develops severe hypoxemia. High concentrations of oxygen in the lung increase the injury induced by paraquat and possibly by diquat as well. There may be some advantage in placing the patient in a moderately hypoxic environment, *i.e.,* 15%-16% oxygen, although the benefit of this treatment has not been established empirically in human poisonings. Inhalation of nitric oxide has been suggested as a method to maintain tissue oxygenation at low inspired oxygen concentrations but is of unproven efficacy. When the lung injury is so far advanced that there is no expectation of recovery, oxygen may be given to relieve air hunger.

5. In serious poisonings, provide care in an intensive care setting to allow proper monitoring of body functions and skilled performance of necessary invasive monitoring and procedures.

6. As it is essential to maintain adequate urinary output,[3] administer intravenous fluids: isotonic saline, Ringer's solution or 5% glucose in water. This is highly advantageous early in poisonings as a means of correcting dehydration, accelerating toxicant excretion, reducing tubular fluid concentrations of paraquat and correcting metabolic acidosis. However, fluid balance must be monitored carefully to forestall fluid overload if renal failure develops. Monitor the urine regularly for protein and cells to warn of impending tubular necrosis. Intravenous infusions must be stopped if renal failure occurs, and extracorporeal hemodialysis is indicated. Hemodialysis is not effective in clearing paraquat or diquat from the blood and tissues.

7. Consider hemoperfusion over cellophane-coated activated charcoal. The procedure has been used in many paraquat poisonings because the adsorbent does efficiently remove paraquat from the perfused blood. However, recent reviews of effectiveness have failed to show any reduction in mortality as a result of hemoperfusion.[2,3,22] The apparent reason for this is the very small proportion of paraquat body burden carried in the circulating blood even when only a few hours have elapsed after ingestion. Theoretically, a patient who can be hemoperfused within 10 hours of paraquat ingestion may derive some marginal benefit, but this has not been demonstrated. If hemoperfusion is attempted, blood calcium and platelet concentrations must be monitored. Calcium and platelets must be replenished if these constituents are depleted by the procedure.

8. Control seizures following procedure in **Chapter 3**.

   CAUTION: ***Be prepared to assist ventilation mechanically if respiration is depressed, to intubate the trachea if laryngospasm occurs and to counteract hypotensive reactions.***

9. Consider administering cyclophosphamide and methylprednisolone. Many drugs have been tested in animals or given in human dipyridyl poisonings without clear evidence of benefit or harm: corticosteroids, superoxide dismutase,

propranolol, cyclophosphamide, vitamin E, riboflavin, niacin, ascorbic acid, clofibrate, desferrioxamine, acetylcysteine, terpin hydrate and melatonin.[23] However, recent evidence regarding the use of cyclophosphamide and methylprednisolone shows that they may be effective in reducing the mortality associated with moderate-to-severe paraquat poisoning. Two studies found a reduced mortality associated with the treatment, while one study found no difference.[24] The dosages used for cyclophosphamide and methylprednisolone were 1 gram daily for 2 days and 1 gram daily for 3 days, respectively, given after the hemoperfusion. Each drug was administered as a 2-hour infusion; white cell counts, serum creatinine levels, chest radiography and liver function tests were monitored.[24] Two controlled trials seem to have confirmed benefit from cyclophosphamide and methylprednisolone therapy with reduction of mortality from 81% to 33% in one study and 86% to 31% in another.[24,25] The protocols for administration of the drugs were similar but not identical.

10. Manage pain with morphine sulfate. Morphine sulfate is usually required to control the pain associated with deep mucosal erosions of the mouth, pharynx and esophagus, as well as abdominal pain from pancreatitis and enteritis.

**Dosage for Morphine Sulfate**

- ***Adults and children over 12 years: 10 - 15 mg subcutaneously every 4 hours.***
- ***Children under 12 years: 0.1 - 0.2 mg /kg body weight every 4 hours.***

Mouthwashes, cold fluids, ice cream or anesthetic lozenges may help to relieve pain in the mouth and throat.

With severe pulmonary toxicity, recovery of the patient may only be accomplished by lung transplantation. However, the transplanted lung is susceptible to subsequent damage due to redistribution of paraquat.[26]

## References

1. Giulivi C, Lavagno CC, Lucesoli F, Bermudez MJ, Boveris A. Lung damage in paraquat poisoning and hyperbaric oxygen exposure: superoxide-mediated inhibition of phospholipase A2. *Free Radic Biol Med.* Feb 1995;18(2):203-213.
2. Pond SM. Manifestations and management of paraquat poisoning. *Med J Aust.* Mar 5 1990;152(5):256-259.
3. Honore P, Hantson P, Fauville JP, Peeters A, Manieu P. Paraquat poisoning. "State of the art". *Acta Clin Belg.* 1994;49(5):220-228.
4. Nordquist RE, Nguyen H, Poyer JL, Carubelli R. The role of free radicals in paraquat-induced corneal lesions. *Free Radic Res.* Jul 1995;23(1):61-71.
5. Bismuth C, Garnier R, Dally S, Fournier PE, Scherrmann JM. Prognosis and treatment of paraquat poisoning: a review of 28 cases. *J Toxicol Clin Toxicol.* Jul 1982;19(5):461-474.
6. Harsanyi L, Nemeth A, Lang A. Paraquat (gramoxone) poisoning in south-west Hungary, 1977-1984. Toxicological and histopathological aspects of group intoxication cases. *Am J Forensic Med Pathol.* Jun 1987;8(2):131-134.

7. Lee CC, Lin JL, Liu L. Recovery of respiratory function in survivors with paraquat intoxication. *Annals of Emergency Medicine.* 1995;26(2):721-722.

8. Tungsanga K, Chusilp S, Israsena S, Sitprija V. Paraquat poisoning: evidence of systemic toxicity after dermal exposure. *Postgrad Med J.* May 1983;59(691):338-339.

9. Vale JA, Meredith TJ, Buckley BM. Paraquat poisoning: clinical features and immediate general management. *Hum Toxicol.* Jan 1987;6(1):41-47.

10. Jones GM, Vale JA. Mechanisms of toxicity, clinical features, and management of diquat poisoning: a review. *J Toxicol Clin Toxicol.* 2000;38(2):123-128.

11. Hughes JT. Brain damage due to paraquat poisoning: a fatal case with neuropathological examination of the brain. *Neurotoxicology.* Summer 1988;9(2):243-248.

12. McKeag D, Maini R, Taylor HR. The ocular surface toxicity of paraquat. *Br J Ophthalmol.* Mar 2002;86(3):350-351.

13. Grant WM, Schuman JS. *Toxicology of the Eye*. 4th ed. Springfield: Charles C Thomas Publisher Ltd; 1993.

14. Olson KR. Paraquat and diquat. *Poisoning and drug overdose*. 2nd ed. Norwalk: Appelton and Lange; 1994:245-246.

15. Vanholder R, Colardyn F, De Reuck J, Praet M, Lameire N, Ringoir S. Diquat intoxication: report of two cases and review of the literature. *Am J Med.* Jun 1981;70(6):1267-1271.

16. Lam HF, Takezawa J, Gupta BN, van Stee EW. A comparison of the effects of paraquat and diquat on lung compliance, lung volumes and single breath diffusing capacity in the rat. *Toxicology.* 1980;18(2):111-123.

17. Sechi GP, Agnetti V, Piredda M, et al. Acute and persistent parkinsonism after use of diquat. *Neurology.* Jan 1992;42(1):261-263.

18. Koo JR, Yoon JW, Han SJ, et al. Rapid analysis of plasma paraquat using sodium dithionite as a predictor of outcome in acute paraquat poisoning. *Am J Med Sci.* Nov 2009;338(5):373-377.

19. Proudfoot AT, Stewart MS, Levitt T, Widdop B. Paraquat poisoning: significance of plasma-paraquat concentrations. *Lancet.* Aug 18 1979;2(8138):330-332.

20. Senarathna L, Eddleston M, Wilks MF, et al. Prediction of outcome after paraquat poisoning by measurement of the plasma paraquat concentration. *QJM.* Apr 2009;102(4):251-259.

21. Kim YT, Jou SS, Lee HS, et al. The area of ground glass opacities of the lungs as a predictive factor in acute paraquat intoxication. *J Korean Med Sci.* Aug 2009;24(4):636-640.

22. Feinfeld DA, Rosenberg JW, Winchester JF. Three controversial issues in extracorporeal toxin removal. *Semin Dial.* Sep-Oct 2006;19(5):358-362.

23. Suntres ZE. Role of antioxidants in paraquat toxicity. *Toxicology.* Oct 30 2002;180(1):65-77.

24. Lin JL, Wei MC, Liu YC. Pulse therapy with cyclophosphamide and methylprednisolone in patients with moderate to severe paraquat poisoning: a preliminary report. *Thorax.* Jul 1996;51(7):661-663.

25. Afzali S, Gholyaf M. The effectiveness of combined treatment with methylprednisolone and cyclophosphamide in oral paraquat poisoning. *Arch Iran Med.* Jul 2008;11(4):387-391.

26. Sequential bilateral lung transplantation for paraquat poisoning. A case report. The Toronto Lung Transplant group. *J Thorac Cardiovasc Surg.* May 1985;89(5):734-742.

*Phosphonate Herbicides*

**COMMERCIAL PRODUCTS**

fosamine ammonium

glyphosate (Brands include Round-Up and Glyfonox)

**HIGHLIGHTS**

Most commonly used herbicide in U.S.

Many reported poisonings

Actual toxicity likely from surfactant

Read label to ascertain possible additional ingredients

**SIGNS & SYMPTOMS**

GI symptoms predominate

Cardiovascular, respiratory, renal systems may be affected

Can be measured in plasma

**TREATMENT**

Decontaminate skin and eyes

Consider GI decontamination

Control seizures

Consider hemodialysis in cases of renal failure

No known antidote

CHAPTER 13

# Other Herbicides

Many herbicides are now available for use in agriculture and for lawn and garden weed control. This chapter discusses herbicides other than the chlorophenoxy compounds, nitro- and chloro-phenols, arsenicals and dipyridyls, which are subjects of separate chapters. Many modern herbicides kill weeds selectively by impairing metabolic processes that are unique to plant life. For this reason, systemic toxicity in mammals is generally low. Nonetheless, some pose a significant risk of poisoning if not handled appropriately, and may result in eye, skin and mucous membrane irritation.

Herbicides mentioned in this chapter should be handled and applied only with proper personal protective equipment and careful attention to hygienic measures that minimize personal contact. Many formulations contain adjuvants (stabilizers, penetrants, surfactants) that may have significant irritating and toxic effects in addition to the primary herbicide. A number of premixed products may be combination formulations with additional active ingredients that are more toxic than the principal herbicide. Therefore, it is important to read the label to identify each active ingredient and its associated toxicities. Good hygienic practice should not be disregarded because only the primary pesticide is reported to have a high $LD_{50}$ in laboratory animals.

Healthcare professionals should have a general understanding of the metabolism and health effects of these compounds after human exposures. This knowledge is necessary to properly assess acute and chronic exposures. Generally, water-soluble herbicides are not retained in body tissues for long periods of time, as were the previously used lipophilic organochlorine insecticides such as DDT. Most of the water-soluble herbicides are primarily excreted, mainly in the urine, within 1-4 days.

This chapter follows a slightly different format than the other chapters in this book. Glyphosate is discussed separately since it is a widely used herbicide. It has been studied extensively and is the subject of numerous publications in the medical literature. The remaining herbicides in the chapter are summarized in a table. A notable inclusion in the table is propanil, an anilide herbicide. Propanil was previously described as having low toxicity; however, data from Sri Lanka have documented significant acute toxicity with the development of methemoglobinemia, including several fatalities.[1,2]

The rat acute oral $LD_{50}$ is given as a rough index of potential lethal toxicity (If several values are reported by various sources, the lowest is recorded here). Adverse effect information given is drawn from many sources, including reregistration eligibility decisions (REDs), product labels, textbooks and published reports. The listing cannot be considered inclusive, either of herbicide products or of effects.

## PHOSPHONATE HERBICIDES

Glyphosate is the most commonly used herbicide in the United States; it is used as weed control on numerous agricultural crops and is also registered for home use.[3] The advent of genetically modified seed producing plants resistant to glyphosate allows the planting of crops such as corn that can tolerate widespread application of glyphosate. The National Poison Data System (NPDS) uses 63 generic categories to classify pesticides. In 2010, among reported human exposures to pesticides, glyphosate ranked eighth.[4] Although Round-Up is the most well-known brand of glyphosate, note that some products with the same brand name may include additional active ingredients. Always read labels carefully.

## Toxicology

**Glyphosate** is marketed in the United States as **isopropylamine salt**. Glyphosate and related compounds have a specific mechanism of action inhibiting the enzyme responsible for synthesizing phenylalanine, tyrosine and tryptophan, which is an enzyme system that is not present in humans.[5,6] Given the plant-specific mechanism of action, there is theoretically a low risk for acute human toxicity. Indeed, glyphosate has low acute toxicity in mammals, with a rat $LD_{50}$ in the range of 4,300 mg/kg. Despite this, there have been a number of reports in the medical literature of acute glyphosate-related poisoning. Most, if not all, of the symptoms may actually be related to the organic surfactant with which glyphosate is combined. Most moderate to severe symptomatic cases have been associated with intentional (suicidal) ingestion.[3,7,8]

Another formulation of glyphosate is **glyphosate-trimesium**, which is not marketed in the United States. Two fatalities have been reported associated with glyphosate-trimesium exposure.[9]

## Signs and Symptoms of Poisoning

Gastrointestinal symptoms predominate, including mouth and throat pain, nausea, vomiting, diarrhea and abdominal discomfort, and are usually self limiting. More severe signs and symptoms may be seen in cases of intentional oral exposures. Cardiovascular, respiratory and renal systems may be affected; and signs and symptoms include tachypnea, dysrhythmias, hypotension, non-cardiogenic pulmonary edema, hypovolemic shock, oliguria and respiratory failure. Seizures and depressed level of consciousness may also occur. Death was often caused by severe hypotension and respiratory failure.[3,8] Hyperkalemia may occur as a complication of renal failure.[3,7]

One study assessed patients prospectively following a report of glyphosate ingestion. Of the 601 cases, most were either asymptomatic (27%) or with minor symptoms (64%). Approximately 5.5% had moderate to severe poisoning, and 3.2% of the patients died.[8] In another series of acutely poisoned patients, 42% had medical complications of some type, with metabolic acidosis (37%) and respiratory failure (28%) being the most common. A late complication in 12% of patients was pancreatitis.[3]

## Confirmation of Poisoning

Glyphosate can be measured in the plasma, with levels above 734 μg/mL being measured in fatal cases.[8]

## Treatment of Glyphosate Toxicosis

1. Provide supportive treatment, as there is no known antidote.

2. Decontaminate the skin with soap and water as outlined in **Chapter 3, *General Principles***. Treat eye contamination by irrigating the exposed eye(s) with copious amounts of clean water or saline for at least 15 minutes. Remove contact lenses, if present, prior to irrigation. If irritation persists after irrigation, specialized medical treatment in a healthcare facility is indicated.

3. If ingested, consider gastrointestinal decontamination as outlined in **Chapter 3**.
4. Control seizures using benzodiazepines. See **Chapter 3** for specific medications and dosages.
5. In cases of severe poisoning resulting in acute renal failure, consider hemodialysis to correct acidosis and hyperkalemia.[7]

## Potential Effects of Other Herbicides

The potential effects of a variety of other herbicides are summarized in the following multi-page table.

| POTENTIAL EFFECTS OF OTHER HERBICIDES | | | | |
|---|---|---|---|---|
| **Chemical Class** | **Generic Name** | **Examples of Proprietary Names** | **Acute Oral $LD_{50}$ mg/kg** | **Known or Suspected Adverse Effects** |
| Acetamides | Metolachlor | Dual, Pennant | 2,780 | Irritating to eyes and skin. Methemoglobinemia has been reported in a mixed herbicide ingestion with the urea derivative metobromuron; however, it is likely that the latter was the cause of the methemoglobinemia. |
| Anilides | Alachlor | Lasso, Alanox | 1,800 | Mild irritant, vomiting. Occasionally hypotension and CNS depression.[10] |
| | Propachlor | Ramrod, Bexton, Prolex | 710 | Dermal irritant and sensitizer. |
| | Propanil | DPA, Chem Rice, Propanex, Riselect, Stam, Stampede | >2,500 | Despite the relatively high $LD_{50}$ in rats, this compound has caused significant methemoglobinemia, reduced consciousness and respiratory depression.[1,2] |
| Benzamide | Pronamide | Kerb, Rapier | 8,350 | Moderately irritating to eyes. |

*continued next page*

| POTENTIAL EFFECTS OF OTHER HERBICIDES, continued | | | | |
|---|---|---|---|---|
| **Chemical Class** | **Generic Name** | **Examples of Proprietary Names** | **Acute Oral $LD_{50}$ mg/kg** | **Known or Suspected Adverse Effects** |
| Benzoic, Anisic Acid derivatives | Trichloro-benzoic acid | TCBA, Tribac, 2,3,6-TBA | 1,500 | Moderately irritating to skin and respiratory tract. |
| | Dicamba | Banvel | 2,700 | |
| Benzonitriles | Dichlobenil | Casoron, Dyclomec, Barrier | >4,460 | Minimal toxic, irritant effects. |
| Benzothiadiazinone dioxide | Bentazone | Basagran | >1,000 | Generally described as irritating to eyes, GI tract and respiratory tract. Some reports of acute renal failure and respiratory failure have been reported with ingestion of large amounts.[11,12] |
| Carbamates and Thiocarbamates (herbicidal) | Asulam | Asulox | >5,000 | Some are irritating to eyes, skin, and respiratory tract, particularly in concentrated form. Some may be weak inhibitors of cholinesterase. |
| | Terbucarb | Azac, Azar | >34,000 | |
| | Butylate | Sutan | 3,500 | |
| | Cycloate | Ro-Neet | 2,000 | |
| | Pebulate | Tillam, PEBC | 921 | |
| | EPTC | Eptam, Eradicane | 1,630 | |
| | Diallate | Di-allate | 395 | |
| | Triallate | Far-go | 1,675 | |
| | Thiobencarb | Bolero, Saturn | 1,300 | |
| Chloropyridinyl | Triclopyr | Garlon, Turflon | 630 | Irritating to skin and eyes. |

*continued next page*

| POTENTIAL EFFECTS OF OTHER HERBICIDES, continued | | | | |
|---|---|---|---|---|
| **Chemical Class** | **Generic Name** | **Examples of Proprietary Names** | **Acute Oral $LD_{50}$ mg/kg** | **Known or Suspected Adverse Effects** |
| **Cyclohexenone derivative** | Sethoxydim | Poast | 3,125 | Irritating to skin and eyes. |
| **Dinitroaminobenzene derivative** | Butralin | Amex, Tamex | 12,600<br>>5,000 | May be moderately irritating, particularly to the GI tract following ingestion.[13] These herbicides do not uncouple oxidative phosphorylation or generate methemoglobin. |
| | Pendi-methalin | Prowl, Stomp, Accotab, Herbodox, Go-Go-San, Wax Up | 2,250 | |
| | Oryzalin | Surflan, Dirimal | >10,000 | |
| **Fluorodinitrotoluidine compounds** | Benfluralin | Benefin, Balan, Balfin, Quilan | >10,000 | May be mildly irritating. These herbicides do not uncouple oxidative phosphorylation or generate methemoglobin. |
| | Ethalfluralin | Sonalan | >10,000 | |
| | Fluchloralin | Basalin | 1,550 | |
| | Trifluralin | Treflan | >10,000 | |
| **Nicotinic idisopropylamine derivative** | Imazapyr | Arsenal | >5,000 | Irritating to eyes and skin. Impaired consciousness, respiratory distress and severe vomiting occurs with large quantity (>100 mL) ingestion.[14] Does not contain arsenic. |
| **Oxadiazolinone** | Oxadiazon | Ronstar | >3,500 | Minimal toxic and irritant effects. |
| **Picolinic acid compound** | Picloram | Tordon, Pinene | 8,200 | Irritating to skin, eyes, and respiratory tract. Low systemic toxicity. |

*continued next page*

| POTENTIAL EFFECTS OF OTHER HERBICIDES, continued | | | | |
|---|---|---|---|---|
| **Chemical Class** | **Generic Name** | **Examples of Proprietary Names** | **Acute Oral $LD_{50}$ mg/kg** | **Known or Suspected Adverse Effects** |
| Triazines | Ametryn | Ametrex, Evik, Gesapax | 1,750 | Systemic toxicity is unlikely unless large amounts have been ingested. There is one report in the literature of metabolic acidosis following massive ingestion of prometryn.[15] Some triazines are moderately irritating to the eyes, skin and respiratory tract. |
| | Atrazine | Aatrex, Atranex, Crisazina | 1,780 | |
| | Desmetryn | Semeron | 1,390 | |
| | Metribuzin | Sencor, Lexone, Sencoral, Sencorex | 1,100 | |
| | Prometryn | Caparol, Gesagard, Prometrex | 5,235 | |
| | Propazine | Milo-Pro, Primatol, Prozinex | >7,000 | |
| | Simazine | Gesatop, Princep, Caliber 90 | >5,000 | |
| | Terbuthy-lazine | Gardoprim, Primatol M | 2,000 | |
| | Tertutryn | Ternit, Prebane, Terbutrex | 2,500 | |
| | Prometon | Gesafram 50, Pramitol 25E | 2,980 | Some formulations of prometon are strongly irritating to eyes, skin and respiratory tract. |
| Triazole | Amitrole, aminotriazole | Amerol, Azolan, Azole, Weedazol | >10,000 | Minimal systemic toxicity. Slight irritant effect. |
| Uracils | Bromacil | Hyvar | 5,200 | Irritant to skin, eyes and respiratory tract. |
| | Lenacil | Venzar | >11,000 | Moderately irritating. |
| | Terbacil | Sinbar | >5,000 | |

*continued next page*

| POTENTIAL EFFECTS OF OTHER HERBICIDES, continued | | | | |
|---|---|---|---|---|
| **Chemical Class** | **Generic Name** | **Examples of Proprietary Names** | **Acute Oral $LD_{50}$ mg/kg** | **Known or Suspected Adverse Effects** |
| Urea derivatives | Chlorimuron ethyl | Classic | >4,000 | Systemic toxicity is unlikely unless large amounts have been ingested. |
| | Chlorotoluron | Dicuran, Tolurex | >10,000 | Chlorimuron ethyl has been associated with asthma.[16] |
| | Diuron | Cekiuron, Crisuron, Dailon, Direx, Diurex, Diuron, Karmex, Unidron, Vonduron | >5,000 | Many substituted ureas are irritating to eyes, skin and mucous membranes. |
| | Ebuthiuron | Spike, Tebusan | 644 | |
| | Flumeturon | Cotoran, Cottonex | 8,900 | |
| | Isoproturon | Alon, Arelon, IP50, Tolkan | 1,826 | |
| | Linuron | Afalon, Linex, Linorox, Linurex, Lorox, Sarclex | 1,500 | |
| | Methabenz-thiazuron | Tribunil | 5,000 | |
| | Metobromuron | Pattonex | 2,000 | Metobromuron has been associated with methemoglobinemia.[17] |
| | Metoxuron | Deftor, Dosaflo, Purivel, Sulerex | 3,200 | |
| | Monolinuron | Aresin | 2,100 | |
| | Monuron | Monuron | 3,600 | |
| | Neburon | Granurex, Neburex | >11,000 | |
| | Siduron | Tupersan | >7,500 | |
| | Sulfometuron-methyl | Oust | >5,000 | |

## Confirmation of Poisoning

Although there are analytical methods for residues of many of the herbicides mentioned in this chapter, and for some of the mammalian metabolites generated from them, these procedures are not generally available to confirm human absorption of the chemicals. Prior exposure must be determined from a recent history of occupational exposure or accidental or deliberate ingestion.

## Treatment of Toxicosis from Other Herbicides

1. Decontaminate skin promptly by washing with soap and water. Treat contamination of the eyes immediately by prolonged flushing with copious amounts of clean water. If dermal or ocular irritation persists, medical attention should be obtained without delay.

2. Ingestions of these herbicides are likely to be followed by vomiting and diarrhea because of the irritant properties of most of the toxicants. Management depends on: (a) the best estimate of quantity originally ingested, (b) the time elapsed since ingestion and (c) the clinical status of the subject.

   If large amounts of herbicide have been ingested and the patient is seen within an hour of the ingestion, consider gastrointestinal decontamination as outlined in **Chapter 3, *General Principles*.** GI decontamination may be effective in limiting irritant effects and reducing absorption of most or all of these herbicides.

3. If serious dehydration and electrolyte depletion have occurred as a result of vomiting and diarrhea, monitor blood electrolytes and fluid balance and administer intravenous infusions of glucose, normal saline, Ringer's solution or Ringer's lactate to restore extracellular fluid volume and electrolytes. Follow this with oral nutrients as soon as fluids can be retained.

4. Use supportive measures to manage excessive exposures to these herbicides. With the exception of treating methemoglobinemia associated with some of these herbicides, there are no specific antidotes for poisoning by most of these compounds. In the case of suicidal ingestions, particularly, the possibility must always be kept in mind that multiple toxic substances may have been swallowed, especially if the patient's condition deteriorates in spite of good supportive care.

5. Antidotal therapy for methemoglobinemia is methylene blue.

### Dosage of Methylene Blue

- ***1-2 mg/kg of 1% methylene blue, slow IV, in symptomatic patients. Additional doses may be required.***

*Other Herbicides*

**TREATMENT**

Decontaminate skin and eyes

GI decontamination if ingestion within 1 hour

Administer IV fluids and electrolytes as appropriate

Consider multiple substance ingestion

Methylene blue for methemoglobinemia

## References

1. Eddleston M, Rajapakshe M, Roberts D, et al. Severe propanil [N-(3,4-dichlorophenyl) propanamide] pesticide self-poisoning. *J Toxicol Clin Toxicol.* 2002;40(7):847-854.

2. Roberts DM, Heilmair R, Buckley NA, et al. Clinical outcomes and kinetics of propanil following acute self-poisoning: a prospective case series. *BMC Clin Pharmacol.* 2009;9:3.

3. Moon JM, Chun BJ. Predicting acute complicated glyphosate intoxication in the emergency department. *Clin Toxicol (Phila).* Aug 2010;48(7):718-724.

4. Bronstein AC, et al. 2010 Annual Report of the American Association of Poison Control Centers' National Poison Data System (NPDS): 28th Annual Report. *Clin Toxicol.* 2011;49:910-941.

5. Steinrücken HC, Amrhem N. The herbicide glyphosate is a potent inhibitor of 5-enolpyruvylshikimic acid-3-phosphate synthase. *Biochem Biophy Res Commun.* 30 Jun 1980;94:1207-12.

6. Bradberry SM, Proudfoot AJ, Vale JA. Glyphosate poisoning. *Toxicol Rev.* 2004;23(3):159-167.

7. Moon JM, Min YI, Chun BJ. Can early hemodialysis affect the outcome of the ingestion of glyphosate herbicide? *Clin Toxicol (Phila).* 2006;44(3):329-332.

8. Roberts DM, Buckley NA, Mohamed F, et al. A prospective observational study of the clinical toxicology of glyphosate-containing herbicides in adults with acute self-poisoning. *Clin Toxicol (Phila).* Feb 2010;48(2):129-136.

9. Sorensen FW, Gregersen M. Rapid lethal intoxication caused by the herbicide glyphosate-trimesium (Touchdown). *Hum Exp Toxicol.* Dec 1999;18(12):735-737.

10. Lo YC, Yang CC, Deng JF. Acute alachlor and butachlor herbicide poisoning. *Clin Toxicol (Phila).* Sep 2008;46(8):716-721.

11. Turcant A, Harry P, Cailleux A, et al. Fatal acute poisoning by bentazon. *J Anal Toxicol.* Mar 2003;27(2):113-117.

12. Wu IW, Wu MS, Lin JL. Acute renal failure induced by bentazone: 2 case reports and a comprehensive review. *J Nephrol.* Mar-Apr 2008;21(2):256-260.

13. Chuang CC, Wang ST, Yang CC, Deng JF. Clinical experience with pendimethalin (STOMP) poisoning in Taiwan. *Vet Hum Toxicol.* Jun 1998;40(3):149-150.

14. Lee HL, Chen KW, Wu MH. Acute poisoning with a herbicide containing imazapyr (Arsenal): a report of six cases. *J Toxicol Clin Toxicol.* 1999;37(1):83-89.

15. Brvar M, Okrajsek R, Kosmina P, et al. Metabolic acidosis in prometryn (triazine herbicide) self-poisoning. *Clin Toxicol (Phila).* Mar 2008;46(3):270-273.

16. Hoppin JA, Umbach DM, London SJ, Lynch CF, Alavanja MC, Sandler DP. Pesticides associated with wheeze among commercial pesticide applicators in the Agricultural Health Study. *Am J Epidemiol.* Jun 15 2006;163(12):1129-1137.

17. Turcant A, Cailleux A, Le Bouil A, Allain P, Harry P, Renault A. Acute metobromuron poisoning with severe associated methemoglobinemia. Identification of four metabolites in plasma and urine by LC-DAD, LC-ESI-MS, and LC-ESI-MS-MS. *J Anal Toxicol.* Apr 2000;24(3):157-164.

*Section IV*

# OTHER PESTICIDES

*N,N-Diethyl-3-methylbenzamide (DEET)*
**HIGHLIGHTS**

Few cases of toxicity when used properly

Skin, GI absorption

**SIGNS & SYMPTOMS**

Skin irritation, contact dermatitis, urticarial

Rarely: headache, restlessness, irritability, ataxia, unconsciousness, hypotension, seizures

**TREATMENT**

Decontaminate skin, eyes

Consider topical steroids, oral antihistamines for severe skin reactions

Consider GI decontamination following substantial ingestion

Control seizures with anticonvulsants

**CONTRAINDICATED**

Induced emesis



CHAPTER 14

# Insect Repellents

Insect repellents are by nature different from all other pesticides because they are the one class of chemicals applied purposefully to humans. The exceptions to this rule are several insecticides (permethrin, lindane and malathion) that may be applied purposefully to human skin or hair to treat scabies and lice. Repellents are not insecticidal; rather they mask the human skin to detection by insects and arthropods (mosquitoes, gnats, ticks).

Hundreds of insect repellents are marketed in the United States. The primary synthetic insect repellents used in the U.S. are N,N-diethyl-3-methylbenzamide (also and formerly known as N,N-diethyl-m-toluamide, DEET) and KBR 3023 (picaridin). DEET was developed by the military around the time of World War II and has long been considered the gold standard of insect repellents. Picaridin was developed in the late 1990s and, prior to being marketed in the United States in the mid 2000s, was available in Europe and Australia. Several natural products have been used as insect repellents and are listed by the EPA as minimum risk pesticides, making them exempt from federal regulation. These include oil of citronella, cedar oil, lemongrass oil and others that are available in the retail market.[1] Oil of lemon eucalyptus, its synthetic analog PMD, picaridin, and IR3535 are recommended by the Centers for Disease Control and Prevention (CDC) as alternatives to DEET to control mosquitoes that carry West Nile virus.[2] Picaridin has a duration of action comparable to some formulations with 20%-30% DEET and other repellents.[3] It has not been approved for ticks.

## N,N-DIETHYL-3-METHYLBENZAMIDE (DEET)

This chemical is a widely used liquid insect repellent, suitable for application to skin or to fabrics. It comes in a wide range of concentrations from 5% (Off! Skintastic for Kids) to 100% (Muskol). Despite the widespread use of the product, there are relatively few cases of toxicity reported in the literature.[4] Improper use, ingestion and high-concentration usage on children are risk factors for the rarely observed severe toxicity.[5]

### Toxicology

The toxicokinetics has been well studied in animal models and humans. Approximately 19%-48% of **DEET** penetrates the epidermis in about 6 hours in guinea pigs. DEET can be detected within the blood and other tissues in mice within 2 hours of application, and excretion in the form of inactive metabolites is the primary mode of elimination.[6,7] Similar absorption, distribution, metabolism and excretion are found in humans.[8] DEET is efficiently absorbed across the skin and by the gastrointestinal tract. Blood concentrations of about 3 mg/liter have been reported several hours after dermal application in the prescribed fashion.[9] DEET is rapidly absorbed, peaks at around 6 hours, and within 24 hours its metabolites are completely excreted.[8] Skin permeability for DEET is enhanced by an ethanol substrate, which is how most formulations are prepared.[10] Human skin permeability of DEET is decreased using a polyethylene glycol solvent.[11,12]

For many years, DEET has been effective and generally well tolerated as an insect repellent applied to human skin, although tingling, mild irritation and sometimes desquamation have followed repeated application. It should be pointed out that the label recommends "to avoid over application" and that the product should be washed off upon returning indoors. The chemical tends to leave an oily residue on the skin, and may dissolve plastic or other synthetic materials such as clothing, wrist watches and other objects.

## Signs and Symptoms of Poisoning

Most reports of adverse events following DEET exposure are skin-related findings. These include mild skin irritation, contact dermatitis, exacerbation of preexisting skin disease as well as generalized urticaria.[13,14] DEET is very irritating to the eyes but not corrosive.

Serious adverse cutaneous effects have occurred in tropical conditions, when applied to areas of skin that were occluded during sleep (mainly the antecubital and popliteal fossae). Under these conditions, the skin became red and tender and then exhibited blistering and erosion, leaving painful, weeping, denuded areas that were slow to heal. Severe scarring occasionally resulted from some of these severe reactions.[15]

Toxic encephalopathic reactions have been reported in rare instances following ingestion or dermal application. Manifestations of toxic encephalopathy have been headache, restlessness, irritability, ataxia, rapid loss of consciousness, hypotension and seizures. Some cases have shown flaccid paralysis and areflexia. Deaths have occurred following very large doses.[4,5,16] Plasma levels of DEET found in fatal systemic poisonings have ranged for 168 to 240 milligrams per liter.[5] One well-documented case of anaphylactic reaction to DEET has been reported.[17] One fatal case of encephalopathy in a child heterozygous for ornithine carbamoyl transferase deficiency resembled Reye's syndrome, but the postmortem appearance of the liver was not characteristic of the syndrome.[18]

While most severe toxicity reports relate to multiple dermal applications of various concentrations including as low as 10%,[16,19,20,21] seizures following less frequent dermal exposure has also been reported.[22,23] A summary of the 22 cases reported in the medical literature was reviewed in *Pediatric Annals*.[24] The difficulty of such case reports is that exposure details cannot always be ascertained. The more severe cases of systemic toxicity have often occurred following ingestion.[4] No dose-response patterns appear to exist among the small number of human toxicity reports.

The National Poison Data System, formerly known as the Toxic Exposure Surveillance System, is used by the American Association of Poison Control Centers (AAPCC) and allows a retrospective review of reports to poison control centers. Every year in the annual report of poison control centers (PCCs), reports of adverse events following insect repellents number in the hundreds, but most are listed as mild-to-moderate effects, and no details are given as to the nature of the symptoms. Generally, data are based on reports to PCCs, sometimes with or without follow-up information, so the data are limited by what is reported and collected. Greater detail can be found in a review published in 2002, based on 1993-1997 reports to poison control. Of 20,764 exposures reported to PCCs, information on outcomes was reported for 11,600. Of these, 11,159 (96.2%) were considered minor, and 437 (3.8%) were classified as moderate (409), major (26) or fatal (2). The two deaths occurred in adults, both following exposure to a concentration greater than 50% DEET. Of the 26 cases with major effects, half were adults and half were 0-19 years of age. Two patients were exposed to less than 11% DEET, one of whom had neurological effects and the other was admitted to critical care although symptoms were not reported. Thirteen of the 26 cases did not have DEET concentrations available,

***N,N-Diethyl-3-methylbenzamide (DEET)***

**COMMERCIAL PRODUCTS**

Auton
Cutter
Detamide
Metadelphene
MGK
Muskol
OFF!
Sawyer
Skeeter Beater
Skeeter Cheater

and 7 did not have symptom data available. Of those with symptom data available (17), 11 reported neurological symptoms.[25]

Discretion should be exercised in recommending DEET for persons who have acne, psoriasis, an atopic predisposition or other chronic skin condition. According to the label, it should not be applied over cuts, wounds or irritated skin. In addition, it should not be applied to any skin area that is likely to be opposed to another skin surface for a significant period of time (antecubital and popliteal fossae, inguinal areas).[15]

Great caution should be exercised in using DEET on children. A wide variation in applied concentrations has been associated with the reported cases of pediatric seizures or major effects. Care should be taken to balance the risks of prevention of arthropod-borne diseases, possible adverse effects and the duration of time the child may be exposed. The adverse event reports suggest that multiple applications can play a role in toxicity and reinforce the need to follow the product label on reapplication. Specifically, the label says "Avoid over application. Frequent reapplication and saturation are unnecessary." The product should also be washed off after returning indoors. To avoid multiple applications, use the concentration that best fits the duration of possible exposure. If the child is going to be outside for 1-2 hours, a product containing 10% DEET is likely to be effective. If the exposure time is longer, a product with 5-7 hours of protection time (such as 25%-30% DEET) may be more appropriate.[3] The application should be limited to exposed areas of skin, using as little repellent as possible. If headache or any emotional or behavioral change occurs, use of DEET should be discontinued immediately. The American Academy of Pediatrics does not recommend using products that contain of DEET on infants less than 2 months of age.

## Confirmation of Poisoning

Methods exist for measurement of DEET in plasma and tissues and of DEET metabolites in urine, but these are not widely available. The Centers for Disease Control and Prevention (CDC) has developed a method for measuring DEET (not the metabolite) in the urine. In the nationally representative sample of U.S. residents conducted by CDC for the years 2001 and 2002, DEET was detected in approximately 10% of the U.S. population, with a $90^{th}$ percentile geometric mean of 0.100 µg/l, and a $95^{th}$ percentile geometric mean of 0.170 µg/l.[26] Because the parent compound is excreted within 24 hours of exposure, this likely reflects individuals with recent exposure.

## Treatment of DEET Toxicosis

1. Decontaminate the skin with soap and water as outlined in **Chapter 3, *General Principles***. Eye contamination should be removed by prolonged flushing of the eye with copious amounts of clean water or saline. If irritation persists, specialized medical treatment should be obtained. Topical steroids and oral antihistamines have been used for severe skin reactions that occasionally follow application of DEET.[15]

2. If a substantial amount of DEET has been ingested within an hour of treatment, consider gastrointestinal decontamination as outlined in **Chapter 3**. Induced emesis is usually considered contraindicated in these poisonings because of the rapid onset of seizures.

3. Provide supportive treatment, controlling seizures with anticonvulsants as outlined in **Chapter 3**. Persons surviving poisoning by ingestion of DEET have usually recovered within 36 hours or less.[4,5]

## PICARIDIN

Picaridin, also known as KBR 3023, is a relatively new synthetic insect repellent. Marketed in Europe in the 1990s, it was introduced to the U.S. market in 2005. It has been shown to have relatively similar protection time as DEET when tested in similar concentrations. It tends to be less oily, better tolerated and less pungent than DEET and, unlike DEET, does not damage plastics and synthetic fabrics.[27]

### Toxicology

The mechanism of action of **picaridin** is unknown.[28] Animal studies have not demonstrated dermal, organ-specific or reproductive toxicity.[29] Duration of protection from mosquito bites depends on its concentration. Animal studies did not reveal acute toxicity in doses as high as 200 mg/kg body weight.[30] Likewise, animal studies did not demonstrate any teratologic, developmental or neoplastic abnormalities.[30,31] Most commercially available products contain 7.5%, 10% and 15% picaridin, with the lower-concentration product lasting approximately 2 hours and the 15% picaridin lasting approximately 4-6 hours. Protection appears comparable to DEET, and this agent seems better tolerated.[27,32,33,34,35]

### Signs and Symptoms of Poisoning

Allergic contact dermatitis has been reported in a human following routine application and produced erythema and pruritis. It is not clear whether the solvent methyl glucose-dioleate had a causative or additive effect.[36] Other than the skin irritation, there are no additional reports of toxic effects in humans.

### Treatment of Picaridin Toxicosis

1. Treat skin irritation with oral antihistamines and topical steroids.
2. For eye exposure, irrigate eyes with copious amounts of water or normal saline. If contact lenses are present, they should be removed.
3. Otherwise, provide supportive treatment.

*Picaridin*

### HIGHLIGHTS

Similar protection time as DEET with fewer undesirable impacts

Unknown mechanism of action

### SIGNS & SYMPTOMS

Skin irritation

### TREATMENT

Oral antihistamine, topical steroid

Irrigate eyes if exposed

## ESSENTIAL OILS

Numerous natural or essential-oil-based products are in use as insect repellents. The repellency activity is thought to derive from camphor in some of the products, but other activity is unknown. There is marked variability of the ingredients of oils in various repellents and their efficacy, with some results supporting an essential oil as effective or almost as effective as DEET,[37,38,39] while other results do not support these efficacy findings.[3,40] Of the oils, oil of lemon eucalyptus is the only one that has been recommended by the CDC as being an effective alternative to DEET.[41] Several essential oils are considered by the EPA to be minimum risk pesticides and are not subject to federal registration requirements.

### Toxicology

**Oil of lemon eucalyptus** is colorless to pale yellow in color and has an aromatic odor and pungent taste. It primarily contains 1,8 cinole, along with a small amount of

*Essential Oils*
## COMMERCIAL PRODUCTS

cedar oil

lemongrass oil

oil of citronella

oil of lemon eucalyptus

## HIGHLIGHTS

Variability of ingredients and efficacy

CNS impacts from ingestion

## SIGNS & SYMPTOMS

Most toxicities from ingestion

Vomiting, lethargy, coma, seizures

Dermal irritation possible

## TREATMENT

Supportive

Consider GI decontamination

other compounds, including hydrocyanic acid, which is thought to be the source of its toxicity. Ingestion of eucalyptus oil is known to cause significant neurological toxicity, and an adult fatality has been reported following ingestion of as little as 3.5 mL.[42,43,44,45]

## Signs and Symptoms of Poisoning

Most reports of toxicity from eucalyptus oil have arisen from ingestion.[43,44,45,46] Most preparations include a combination of camphor, eucalyptus oil and menthol, such as those used in a vaporizer solution or other medicinal purposes.[47,48] The main symptoms reported include vomiting, lethargy, coma and seizures.[43,44]

One published case report of systemic toxicity from essential oils related to dermal application.[42] In this case, a 6-year-old girl had numerous applications of occlusive bandages soaked in a homemade solution including eucalyptus oil. Approximately 24 hours after the applications were initiated, she appeared intoxicated and progressed to complete loss of consciousness. Removal of the exposure and rinsing her skin with water resulted in a full recovery within 24 hours.[42] Several cases of irritant dermatitis have also been reported, with signs and symptoms including erythema, pruritis and a burning sensation.[45,49]

## Treatment of Essential Oil Toxicosis

1. Provide supportive care, as there is no antidote.

2. If the patient is symptomatic or has ingested a large amount of essential oils, consider GI decontamination.[45] For specific information on GI decontamination, please see **Chapter 3, *General Principles*.**

## References

1. United States Environmental Protection Agency. Minimum Risk Pesticides. 2010. http://www.epa.gov/oppbppd1/biopesticides/regtools/25b_list.htm. Accessed December 30, 2012.

2. Center for Disease Control and Prevention. Updated Information regarding Insect Repellents. 2009. http://www.cdc.gov/ncidod/dvbid/westnile/repellentupdates.htm. Accessed December 30, 2012.

3. Fradin MS, Day JF. Comparative efficacy of insect repellents against mosquito bites. *N Engl J Med.* Jul 4 2002;347(1):13-18.

4. Veltri JC, Osimitz TG, Bradford DC, Page BC. Retrospective analysis of calls to poison control centers resulting from exposure to the insect repellent N,N-diethyl-m-toluamide (DEET) from 1985-1989. *J Toxicol Clin Toxicol.* 1994;32(1):1-16.

5. Tenenbein M. Severe toxic reactions and death following the ingestion of diethyltoluamide-containing insect repellents. *JAMA.* Sep 18 1987;258(11):1509-1511.

6. Blomquist L, Thorsell W. Distribution and fate of the insect repellent 14C-N, N-diethyl-m-toluamide in the animal body. II. Distribution and excretion after cutaneous application. *Acta Pharmacol Toxicol (Copenh).* Sep 1977;41(3):235-243.

7. Robbins PJ, Cherniack MG. Review of the biodistribution and toxicity of the insect repellent N,N-diethyl-m-toluamide (DEET). *J Toxicol Environ Health.* 1986;18(4):503-525.

8. Selim S, Hartnagel RE, Jr., Osimitz TG, Gabriel KL, Schoenig GP. Absorption, metabolism, and excretion of N,N-diethyl-m-toluamide following dermal application to human volunteers. *Fundam Appl Toxicol.* Apr 1995;25(1):95-100.

9. Wu A, Pearson ML, Shekoski DL, Soto RJ, Stewart RD. High resolution gas chromatography/mass spectrometric characterization of urinary metabolites of N,N-diethyl-m-toluamide (DEET) in man. *J High Resolution Chromatogr.* 1979;2(9):558-562.

10. Stinecipher J, Shah J. Percutaneous permeation of N,N-diethyl-m-toluamide (DEET) from commercial mosquito repellents and the effect of solvent. *J Toxicol Environ Health.* Oct 10 1997;52(2):119-135.

11. Qiu H, Jun HW, Dzimianski M, McCall J. Reduced transdermal absorption of N,N-diethyl-m-toluamide from a new topical insect repellent formulation. *Pharm Dev Technol.* Feb 1997;2(1):33-42.

12. Ross J, Shah J. Reduction in skin permeation of *N,N*-diethyl-*m*-toluamide (DEET) by altering the skin / vehicle partition coefficient. *Journal of Controlled Release.* 2000;67:211-221.

13. Maibach HI, Johnson HL. Contact urticaria syndrome. Contact urticaria to diethyltoluamide (immediate-type hypersensitivity). *Arch Dermatol.* Jun 1975;111(6):726-730.

14. Wantke F, Focke M, Hemmer W, Gotz M, Jarisch R. Generalized urticaria induced by a diethyltoluamide-containing insect repellent in a child. *Contact Dermatitis.* Sep 1996;35(3):186-187.

15. Reuveni H, Yagupsky P. Diethyltoluamide-containing insect repellent: adverse effects in worldwide use. *Arch Dermatol.* Aug 1982;118(8):582-583.

16. Lipscomb JW, Kramer JE, Leikin JB. Seizure following brief exposure to the insect repellent N,N-diethyl-m-toluamide. *Ann Emerg Med.* Mar 1992;21(3):315-317.

17. Miller JD. Anaphylaxis associated with insect repellent. *N Engl J Med.* Nov 18 1982;307(21):1341-1342.

18. Heick HM, Shipman RT, Norman MG, James W. Reye-like syndrome associated with use of insect repellent in a presumed heterozygote for ornithine carbamoyl transferase deficiency. *J Pediatr.* Sep 1980;97(3):471-473.

19. de Garbino P, Laborde A. Toxicity of an insect repellent: N-N-diethyltoluamide. *Vet Hum Toxicol.* 1983;25(6):422-423.

20. Hampers LC, Oker E, Leikin JB. Topical use of DEET insect repellent as a cause of severe encephalopathy in a healthy adult male. *Acad Emerg Med.* Dec 1999;6(12):1295-1297.

21. Zadikoff CM. Toxic encephalopathy associated with use of insect repellent. *J Pediatr.* Jul 1979;95(1):140-142.

22. Briassoulis G, Narlioglou M, Hatzis T. Toxic encephalopathy associated with use of DEET insect repellents: a case analysis of its toxicity in children. *Hum Exp Toxicol.* Jan 2001;20(1):8-14.

23. Seizures temporally associated with use of DEET insect repellent--New York and Connecticut. *MMWR Morb Mortal Wkly Rep.* Oct 6 1989;38(39):678-680.

24. Roberts JR, Reigart JR. Does anything beat DEET? *Pediatr Ann.* Jul 2004;33(7):443-453.

25. Bell JW, Veltri JC, Page BC. Human Exposures to N,N-diethyl-m-toluamide insect repellents reported to the American Association of Poison Control Centers 1993-1997. *Int J Toxicol.* Sep-Oct 2002;21(5):341-352.

26. Department of Health and Human Services Centers for Disease Control and Prevention. *Third National Report on Human Exposure to Environmental Chemicals.* 2005.

27. Katz TM, Miller JH, Hebert AA. Insect repellents: historical perspectives and new developments. *J Am Acad Dermatol.* May 2008;58(5):865-871.

28. Kendrick DB. Mosquito repellents and superwarfarin rodenticides--are they really toxic in children? *Curr Opin Pediatr.* Apr 2006;18(2):180-183.

29. Picaridin - A New Insect Repellent. *The Medical Letter on Drugs and Therapeutics.* 2005;47(1210):46-47.

30. Astroff AB, Freshwater KJ, Young AD, Stuart BP, Sangha GK, Thyssen JH. The conduct of a two-generation reproductive toxicity study via dermal exposure in the Sprague-Dawley rat--a case study with KBR 3023 (a prospective insect repellent). *Reprod Toxicol.* May-Jun 1999;13(3):223-232.

31. Wahle BS, Sangha GK, Lake SG, Sheets LP, Croutch C, Christenson WR. Chronic toxicity and carcinogenicity testing in the Sprague-Dawley rat of a prospective insect repellent (KBR 3023) using the dermal route of exposure. *Toxicology.* Dec 20 1999;142(1):41-56.

32. Costantini C, Badolo A, Ilboudo-Sanogo E. Field evaluation of the efficacy and persistence of insect repellents DEET, IR3535, and KBR 3023 against *Anopheles gambiae* complex and other Afrotropical vector mosquitoes. *Trans R Soc Trop Med Hyg.* Nov 2004;98(11):644-652.

33. Frances SP, Van Dung N, Beebe NW, Debboun M. Field evaluation of repellent formulations against daytime and nighttime biting mosquitoes in a tropical rainforest in northern Australia. *J Med Entomol.* May 2002;39(3):541-544.

34. Frances SP, Waterson DG, Beebe NW, Cooper RD. Field evaluation of commercial repellent formulations against mosquitoes (Diptera: Culicidae) in Northern Territory, Australia. *J Am Mosq Control Assoc.* Dec 2005;21(4):480-482.

35. Scheinfeld N. Picaridin: a new insect repellent. *J Drugs Dermatol.* Jan-Feb 2004;3(1):59-60.

36. Corazza M, Borghi A, Zampino MR, Virgili A. Allergic contact dermatitis due to an insect repellent: double sensitization to picaridin and methyl glucose dioleate. *Acta Derm Venereol.* 2005;85(3):264-265.

37. Barnard DR, Xue RD. Laboratory evaluation of mosquito repellents against *Aedes albopictus, Culex nigripalpus,* and *Ochierotatus triseriatus* (Diptera: Culicidae). *J Med Entomol.* Jul 2004;41(4):726-730.

38. Choi WS, Park BS, Ku SK, Lee SE. Repellent activities of essential oils and monoterpenes against *Culex pipiens pallens. J Am Mosq Control Assoc.* Dec 2002;18(4):348-351.

39. Tawatsin A, Thavara U, Chansang U, et al. Field evaluation of DEET, Repel Care, and three plant based essential oil repellents against mosquitoes, black flies (Diptera: Simuliidae) and land leeches (Arhynchobdellida: Haemadipsidae) in Thailand. *J Am Mosq Control Assoc.* Jun 2006;22(2):306-313.

40. Sfara V, Zerba EN, Alzogaray RA. Fumigant insecticidal activity and repellent effect of five essential oils and seven monoterpenes on first-instar nymphs of *Rhodnius prolixus. J Med Entomol.* May 2009;46(3):511-515.

41. Kuehn BM. CDC: new repellents for West Nile fight. *JAMA.* Jun 1 2005;293(21):2583.

42. Darben T, Cominos B, Lee CT. Topical eucalyptus oil poisoning. *Australas J Dermatol.* Nov 1998;39(4):265-267.

43. Hindle RC. Eucalyptus oil ingestion. *N Z Med J.* May 11 1994;107(977):185-186.

44. Patel S, Wiggins J. Eucalyptus oil poisoning. *Arch Dis Child.* May 1980;55(5):405-406.

45. Webb NJ, Pitt WR. Eucalyptus oil poisoning in childhood: 41 cases in south-east Queensland. *J Paediatr Child Health.* Oct 1993;29(5):368-371.

46. Orr J. Eucalyptus Poisoning. *Br Med J.* May 12 1906;1(2367):1085.

47. Day LM, Ozanne-Smith J, Parsons BJ, Dobbin M, Tibballs J. Eucalyptus oil poisoning among young children: mechanisms of access and the potential for prevention. *Aust N Z J Public Health.* Jun 1997;21(3):297-302.

48. Flaman Z, Pellechia-Clarke S, Bailey B, McGuigan M. Unintentional exposure of young children to camphor and eucalyptus oils. *Paediatr Child Health.* Feb 2001;6(2):80-83.

49. Schaller M, Korting HC. Allergic airborne contact dermatitis from essential oils used in aromatherapy. *Clin Exp Dermatol.* 1995;20(2):143-145.

CHAPTER 15

# Arsenical Pesticides

Many arsenic compounds have been discontinued in the United States as a result of government regulations. However, they are still widely available in some countries, and many homes and farms have leftover supplies that continue to present risk. Arsenic trioxide is still used in some ant bait stations, which have been a source for childhood exposure via ingestion in recent years.[1] Another arsenic compound, arsine gas, is not a pesticide but is released as a byproduct in pesticide manufacturing and metal refining operations and is the most toxic of all forms of arsenic. It is discussed separately in this chapter.

## Toxicology

**Arsenic** is a natural element having both metal and nonmetal physical/chemical properties. In one respect or another, it resembles nitrogen, phosphorus, antimony and bismuth in its chemical behavior. In nature it exists in elemental, trivalent (-3 or +3) and pentavalent (+5) states. It binds covalently with most nonmetals (notably oxygen and sulfur) and with metals (*e.g.,* calcium and lead). It forms stable trivalent and pentavalent organic compounds. In biochemical behavior, it resembles phosphorus, competing with phosphorus analogs for chemical binding sites. Toxicity of the various arsenic compounds in mammals extends over a wide range, determined, in part, by the unique biochemical actions of each compound, but also by absorbability and efficiency of biotransformation and disposition. After arsine gas, arsenites (inorganic trivalent compounds) represent the next most toxic hazard of arsenic compounds. Doses of 78-180 mg of arsenic trioxide (~1-2.5 mg/kg in a child) are considered high enough to be lethal.[2] Inorganic pentavalent compounds (arsenates) are somewhat less toxic than arsenites, while the organic (methylated) pentavalent compounds (arsonates) incur the least hazard of the arsenicals that are used as pesticides.[3]

The pentavalent arsenicals are relatively water soluble and absorbable across mucous membranes, while trivalent arsenicals, having greater lipid solubility, are more readily absorbed across the skin.[4] However, acute, systemic poisonings that arise following dermal absorption of either form have been extremely rare. There are numerous dermal manifestations of arsenic poisoning, which will be discussed later in the chapter. Ingestion has been the usual basis of poisoning; gut absorption efficiency depends on the physical form of the compound, its solubility characteristics, the gastric pH, gastrointestinal motility and gut microbial transformation. Inhalation is the major route of arsine exposure; toxic effects may also occur with other arsenicals through inhalation of aerosols.

Once absorbed, many arsenicals cause toxic injury to cells of the nervous system, blood vessels, liver, kidney and other tissues. Two biochemical mechanisms of toxicity are recognized: (1) reversible combination with thiol groups contained in tissue proteins and enzymes and (2) substitution of arsenic anions for phosphate in many reactions, including those critical to oxidative phosphorylation.[5,6] Arsenic is readily metabolized in the liver to a methylated form, which is much less toxic and easily excreted. However, it is prudent to manage cases of arsenical pesticide ingestion as though all are highly toxic.

***Arsenic Compounds***
### HIGHLIGHTS

Life-threatening effects on CNS, blood vessels, kidney, liver

### SIGNS & SYMPTOMS

Acute cases

Garlic odor of the breath and feces

Metallic taste in mouth

Adverse GI symptoms

Also CNS, renal & cardiovascular symptoms

Jaundice

Chronic cases

Muscle weakness

Fatigue

Weight loss

Hyperpigmentation

Hyperkeratosis

Mees lines

### TREATMENT

Skin, eye, GI decontamination

IV hydration

Chelation therapy with BAL, DMSA, or d-penicillamine

Consider hemodialysis

## Inorganic Trivalent

***Arsenic trioxide***
"White arsenic." Arsenous oxide. Has been discontinued but still stocks may still be on hand from prior registrations.

***Sodium arsenite***
Sodanit, Prodalumnol Double. Used as a fungicide in vineyards.

***Calcium arsenite***
Mono-calcium arsenite. Flowable powder for insecticidal use on fruit.

***Copper arsenite (Acid copper arsenite)***
Wettable powder, for use as insecticide, wood preservative

***Copper acetoarsenite***
Insecticide. Paris Green, Schweinfurt green, emerald green, French green, mitis green. No longer used in the United States; still used outside the United States.

***Arsine***
Not a pesticide.

Occasionally generated during manufacture of arsenicals.

See separate discussion in subsection on p. 140.

## Signs and Symptoms of Poisoning

Manifestations of **acute poisoning** (large amount absorbed over a short time) are distinguishable from those of chronic poisoning (lesser doses absorbed over a longer time interval).

The symptoms and signs of acute arsenic poisoning usually appear within 1 hour after ingestion, but may be delayed several hours. A garlic odor to the breath and feces may help to identify the responsible toxicant in a severely poisoned patient. There is often a metallic taste in the mouth. Gastrointestinal (GI) adverse effects predominate, with vomiting, abdominal pain and rice-water or bloody diarrhea being the most common.[1,3,7,8] Other GI effects include inflammation, vesicle formation and eventual sloughing of the mucosa in the mouth, pharynx and esophagus.[7] These effects result from the action of an arsenical metabolite on blood vessels generally, but the splanchnic vasculature particularly, causing dilation and increased capillary permeability.

The central nervous system is another system commonly affected during acute poisoning. Symptoms may begin with headache, dizziness, drowsiness and confusion. Symptoms may progress to include muscle weakness and spasms, hypothermia, lethargy, delirium, coma and convulsions.[3] Renal injury is manifest as proteinuria, hematuria, glycosuria, oliguria, casts in the urine and, in severe poisoning, acute tubular necrosis. Cardiovascular manifestations include shock, cyanosis and cardiac arrhythmia,[9,10] which are due to direct toxic action and electrolyte disturbances. Liver damage may manifest as elevated liver enzymes and jaundice. Injury to blood-forming tissues may cause anemia, leukopenia and thrombocytopenia. In lethal exposures death usually occurs 1-3 days following symptom onset and is often the result of circulatory failure, although renal failure may also contribute.[3] If the patient survives, painful paresthesias, tingling and numbness in the hands and feet may be experienced as delayed sequelae of acute exposure. This sensorimotor peripheral neuropathy, which may include muscle weakness and spasms, typically begins 1-3 weeks after exposure.[11] The muscle weakness may be confused with Guillain-Barre syndrome.[12]

Other organ systems are affected with arsenic toxicity. Liver injury reflected in hepatomegaly and jaundice may progress to cirrhosis, portal hypertension and ascites. Arsenic has direct glomerular and tubular toxicity resulting in oliguria, proteinuria and hematuria. Electrocardiographic abnormalities (prolongation of the QTc interval and torsades de pointes) and peripheral vascular disease have been reported. The latter includes acrocyanosis, Raynaud's phenomenon and frank gangrene.[3,13] Hematologic abnormalities include anemia, leukopenia and thrombocytopenia.[3] Late sequelae of protracted high intakes of arsenic include skin cancer and an increased risk of lung cancer.[3,14]

Numerous chronic effects are associated with arsenic. Most uses of arsenic as a pesticide, as previously noted, have been discontinued, and most arsenic exposure today is due to naturally occurring arsenic found in shallow well water. Several review articles summarize the evidence of chronic arsenic toxicity.[6,15,16] Repeated absorption of subacutely toxic amounts of arsenic generally has an insidious onset of clinical effects and may be difficult to diagnose. Neurologic, dermal and nonspecific manifestations are usually more prominent than the gastrointestinal effects that characterize acute poisoning. Muscle weakness and fatigue can occur, as can anorexia and weight loss. Hyperpigmentation is a common sign and tends to be accentuated in areas that are already more pigmented such as the groin and areola. Hyperkeratosis is another common sign, especially on the palms and soles.[14,17] Subcutaneous edema of the face, eyelids and ankles; stomatitis; white striations across the nails (Mees lines) and loss of nails or hair are other signs of chronic, continuous exposure.[3,17] Chronic neurologic effects and carcinogenic risks are discussed in **Chapter 21, *Chronic Effects***.

## Confirmation of Poisoning

Measurement of 24-hour urinary excretion of arsenic (micrograms per day) is the most common way to confirm excessive absorption and is the preferred method to follow serial levels and evaluate chronic exposure.[3,18] Spot urine arsenic analysis expressed as a ratio with urinary creatinine is the recommended method to evaluate occupational exposures.[19] Methods to determine blood arsenic concentration are available; however, blood levels tend to poorly correlate with exposure or effect except in the initial acute phase.[18,20] Special metal-free, acid-washed containers should be used for sample collection. Arsenic excretion above 100 µg per day should be considered abnormal. Excretions above 200 µg per day reflect a toxic intake, unless seafood was ingested.[18,20,21,22,22] Diets rich in seafood, primarily shellfish eaten in the previous 48 hours, may generate 24-hour urine excretion levels as high as 200 µg/day and sometimes more.[1,7,22] In some labs and reports, urinary arsenic levels are expressed as µg/l. Normal values are 0-50 µg/l for a 24-hour urine level.[1,23]

The majority of marine arsenic that is excreted is in the methylated form (arsenobetaine) and not considered acutely toxic. However, some of the arsenic released from mussels may contain higher amounts of arsenic trioxide than previously thought.[22] Urinary arsenic should be speciated into inorganic and organic fractions to help determine the source of the exposure and to help guide treatment.

Concentrations of arsenic in blood, urine or other biologic materials can be measured by either wet or dry ashing, followed by colorimetric or atomic absorption spectrometric analysis. This latter method is preferred. Arsenic can be measured in human urine by an inductively coupled plasma mass spectrometry (ICP-MS) method. Blood concentrations in excess of about 100 µg per liter probably indicate excessive intake or occupational exposure, provided that seafood was not ingested before the sample was taken.[7,18,20,21] Blood samples tend to correlate with urine samples during the early stages of acute ingestion,[18] but because arsenic is rapidly cleared from the blood, the 24-hour urine sample remains the preferred method for detection and for ongoing monitoring.[3,18,20]

Hair has been used for evaluation for chronic exposure. Levels in unexposed people are usually less than 1 mg/kg and levels in individuals with chronic poisoning range between 1 and 5 mg/kg.[21] Hair samples should be viewed with caution because external environmental contamination such as air pollution may artificially elevate arsenic levels. Additionally, commercial laboratories have not been shown to have reliably consistent results.[24] Therefore, hair arsenic may be a reasonable tool for use in research but not in the assessment of an acutely poisoned patient.

Special tests for arsine toxicosis are described in the *Arsine Gas* subsection beginning on p. 140.

## Treatment of Arsenic Compound Toxicosis

The following discussion applies principally to poisonings by arsenicals in solid or dissolved form. Treatment of poisoning by arsine gas requires special measures described in the *Arsine Gas* subsection beginning on p. 140.

1. **Skin decontamination.** Wash arsenical pesticide from skin and hair with copious amounts of soap and water.

2. If a high-concentration solution is in contact with the eyes, wash eyes with a profuse amount of water and examine the corneas carefully. If burns have occurred, appropriate ophthalmologic care should be provided. See **Chapter 3, *General Principles*.**

### Inorganic Pentavalent

***Arsenic acid***
Hi-Yield Desiccant H-10, Zotox. Water solutions used as defoliants, herbicides.

***Sodium arsenate***
Disodium arsenate, Jones Ant Killer, Terro Ant Killer. All discontinued, but may still be encountered from old registration.

***Calcium arsenate***
Tricalcium arsenate, Spra-cal, Turf-Cal. Flowable powder formulations used against weeds, grubs.

***Lead arsenate***
Gypsine, Soprabel. Limited use in the United States; wettable powder used as insecticide outside the United States.

***Zinc arsenate***
Powder once used in the United States as insecticide on potatoes and tomatoes.

## Organic (Pentavalent)

***Cacodylic acid (sodium cacodylate)***
Non-selective herbicide, defoliant, silvicide.

Bolate, Bolls-Eye, Bophy, Dilie, Kack, Phytar 560, Rad-E-Cate 25, Salvo.

***Methane arsonic acid***
MAA. Non-selective herbicide.

***Monosodium methane arsonate***
MSMA. Non-selective herbicide, defoliant, silvicide.

Ansar 170, Arsonate Liquid, Bueno 6, Daconate 6, Dal-E-Rad, Drexar 530, Herbi-All, Merge 823, Mesamate, Target MSMA, Trans-Vert, Weed-E-Rad, Weed-Hoe.

***Disodium methane arsonate***
DSMA. Selective post-emergence herbicide, silvicide.

Ansar 8100, Arrhenal, Arsinyl, Crab-E-Rad, Di-Tac, DMA, Methar 30, Sodar, Weed-E-Rad 360.

***Monoammonium methane arsonate***
MAMA. Selective post-emergence herbicide.

***Calcium acid methane arsonate***
CAMA. Selective post-emergence herbicide.

Calar, Super Crab-E-Rad-Calar, Super Dal-E-Rad.

3. **Gastrointestinal decontamination**. If an arsenical pesticide has been ingested within an hour of treatment, consider GI decontamination, as outlined in **Chapter 3**. Because poisoning by ingested arsenic often results in profuse diarrhea, it is generally not appropriate to administer a cathartic. Although it is not clear how well arsenic is absorbed by charcoal, charcoal and whole bowel irrigation were used in recent case reports.[1,8,25] Gastric lavage is also recommended, especially if there are visible opacities on abdominal X-rays.[8,26]

4. **Intravenous fluids**. Administer intravenous fluids to restore adequate hydration, support urine flow and correct electrolyte imbalances. Aggressive rehydration is needed to correct the significant amount of fluid lost from the GI tract. Serum electrolytes including magnesium and calcium should be monitored. Monitor intake/output continuously to guard against fluid overload. If acute renal failure occurs, monitor blood electrolytes regularly.

5. **Hypovolemic shock**. As above, use isotonic fluids (normal saline or lactated ringers) to treat hypovolemia and hypotension associated with shock. Dopamine and/or norepinepherine may be needed.

6. **Cardiac monitoring**. Obtain an electrocardiogram (ECG) to detect ventricular arrhythmias, including prolonged Q-T interval and ventricular tachycardia and toxic myocardiopathy (T wave inversion, long S-T interval).

7. **Chelation therapy**. Use chelation for severe poisoning, including symptomatic poisoning or someone with a recent significant exposure. While there is not a definitive cut-off value at which an asymptomatic patient should be chelated, a urine arsenic level of 200 µg/liter has been suggested.[1,27] Administration of dimercaprol (BAL) has long been the chelator of choice in symptomatic arsenic poisonings.[8,28] However, it is given as painful and frequent intramuscular injections. Oral agents, such as dimercaptosuccinic acid (DMSA, also known as succimer) or d-penicillamine, have also been used more frequently in individual cases;[1,25,29,30,31,32] however, neither has been approved by the FDA for arsenic toxicity. (DMSA is FDA-approved for lead toxicity.) DMSA and d-penicillamine are discussed in greater detail below. The following dosage schedules have proven to be effective in accelerating arsenic excretion.

### Dosage of Dimercaprol (BAL)

- ***Adults: BAL is provided as a 100 mg/mL solution in peanut oil. The dosage is 3-5 mg/kg q 4-12 hours.***
- ***Children: Dosages are similar, but may start with 2.5-3.0 mg/kg.***[7,8,28]

CAUTION: *□isa□reea□le si□e e□□ec□s □□en acc□□pan□ □□e □se □□□□□ na□sea□□ea□ac□e□□□rnin□ an□ □in□lin□ sensa□i□ns□s□ea□in□ pain in □□e □ac□ an□ a□□□en□re□ □r□res□lessness□ □ac□□car□ia an□ □□per□ensi□n□ □c□e s□□p□□s □s□all□ s□□si□e in □□□□ □in□es□□n□i□is a□ine □r□□s □a□pr□i□e relie□□especiall□ i□□i□en pri□r □□ □□□□□□□ □a□p□en□iall□ □a□e □□er a□□erse e□□ec□s □n ra□□i□s □rea□□en□□ arseni□e e□p□s□re □i□□ □□ □increase□ □rain arsenic le□els□ □eca□se □□□ese si□e e□□ec□s c□nsi□er an □ral c□ela□i□n a□en□ i□ □□lera□e□□*

After the gastrointestinal tract is reasonably free of arsenic or if the patient can tolerate oral chelation from the outset, consider an oral chelating agent. D-penicillamine has been suggested; however, its effectiveness for arsenic exposure has been questioned in experimental models, though it has been used with some success in earlier human case reports.[25,30,31,32,34]

### Dosage of D-penicillamine

- ***Adults and children over 12 years: 0.5 gm every 6 hours, given 30-60 minutes before meals and at bedtime for about 5 days.***
- ***Children under 12 years: 0.1 gm/kg body weight, not exceeding 1.0 gm per day, every 6 hours, 30-60 minutes before meals and at bedtime for about 5 days.***

CAUTION: *□□erse reac□i□ns □□s□□r□□er□ □□erap□are rare□□□□e□er□ pers□ns aller□ic □□penicillin □a□s□□□er aller□ic reac□i□ns □□ □penicilla□ □ine□□□e□s□□□l□n□recei□e □penicilla□ine□*

DMSA (succimer) has also been shown to be an effective chelator of arsenic, though it is not labeled for this indication.[29] In light of the lack of effectiveness of d-penicillamine, coupled with the low toxicity and high therapeutic index of DMSA, it appears that the latter agent may be the preferred method for chronic toxicity or when oral chelation is acceptable.[1,25,29,34]

### Dosage of DMSA (Succimer)

- ***Adults and Children: 10 mg/kg every 8 hours for 5 days, followed by 10 mg/kg every 12 hours for an additional 14 days. (Maximum 500 mg per dose). It should be given with food.***

8. **Extracorporeal hemodialysis**. Consider whether to use extracorporeal hemodialysis. When used in combination with BAL therapy, hemodialysis has limited effectiveness in removing arsenic from the blood. Hemodialysis may be indicated early in the course of poisoning to enhance arsenic elimination and maintain extracellular fluid composition if severe acute renal failure occurs.[35] A recent case with acute renal failure resolved without the need for dialysis.[8]

9. **Urinary arsenic excretion.** Monitor urinary arsenic excretion while any chelating agent is being administered. Once arsenic levels fall into the normal reference range of 0-50 µg/l or less than 50 µg/day, it is reasonable to discontinue chelation therapy.[1]

*Arsine Gas*
**HIGHLIGHTS**

Powerful hemolysin

**SIGNS & SYMPTOMS**

Fatigue, headaches, malaise, dizziness, nausea, abdominal pain

Hemoglobinuria, jaundice, very dark plasma

**TREATMENT**

Fresh air

IV fluids

Consider red blood cell or plasma exchange transfusion

## ARSINE GAS

Arsine is not used as a pesticide. However, some poisonings by arsine have occurred in pesticide manufacturing plants and metal-refining operations when arsenicals came into contact with mineral acids or strong reducing agents.[36,37] Arsine may also be released following poisoning by arsenic trioxide.[38]

### Toxicology and Signs and Symptoms of Arsine Poisoning

**Arsine** is a powerful hemolysin, a toxic action not exhibited by other arsenicals. Arsine exposure occurs through inhalation with very little exposure required to cause a serious hemolytic reaction. Death is due to hemolysis and secondary renal failure. Exposure times of 30 minutes at 25-50 parts per million are considered lethal.[39]

Symptoms of poisoning usually appear 30-60 minutes after exposure but may be delayed for up to 24 hours. Patients may exhibit a garlic odor to their breath. Signs and symptoms are the result of sometimes profound hemolysis leading to hemolytic anemia and include fatigue, headache, malaise, weakness, dizziness, dyspnea, nausea, abdominal pain and vomiting. Red staining of the conjunctiva may be present. Free hemoglobin may be present in plasma. Basophilic stippling of red cells, red cell fragments, and ghosts are seen in the peripheral blood smear. Plasma will appear very dark, almost black, resulting from elevated level of unconjugated bilirubin.[36] Hyperkalemia secondary to hemolysis is possible.

Elevated concentrations of arsenic are found in the urine, but these are not nearly as high as are found in poisonings by solid arsenicals. Dark red urine (hemoglobinuria) is often passed 4-6 hours after exposure. Usually 1-2 days after hemoglobinuria appears, jaundice and bronzing of the skin may be evident. Abdominal tenderness, hepatomegaly, and elevated hepatic enzymes all may occur.[36,40]

Renal failure due to direct toxic action of arsine and to products of hemolysis represents the chief threat to life in arsine poisoning.[41]

Polyneuropathy and a mild psycho-organic syndrome are reported to have followed arsine intoxication after a latency of 1-6 months.

### Treatment of Arsine Toxicosis

1. Remove the victim to fresh air.

2. Administer intravenous fluids to keep the urine as dilute as possible and to support excretion of arsenic and products of hemolysis. In the past urinary alkalinization to pH 7.5 has been recommended, but this therapy is not proven.

   ***Monitor fluid balance carefully to avoid fluid overload if renal □ai□re s□per□enes□□ □ni□□r plas□a elec□r□l□es□□□□ an□crea□inine □□□e□ec□ □is□□r□ances □par□ic□larl□□□per□ale□ia□as earl□as p□ssi□le□***

3. Monitor urinary arsenic excretion to assess severity of poisoning. The amount of arsine that must be absorbed to cause poisoning is small, and therefore high levels of urinary arsenic excretion may not always occur, even in the face of significant poisoning.[41,42]

4. If poisoning is severe, consider red blood cell exchange transfusion. It was successful in rescuing one adult victim of arsine poisoning.[36]

5. Consider plasma exchange, which has also been used to treat acute arsine poisoning. A retrospective review study in China reported successful treatment of 12 patients.[40] Another case was treated with a combination of plasma exchange and red blood cell exchange transfusion.[36]

6. Use extracorporeal hemodialysis to maintain normal extracellular fluid composition and to enhance arsenic elimination if renal failure occurs, but it is not very effective in removing arsine carried in the blood.

## References

1. Yarris JP, Caravati EM, Horowitz ZB, Stromness JR, Crouch BI, McKeown NJ. Acute arsenic trioxide ant bait ingestion by toddlers. *Clin Toxicol (Phila).* Nov 2008;46(9):785-789.
2. Hazardous Substance Data Bank: Arsenic trioxide. 2010. http://toxnet.nlm.nih.gov/cgi-bin/sis/search/f?./temp/~zoPqmn:1. Accessed February 23, 2010.
3. Malachowski ME. An update on arsenic. *Clin Lab Med.* Sep 1990;10(3):459-472.
4. Ellenhorn M. Arsenic: Metals and related compounds. In: Schonwald S, Ordog G, Wasserberger J, eds. *Ellenhorn's Medical Toxicology, Diagnosis and Treatment of Human Poisoning.* 2 ed. Baltimore: Williams & Wilkins; 1997:1540.
5. Hughes MF. Arsenic toxicity and potential mechanisms of action. *Toxicol Lett.* Jul 7 2002;133(1):1-16.
6. Vahidnia A, van der Voet GB, de Wolff FA. Arsenic neurotoxicity--a review. *Hum Exp Toxicol.* Oct 2007;26(10):823-832.
7. Campbell JP, Alvarez JA. Acute arsenic intoxication. *Am Fam Physician.* Dec 1989;40(6):93-97.
8. Yilmaz Y, Armagan E, Olmez O, Esen M, Alkis N, Dolar E. Acute arsenic self-poisoning for suicidal purpose in a dentist: a case report. *Hum Exp Toxicol.* Jan 2009;28(1):63-65.
9. Goldsmith S, From AH. Arsenic-induced atypical ventricular tachycardia. *N Engl J Med.* Nov 6 1980;303(19):1096-1098.
10. St Petery J, Gross C, Victorica BE. Ventricular fibrillation caused by arsenic poisoning. *Am J Dis Child.* Oct 1970;120(4):367-371.
11. Heyman A, Pfeiffer JB, Jr., Willett RW, Taylor HM. Peripheral neuropathy caused by arsenical intoxication; a study of 41 cases with observations on the effects of BAL (2, 3, dimercapto-propanol). *N Engl J Med.* Mar 1 1956;254(9):401-409.
12. Donofrio PD, Wilbourn AJ, Albers JW, Rogers L, Salanga V, Greenberg HS. Acute arsenic intoxication presenting as Guillain-Barre-like syndrome. *Muscle Nerve.* Feb 1987;10(2):114-120.
13. Lin TH, Huang YL, Wang MY. Arsenic species in drinking water, hair, fingernails, and urine of patients with blackfoot disease. *J Toxicol Environ Health A.* Jan 23 1998;53(2):85-93.
14. Maloney ME. Arsenic in Dermatology. *Dermatol Surg.* Mar 1996;22(3):301-304.
15. Celik I, Gallicchio L, Boyd K, et al. Arsenic in drinking water and lung cancer: a systematic review. *Environ Res.* Sep 2008;108(1):48-55.
16. Rahman MM, Naidu R, Bhattacharya P. Arsenic contamination in groundwater in the Southeast Asia region. *Environ Geochem Health.* Apr 2009;31 Suppl 1:9-21.
17. Navarro B, Sayas MJ, Atienza A, Leon P. An unhappily married man with thick soles. *Lancet.* Jun 8 1996;347(9015):1596.
18. Fesmire FM, Schauben JL, Roberge RJ. Survival following massive arsenic ingestion. *Am J Emerg Med.* Nov 1988;6(6):602-606.

19. ACGIH. 1997 TLVs and BEIs. Threshold limit values for chemical substances and physical agents. Biological exposure indices. Cincinnati1997.

20. Wagner SL, Weswig P. Arsenic in blood and urine of forest workers as indices of exposure to cacodylic acid. *Arch Environ Health.* Feb 1974;28(2):77-79.

21. Baselt R, Cravey R. Arsenic. *Disposition of Toxic Drugs and Chemicals in Man.* 3 ed. Chicago: Year Book Medical Publishers. 1990:65-69.

22. Buchet JP, Pauwels J, Lauwerys R. Assessment of exposure to inorganic arsenic following ingestion of marine organisms by volunteers. *Environ Res.* Jul 1994;66(1):44-51.

23. Agency for Toxic Substance and Disease Registry. Toxicological Profiles. 2010. http://www.atsdr.cdc.gov/toxprofiles/index.asp. Accessed December 14, 2012.

24. Barrett S. Commercial hair analysis. Science or scam? *JAMA.* Aug 23-30 1985;254(8):1041-1045.

25. Isbister GK, Dawson AH, Whyte IM. Arsenic trioxide poisoning: a description of two acute overdoses. *Hum Exp Toxicol.* Jul 2004;23(7):359-364.

26. Michaux I, Haufroid V, Dive A, et al. Repetitive endoscopy and continuous alkaline gastric irrigation in a case of arsenic poisoning. *J Toxicol Clin Toxicol.* 2000;38(5):471-476.

27. Kersjes MP, Maurer JR, Trestrail JH, McCoy DJ. An analysis of arsenic exposures referred to the Blodgett Regional Poison Center. *Vet Hum Toxicol.* Feb 1987;29(1):75-78.

28. Roses OE, Garcia Fernandez JC, Villaamil EC, et al. Mass poisoning by sodium arsenite. *J Toxicol Clin Toxicol.* 1991;29(2):209-213.

29. Muckter H, Liebl B, Reichl FX, Hunder G, Walther U, Fichtl B. Are we ready to replace dimercaprol (BAL) as an arsenic antidote? *Hum Exp Toxicol.* Aug 1997;16(8):460-465.

30. Kuruvilla A, Bergeson PS, Done AK. Arsenic poisoning in childhood. An unusual case report with special notes on therapy with penicillamine. *Clin Toxicol.* 1975;8(5):535-540.

31. Peterson RG, Rumack BH. d-penicillamine therapy of acute arsenic poisoning. *J Pediatr.* Oct 1977;91(4):661-666.

32. Watson WA, Veltri JC, Metcalf TJ. Acute arsenic exposure treated with oral d-penicillamine. *Vet Hum Toxicol.* 1981;23:164-166.

33. Hoover TD, Aposhian HV. BAL increases the arsenic-74 content of rabbit brain. *Toxicol Appl Pharmacol.* Aug 1983;70(1):160-162.

34. Kreppel H, Reichl FX, Forth W, Fichtl B. Lack of effectiveness of d-penicillamine in experimental arsenic poisoning. *Vet Hum Toxicol.* Feb 1989;31(1):1-5.

35. Blythe D, Joyce DA. Clearance of arsenic by haemodialysis after acute poisoning with arsenic trioxide. *Intensive Care Med.* Jan 2001;27(1):334.

36. Danielson C, Houseworth J, Skipworth E, Smith D, McCarthy L, Nanagas K. Arsine toxicity treated with red blood cell and plasma exchanges. *Transfusion.* Sep 2006;46(9):1576-1579.

37. Pullen-James S, Woods SE. Occupational arsine gas exposure. *J Natl Med Assoc.* Dec 2006;98(12):1998-2001.

38. Kinoshita H, Hirose Y, Tanaka T, Yamazaki Y. Oral arsenic trioxide poisoning and secondary hazard from gastric content. *Ann Emerg Med.* Dec 2004;44(6):625-627.

39. Blackwell M, Robbins A. Arsine (arsenic hydride) poisoning in the workplace. *Am Ind Hyg Assoc J.* Oct 1979;40(10):A56-61.

40. Song Y, Wang D, Li H, Hao F, Ma J, Xia Y. Severe acute arsine poisoning treated by plasma exchange. *Clin Toxicol (Phila).* Sep 2007;45(6):721-727.

41. Fowler BA, Weissberg JB. Arsine poisoning. *N Engl J Med.* Nov 28 1974;291(22):1171-1174.

42. Rathus E, Stinton RG, Putman JL. Arsine poisoning, country style. *Med J Aust.* Mar 10 1979;1(5):163-166.

CHAPTER 16

# Fungicides

Fungicides are extensively used in industry, agriculture and the home and garden for:

1. protection of seed grain during storage, shipment and germination;
2. protection of mature crops, berries, seedlings, flowers and grasses in the field, in storage and during shipment;
3. suppression of mildews that attack painted surfaces;
4. control of slime in paper pulps; and
5. protection of carpet and fabrics in the home.

Approximately 500 million pounds of fungicides are applied worldwide annually (see **Chapter 1,** ***□n□r□□□c□i□n***).

Fungicides vary enormously in their potential for causing adverse effects in humans. Historically, some of the most tragic epidemics of pesticide poisoning occurred by mistaken consumption of seed grain treated with organic mercury or hexachlorobenzene. However, most fungicides currently in use and registered for use in the United States are unlikely to cause frequent or severe acute systemic poisonings for several reasons: (1) many have low inherent toxicity in mammals and are inefficiently absorbed; (2) many are formulated as suspensions of wettable powders or granules, from which rapid, efficient absorption is unlikely; and (3) methods of application are such that relatively few individuals are intensively exposed. Apart from systemic poisonings, fungicides as a class also cause irritant injuries to skin and mucous membranes, as well as some dermal sensitization.

The following discussion considers the recognized adverse effects of widely used fungicides. In the case of those agents that have caused systemic poisoning, some recommendations for management of poisonings and injuries are provided. For those fungicides not known to have caused systemic poisonings in the past, only general guidelines can be offered.

The discussion of fungicide-related adverse effects proceeds in this order:

***Substituted Benzenes***

***Strobilurins***

***Thiocarbamates***

***Ethylene Bis Dithiocarbamates***

***Thiophthalimides***

***Triazoles***

***Copper Compounds***

***Organomercury Compounds***

***Organotin Compounds***

***Cadmium Compounds***

***Miscellaneous Organic Fungicides***

## HIGHLIGHTS

Numerous fungicides in use with varying levels of toxicity

Most are unlikely to cause systemic poisonings, exceptions being

- Organomercury compounds
- Triazoles
- Some copper compounds
- Isolated EBDC exceptions

*Substituted Benzine*

**COMMERCIAL PRODUCTS**

chloroneb (Terraneb SP)

chlorothalonil (Bravo, Clorto Caffaro, Clortosip, Daconil 2787, Exotherm Termil, Tuffcide, others)

dicloran (DCNA, Allisan, Clortran)

hexachlorobenzene (HCB, Anticarie, Ceku C.B., No Bunt)

PCNB, also known as pentachloronitrobenzene (PCNB, quintozene, Avicol, Earthcide, Folosan, Kobu, Kobutol, Pentagen, Tri-PCNB, terraclor, and others)

**HIGHLIGHTS**

No cases of human systemic poisoning have been reported in the medical literature for chloroneb, chlorothalonil, dicloran or PCNB

**TREATMENT**

Decontaminate skin and eyes

In cases where large amounts of HCB have been ingested, consider GI decontamination

## SUBSTITUTED BENZENES

### Toxicology

**Chloroneb** is supplied as wettable powder for treatment of soil and seed. This agent exhibits very low oral toxicity in mammals.[1] It may be moderately irritating to skin and mucous membranes. The metabolite dichloromethoxyphenol is excreted in the urine. No cases of systemic poisoning in humans have been reported.

**Chlorothalonil** is available as wettable powder, water dispersible granules and flowable powders. Chlorothalonil has caused irritation of skin and mucous membranes of the eye and respiratory tract on contact. Cases of allergic contact dermatitis have been reported.[2,3,4,5] There is one report of immediate anaphylactoid reaction to skin contact.[6] Chlorothalonil is poorly absorbed across the skin and the gastrointestinal lining. In a man known to have atopic dermatitis and allergic rhinitis, occupational exposure to chlorothalonil was reported to induce asthma symptoms, which resolved following cessation of exposure.[7] No cases of systemic poisoning in humans have been reported in the published medical literature.

**Dicloran**, also known as DCNA, is formulated as wettable powder, dust, liquid and flowable powder. This broad-spectrum fungicide was widely used to protect perishable produce. It is absorbed through the skin by occupationally exposed workers, but promptly eliminated, at least partly, in the urine. Biotransformation products include dichloroaminophenol, which is an uncoupler of oxidative phosphorylation (enhances heat production). According to the EPA's Registration Eligibility Decision (RED), there is low oral toxicity to mammals (rat $LD_{50}$ is 3,400 mg/kg). There have been no cases of human systemic poisoning reported in the medical literature.

**Hexachlorobenezene (HCB)** is principally formulated as dusts and powders. All registrations in the United States have been canceled. It differs chemically and toxicologically from hexachlorocyclohexane, the gamma isomer of which is also known as lindane, which is still used in limited amounts as an insecticide and as a pharmaceutical agent for the treatment of lice and scabies (see **Chapter 7, *□r□an□c□l□rines***).

Although this seed-protectant fungicide has only slight irritant effects and relatively low single-dose toxicity, long-term ingestion of HCB-treated grain by Turkish farm dwellers in the late 1950s caused several thousand cases of toxic porphyria resembling porphyria cutanea tarda.[8] This condition was due to impaired hemoglobin synthesis, leading to toxic end products (porphyrins) in body tissues. The disease was characterized by excretion of red-tinged (porphyrin-containing) urine, bullous lesions of light-exposed skin, scarring and atrophy of skin with overgrowth of hair, liver enlargement, loss of appetite, arthritic disease and wasting of skeletal muscle mass. Although most adults ultimately recovered after they stopped consuming the HCB-treated grain, some infants nursed by affected mothers died.[8]

Hexachlorobenzene is effectively dechlorinated and oxidized in humans; trichlorophenols are the major urinary excretion products. Disposition is sufficiently prompt that occupationally exposed workers usually show only slight elevation of blood HCB concentrations. HCB is sometimes present in blood specimens from "non-occupationally exposed" persons up to concentrations of about 5 µg per liter. Residues in food are the probable cause. Studies have suggested that adverse neurobehavioral effects in children may occur following exposure to hexachlorobenzene, and these are discussed in **Chapter 21, *□□r□nic □□ec□s***.[9,10]

**PCNB** (also known as Pentachloronitrobenzene) is used to treat seed and soil. Formulations include emulsifiable concentrates, wettable powders and granules. Hexachlorobenzene is a minor contaminant to technical PCNB.

Systemic poisonings have not been reported in humans. Clearance in laboratory animals is chiefly biliary, with some conversion to pentachloroaniline, pentachlorophenol and other metabolites in the liver.[11,12] Although a methemoglobinemic effect is suspected (as from nitrobenzene), this has not been reported in man or animals, nor has toxic porphyria (as from hexachlorobenzene) been reported.

## Confirmation of Poisoning

Chloroneb, chlorothalonil, dicloran, HCB and PCNB all have described methods for analysis by chromatography, but those methods are not widely available. The trichlorophenol metabolites of HCB can be measured in the urine.

Although inherited disease and a number of exogenous agents may cause porphyrins to appear in the urine, a test for porphyrins may be useful for toxicological diagnosis if there has been a known exposure to HCB or if a patient exhibits signs suggestive of porphyria cutanea tarda.

## Treatment of Substituted Benzene Toxicosis

1. Wash off dermal contamination with soap and water. Remove contamination of the eyes by flushing with copious amounts of water. If irritation persists, specialized medical care should be obtained. See **Chapter 3, *General Principles*.**

2. If a large amount of HCB has been ingested in the last few hours, and if copious vomiting has not already occurred, consider GI decontamination as outlined in **Chapter 3**. If contact with the toxicant has been minimal (for example, oral contamination only) promptly flushing out the mouth and observation are probably sufficient.

Persons affected by porphyria should avoid sunlight, which exacerbates the dermal injury by porphyrins.

## STROBILURIN FUNGICIDES

Strobilurin compounds are a relatively newer class of fungicides, discovered in the 1990s and introduced to the market in the late 1990s and early 2000s. They are used in agriculture to kill numerous types of pathogenic fungi including mildews, molds and rusts.

## Toxicology

**Strobilurin** fungicidal activity inhibits mitochondrial respiration by disrupting the cytochrome complex, thus blocking electron transfer.[13] Strobilurin compounds work on a broad range of fungal pests and are now used on a wide range of crops, most notably corn, since 2004.[14]

These compounds have a relatively low acute toxicity; most have a reported $LD_{50}$ oral of over 5,000 mg/kg, except orysastrobin and metominostrobin, with $LD_{50}$ of 356 mg/kg and 708 mg/kg, respectively.[13,15] Few human data are available, though several separate incidents in July 2007 were reported by the CDC. All reports were based on pyraclostrobin. Toxic effects were considered minimal and short term, with resolution after patient was removed from exposure. Symptoms and signs included eye irritation, upper respiratory tract irritation, weakness, dizziness, purities, skin redness and chest pain. In one case, workers in an adjacent corn field were exposed and felt the droplets following aerial spraying, and the major symptoms reported in this incident were upper respiratory tract pain and chest pain.[14]



*Strobilurin*

### COMMERCIAL PRODUCTS

azoxystrobin (Abound, Amistar, Azo-shield, Azotech, Azoxy, Banner Heritage, Dynasty, Dyna-shield, Graduate A+, Heritage, Protégé, Quadris, Quartet, Quilt, Renown, Soygard, Sporgard, Trio, Uniform)

kresoxim-methyl (Allegro, Cygnus, Sovran)

metominostrobin

orysastrobin

picoxystrobin (Benzeneacetic acid, Cygnus, Juwel, Mentor, Ogam, Stroby/Sovran)

pyraclostrobin (Bas, Cabrio, Cornet, Headline, Honor, Insignia, Opera, Pageant, Pristine, Stamina)

trifloxystrobin (Absolute, Armada, Chipco, Compass, Distinguish, Dyna-shield, Flint, Four way, Gem, Prosper, RTU-trifloxystrobin-metalaxyl, Stratego, Tartan, Three way, Trilex, USF)

### HIGHLIGHTS

Widely used on many crops

Low acute toxicity

### SIGNS & SYMPTOMS

Eye & respiratory irritation

Weakness, dizziness

### TREATMENT

Supportive

Consider skin/eye decontamination

### *Thiocarbamate* COMMERCIAL PRODUCTS

thiram (Aules, Chipco Thiram 75, Fermide 850, Fernasan, Hexathir, Mercuram, Nomersam, Polyram-Ultra, Pomarsol forte, Spotrete-F, Spotrete WP 75, Tetrapom, Thimer, Thioknock, Thiotex, Thiramad, Thirasan, Thiuramin, Tirampa, TMTD, Trametan, Tripomol, Tuads)

ziram (Cuman, Hexazir, Mezene, Tricarbamix, Triscabol, Vancide MZ-96, Zincmate, Ziram Technical, Ziram F4, Zirberk, Zirex 90, Ziride, Zitox)

ferbam (Carbamate WDG, Ferbam, Ferberk, Hexaferb, Knockmate, Trifungol)

## Confirmation of Poisoning

Tests to detect these compounds are not readily available.

## Treatment of Strobilurin Toxicosis

Remove the patient from the source of exposure.

Provide supportive treatment directed to symptoms. Significant acute toxicity is not generally expected; therefore, exposure can be asymptomatic and symptoms usually do not warrant medical attention.

Consider skin decontamination as outlined in **Chapter 3, *General Principles*.**

Flush eyes with water or normal saline. If eye irritation, redness or swelling persists for more than 15 minutes, recommend consultation with an ophthalmologist.

# THIOCARBAMATES

Thiocarbamates are commonly formulated as dusts, wettable powders or water suspensions. They are used to protect seeds, seedlings, ornamentals, turf, vegetables and fruit including apples. Unlike the N-methyl carbamates (**Chapter 6**), thiocarbamates have very little insecticidal potency. A few exhibit weak anticholinesterase activity, but most have no significant effect on this enzyme. Overall, they are less of a threat to human health than the insecticidal carbamates. Fungicidal thiocarbamates are discussed in this section, while those used as herbicides are considered in **Chapter 13, *Other Herbicides*.**

Metam-sodium, thiram and ziram and ferbam are the thiocarbamate pesticides. They are discussed individually.

## Metam-sodium

Metam-sodium is formulated in aqueous solutions for application as a soil biocide to kill fungi, bacteria, weed seeds, nematodes and insects. All homeowner uses have been canceled in the United States.

### *Toxicology*

Although animal feeding studies do not indicate high toxicity of **metam-sodium** by ingestion, its decomposition in water yields methyl isothiocyanate, a gas that is extremely irritating to the eyes and to respiratory mucous membranes including the lower respiratory tract/lungs. Inhalation of methyl isothiocyanate may cause pulmonary edema, manifesting with severe respiratory distress and coughing of bloody, frothy sputum. For this reason, metam-sodium must be used outdoors only, and stringent precautions must be taken to avoid inhalation of evolved gas. Metam-sodium can be very irritating to the skin.

Theoretically, exposure to metam-sodium may predispose the individual to "Antabuse" reactions if alcohol is ingested after exposure. Such occurrences have not been reported in the medical literature.

### *Confirmation of Poisoning*

There are no tests for metam-sodium or its breakdown products in body fluids.

### *Treatment of Metam-sodium Toxicosis*

Decontaminate skin and GI tract, as outlined in **Chapter 3, *General Principles*.**

If pulmonary irritation or edema occurs as a result of inhaling methyl isothiocyanate, transport the victim promptly to a medical facility. Treatment for pulmonary edema should proceed as outlined in **Chapter 17, *Fumigants*** in the *Treatment of Fumigant Toxicosis* subsection beginning on page 166.

Metam-sodium is not a cholinesterase inhibitor. Atropine is not antidotal.

## Thiram

**Thiram dust** is moderately irritating to human skin, eyes and respiratory mucous membranes. Contact dermatitis has occurred in occupationally exposed workers. A few individuals have experienced sensitization to thiram.[16] Thiram is a common component of latex and possibly responsible for some of the allergies attributed to latex.

### *Toxicology*

Systemic human poisonings by **thiram** itself have been very few, probably due to limited absorption in most circumstances involving human exposure. Those that have been reported have been similar clinically to toxic reactions to **disulfiram** (Antabuse), the ethyl analogue of thiram that has been extensively used in alcohol aversion therapy.[16] In laboratory animals, thiram at high dosage has effects similar to those of disulfiram (hyperactivity, ataxia, loss of muscle tone, dyspnea and convulsions), but thiram appears to be about 10 times more toxic than disulfiram.

Neither thiram nor disulfiram is a cholinesterase inhibitor. Both, however, inhibit the enzyme acetaldehyde dehydrogenase, which is critical to the conversion of acetaldehyde to acetic acid. This is the basis for the "Antabuse" reaction that occurs when ethanol is consumed by a person on regular disulfiram dosage. The "reaction" includes symptoms of nausea, vomiting, pounding headache, dizziness, faintness, mental confusion, dyspnea, chest and abdominal pain, profuse sweating and skin rash. In rare instances, Antabuse reactions may have occurred following ingestion of beverages containing alcohol among workers previously exposed to thiram.

### *Confirmation of Poisoning*

Urinary xanthurenic acid excretion has been used to monitor workers exposed to thiram, but the test is not generally available.

### *Treatment of Thiram Toxicosis*

Decontaminate skin and GI tract as outlined in **Chapter 3, *General Principles*.**

Infuse appropriate intravenous fluids, especially if vomiting and diarrhea are severe. Monitor serum electrolytes and glucose and replace as needed.

### *Treatment of Acetaldehyde Toxicosis (Antabuse reaction)*

Use oxygen inhalation, trendelenburg positioning and intravenous fluids, which are usually effective in relieving manifestations of "Antabuse" reactions.

*Thiocarbamate*
**HIGHLIGHTS**

Formulated as dusts, wettable powders, water suspensions

Less human health threat than insecticidal carbamates

**SIGNS & SYMPTOMS**

Skin, eye, respiratory irritation

For metam-sodium inhalation, respiratory distress, bloody sputum

May result in Antabuse-like reaction if alcohol is consumed after exposure

**TREATMENT**

Decontaminate skin and GI tract

For metam-sodium, treat pulmonary impacts as for fumigant toxicosis

For thiram, ziram and ferbam, IV fluids as needed

For Antabuse reaction, oxygen, IV fluids and trendelenburg positioning

*EBDC Compound*
**COMMERCIAL PRODUCTS**

maneb (Kypman 80, Manex 80, Maneba, Manex, M-Diphar, Sopranebe, Trimangol)

zineb* (Aspor, Dipher, Hexathane, Kypzin, Parzate C, Tritoftorol, Zebtox)

nabam (Chem Bam, DSE, Parzate, Spring Bak)

mancozeb (Manzeb, Dithane, Mancozin, Manzin, Nemispor, Penncozeb, Ziman-Dithane)

** product canceled*

**HIGHLIGHTS**

Most products no longer in use

Low systemic toxicity

**SIGNS & SYMPTOMS**

Skin, eye, respiratory irritation

Possible behavioral, neurological symptoms

**TREATMENT**

Skin, eye decontamination

Consider GI decontamination

Consider hemodialysis for renal failure

Advise persons who have absorbed any significant amount of thiocarbamates to avoid alcoholic beverages for at least 3 weeks. Disposition of thiocarbamates is slow, and their inhibitory effects on enzymes are slowly reversible.

### Ziram and Ferbam

Ziram and ferbam are formulated as flowable and wettable powders and are used widely on fruit and nut trees, apples, vegetables and tobacco.

#### *Toxicology*

Since **ziram** and **ferbam** are similar to thiram, it is reasonable to assume that similar toxic effects may occur, including irritation to the skin, respiratory tract and eyes. However, there are no reports of human poisoning in the medical literature. If absorbed in sufficient dosage, these metallothiocarbamates may theoretically predispose the patient to an "Antabuse" reaction following ingestion of alcohol. (See thiram.) No occurrences of this have been reported.

#### *Confirmation of Poisoning*

No tests for these fungicides or their breakdown products in body fluids are available.

#### *Treatment of Ziram and Ferbam Toxicosis*

Decontaminate skin and GI tract as needed, as outlined in **Chapter 3,** ***General Principles***.

Treat as for thiram.

## ETHYLENE BIS DITHIOCARBAMATES (EBDC COMPOUNDS)

Maneb and zineb are formulated as wettable and flowable powders. Nabam is provided as a soluble powder and in water solution. Mancozeb is a coordination product of zinc ion and maneb. It is formulated as a dust and as wettable and liquid flowable powders. Although some products, including maneb and mancozeb, were widely used in the 1990s and 2000s, particularly in agricultural settings and on golf courses, most of these products are no longer in use.

### Toxicology

**Maneb, zineb, nabam** and **mancozeb** may cause irritation of the skin, respiratory tract and eyes. Some cases of chronic skin disease in occupationally exposed workers have been attributed to both maneb and zineb, possibly by sensitization.[17,18]

Although marked adverse effects may follow injection of EBDC compounds into animals, systemic toxicity by oral and dermal routes is generally low. Nabam exhibits the greatest toxicity, probably due to its greater water solubility and absorbability. Maneb is moderately soluble in water, but mancozeb and zineb are essentially water insoluble. Absorption of the latter fungicides across skin and mucous membranes is probably very limited. Maneb, mancozeb and metriam all are metabolized to the degradation product ethylene thiourea, which may have toxic properties of its own.[19]

Reports of acute systemic poisonings in humans have been rare. However, zineb precipitated an episode of hemolytic anemia in one worker presumably predisposed because of multiple red cell enzyme deficiencies.[20] Maneb toxicity has been reported in one person who developed acute renal failure and was treated with hemodialysis.[21] Behavioral and neurological symptoms may also occur following maneb poisoning. These include mental status changes, loss of consciousness and tonic-clonic seizures. These appear to improve with supportive care.[22,23] Symptoms similar to Parkinson's disease have also been reported in settings of chronic, occupational exposure, possibly due to the manganese component of maneb.[24] Animal studies suggest that following acute exposure at high doses, chronic symptoms similar to Parkinson's may also occur.[19]

The EBDC compounds are not inhibitors of cholinesterase or of acetaldehyde dehydrogenase. They do not induce cholinergic illness or "Antabuse" reactions.

### Confirmation of Poisoning

No tests for these fungicides or their breakdown products in body fluids are available.

### Treatment of Ethylene Bis Dithiocarbamate Toxicosis

See treatment for substituted benzenes, page 145.

Should severe renal failure occur, consider hemodialysis.

## THIOPHTHALIMIDES

Captan, captafol and folpet are widely used to protect seed, field crops and stored produce. They are formulated as dusts and wettable powders.

### Toxicology

**Captan, captafol** and **folpet** are moderately irritating to the skin, eyes and respiratory tract. Dermal sensitization may occur; captafol has been associated with several episodes of occupational contact dermatitis.[25,26] Very few systemic poisonings by thiophthalimides have been reported in humans. Captafol has been reported to have exacerbated asthma after occupational exposure.[27] A 17-year-old who ingested captafol in a suicide attempt had symptoms including headache, nausea, weakness, numbness of upper limbs and substernal chest pain with an accompanying elevation in creatine kinase and aspartate aminotransferase, and inverted T waves on electrocardiogram. All abnormalities resolved with supportive care over a 72-hour period.[28]

### Confirmation of Poisoning

Following oral exposure, captan fungicides are rapidly metabolized in the body to yield two metabolites that can be measured in the urine: tetrahydrophthalimide (THPI) and thiazolidine-2-thione-4-carboxilic acid (TTCA). Both are considered useful biomarkers for occupational exposure.[29]

### Treatment of Thiophthalimide Toxicosis

See treatment for substituted benzenes, page 145.

*Thiophthalimide*

**COMMERCIAL PRODUCTS**

captan (Captanex, Captaf, Merpan, Orthocide, Vondcaptan)

captafol (Crisfolatan, Difolatan, Foltaf, Haipen, Merpafol, Mycodifol, Sanspor)

folpet (Folpan, Phaltan, Thiophal, Fungitrol II)

**HIGHLIGHTS**

Dusts, wettable powders

Used in seed & field crops, stored produce

Few systemic poisonings reported in the medical literature

**SIGNS & SYMPTOMS**

Skin, eye, respiratory irritation

**TREATMENT**

Skin, eye decontamination

Consider GI decontamination

*Triazole*
**COMMERCIAL PRODUCTS**

triadimefon (Bayleton, Amiral)

myclobutanil

propiconazole (Tilt)

flutriafol

**HIGHLIGHTS**

Moderate acute oral toxicity

Possible chronic effects

Low dermal toxicity

**TREATMENT**

Skin, eye decontamination

Consider GI decontamination

## TRIAZOLE FUNGICIDES

Triazoles are supplied as wettable powder, emulsifiable concentrate, suspension concentrate, paste and dry flowable powder. Most triazoles are used on fruit, cereals, vegetables, coffee, ornamentals, sugarcane, pineapples and turf. (Another compound in this class is fluconazole, a pharmaceutical commonly used to treat fungal infections in humans.) Uses of triadimefon were voluntarily canceled by the registrant in 2008.

### Toxicology

Triazole fungicides – **triadimefon**, **myclobutanil, propiconazole** and **flutriafol** – exhibit moderate acute oral toxicity in laboratory animals, but dermal toxicity is low. All except for triadimefon will cause hepatocyte hypertrophy in mice.[30] Eye exposure may cause irritation. Triadimefon is absorbed across the skin.

Animal data suggest that the triazole fungicides have some central nervous system effects. One study in rats demonstrated that flutriafol induces dopamine release. While this effect is not considered to result in acute toxicity, there is concern for chronic effects.[31] Triadimefon blocks reuptake of dopamine and has demonstrated hyperactivity in mice and rats.[32,33] It is expected that the findings in humans would be similar, although investigation of this has not been reported in the literature.

### Confirmation of Poisoning

No tests for these fungicides or their breakdown products in body fluids are available.

### Treatment of Triazole Toxicosis

See treatment for substituted benzenes, page 145.

## COPPER COMPOUNDS

Insoluble inorganic and organic copper compounds are formulated as wettable powders and dusts. Soluble inorganic and organic copper salts are prepared as aqueous solutions. Some organometallic compounds are soluble in mineral oils.

A great many commercial copper-containing fungicides are available. Some are mixtures of copper compounds. Others include lime, other metals and other fungicides. Compositions of specific products can usually be provided by manufacturers or by poison control centers.

Copper-arsenic compounds such as Paris Green may still be used in agriculture in some countries, but their use has been discontinued in the United States. Toxicity of these is chiefly due to arsenic content (see **Chapter 15, *Arsenicals***). Another copper-arsenic compound, copper chromium arsenate, was formerly used as a wood preservative. That use was discontinued in 2003 on wood being used around the home or on playgrounds.[34]

### Toxicology

The dust and powder preparations of **copper compounds** are irritating to the skin, respiratory tract and particularly the eyes. The soluble copper salts (such as the sulfate and acetate) are corrosive to mucous membranes and the cornea. Limited solubility and absorption probably account for generally low systemic toxicity of most compounds. The more absorbable organic copper compounds exhibit the greatest systemic toxicity in laboratory animals.

Most of what is known about mammalian toxicity of copper compounds has come from veterinary toxicology (livestock seem uniquely vulnerable) and poisonings in man due to deliberate ingestion of copper sulfate or to consumption of water or food that had been contained in copper vessels. The mechanism of toxicity is not clear, although copper appears to release an excess of the cupric ion.[35] This affects enzymes including G6PD and glutathione reductase, which can damage the erythrocyte membrane and produce hemolysis.[36,37] Other enzyme systems may also be affected, including nicotinamide-adenine dinucleotide phosphate (NADPH). Early signs and symptoms of copper poisoning include a metallic taste, nausea, vomiting and epigastric pain. In more severe poisonings, the gastrointestinal irritation will worsen with hemetemesis and melanotic stools. Jaundice and hepatomegaly are common.[38,39] As mentioned above, hemolysis can occur, resulting in circulatory collapse and shock and may be prolonged, particularly in patients with an existing condition such as G6PD deficiency. Methemoglobinemia has been reported in these cases, usually related to copper sulfate.[35,38,40,41] Acute renal failure with oliguria can also occur. Shock is a primary cause of death early in the course, and renal failure and hepatic failure contribute to deaths occurring more than 24 hours after poisoning.[42] A case report from China describes an adult male developing severe hemolysis and methemoglobinemia after ingestion of copper-8-hydroxyquinolate.[35]

## Confirmation of Poisoning

Whole blood and serum copper levels can be measured, with a reported average red blood cell level in normal adults of 89 µg/dL and average serum level of 114 µg/dL.[43] Most reported cases of acute copper poisoning are at levels exceeding 200 µg/dL, and some as high as 1,650 µg/dL.[15,35,41]

## Treatment of Copper Toxicosis

Management of poisonings by ingestion of copper-containing fungicides depends on the chemical nature of the compound: the strongly ionized salts present the greatest hazard; the oxides, hydroxides, oxychloride and oxysulfate are less likely to cause severe systemic poisoning.

Decontaminate skin with soap and water. The eyes should be flushed free of irritating dust, powder or solution, using clean water or saline. If eye or dermal irritation persists, medical treatment should be obtained. Eye irritation may be severe.

Give water or milk as soon as possible to dilute the toxicant and mitigate corrosive action on the mouth, esophagus and gut. Do not be overly aggressive with the dilution to avoid accidental inducement of vomiting.[44] There is not a specific amount that should be given, although the Poisondex Editorial Board consensus is no more than 4 ounces in children and 8 ounces in adults.[42,44]

Do not induce emesis because the corrosive nature of some copper salts can cause further damage to the esophagus, although vomiting is usually spontaneous in acute copper ingestion. Further gastrointestinal decontamination should be determined on a case-by-case basis as outlined in the **Chapter 3, *General Principles***, understanding that gastric lavage may cause further damage.[42] Charcoal's adsorbent effectiveness has not been widely studied in metal poisonings.

CAUTION: ***If corrosive action has been severe, it may be best to avoid gastric intubation, as this may pose a serious risk of esophageal perforation. It may be prudent to consider referral to a gastroenterologist for endoscopy, given the caustic nature of the ingestion.***

*Copper Compound*
### COMMERCIAL PRODUCTS

copper acetate

copper ammonium carbonate

copper carbonate, basic

copper chromium acetate (CCA)

copper hydroxide

copper lime dust

copper oxychloride

copper potassium sulfide

copper silicate

copper sulfate

copper sulfate, tribasic (Bordeaux Mixture)

cupric oxide

cuprous oxide

### SIGNS & SYMPTOMS

Skin, eye, respiratory irritation

### TREATMENT

Skin, eye decontamination

Water or milk for GI dilution

IV fluids with glucose, electrolytes if systemic

Methylene blue for severe methemoglobinemia

Consider BAL

### CONTRAINDICATED

Induced emesis, intubation

*Organomercury*
**COMMERCIAL PRODUCTS**

***Methyl mercury compounds***

methyl mercury hydroxide

acetate

propionate

pentachlorophenate

quinolinolate

**Methoxyethyl mercury compounds**

methoxyethyl mercury acetate (MEMA, Panogen, Panogen M)

methoxyethyl mercury chloride (MEMC, Emisan 6, Ceresan)

**Phenylmercuric acetate**

Agrosan, Shimmer-ex, Tag HL 331, Unisan

Setrete (Gallotox, PMAA) is phenyl mercury ammonium acetate

If indications of systemic illness appear, administer intravenous fluids containing glucose and electrolytes. Monitor fluid balance and correct blood electrolyte concentrations as needed. If shock develops, give blood transfusions and vasopressor amines as required.

Monitor plasma for evidence of hemolysis (free hemoglobin) and the red cells for methemoglobin. If methemoglobinemia is severe (>30%), or the patient is cyanotic, administer methylene blue.

**Dosage of Methylene Blue**

***Adults/children 1-2 mg/kg/dose, given as a slow IV push over a few minutes, every 4 hours as needed.[42]***

In patients with severe methemoglobinemia that is unresponsive to methylene blue, or those with G6PD deficiency, chelation (see below) and plasma exchange may be required.[35]

Administer morphine if the patient is in severe pain.

Consider administering BAL. The value of chelating agents in copper poisoning has not been established.[35,45] However, BAL appears to accelerate copper excretion and may alleviate illness. d-penicillamine is the treatment for Wilson's disease due to chronic copper toxicity, but in the context of severe vomiting and/or mental status changes from an acute ingestion, BAL would be a more likely initial choice.[40,42] For a recommended schedule of dosage for initial therapy with BAL and subsequent d-penicillamine administration, see **Chapter 15**, ***Arsenicals***.

Although hemodialysis is indicated for patients with renal failure, copper is not effectively removed in the dialysate.[35,38]

## ORGANOMERCURY COMPOUNDS

Methyl mercury and methoxyethyl mercury compounds and phenyl mercuric acetate fungicides have been formulated as aqueous solutions and dusts. They have been used chiefly as seed protectants and had historically been added to household paint. Use of alkyl mercury fungicides in the United States has been prohibited since the early 1990s. Phenyl mercuric acetate is no longer manufactured or used in the United States.

### Toxicology

The mercurial fungicides — **methyl mercury** and **methoxyethyl mercury** compounds and **phenyl mercuric acetate** — are among the most toxic pesticides ever developed, in terms of both acute and chronic toxicity. Epidemics of severe, often fatal, neurologic disease have occurred when indigent residents of less developed countries consumed methyl mercury-treated grain intended for planting of crops.[46,47] Poisoning has also occurred from eating meat from animals fed mercury-treated seed.[48] Most of what is known of poisoning by organic mercurial fungicides has come from these occurrences.

Organic mercury compounds are efficiently absorbed across the gut and possibly across the skin. Volatile organic mercury is readily taken up across the pulmonary epithelium. Methyl mercury is selectively concentrated in the tissues of the nervous system and also in red blood cells. Other alkyl mercury compounds are probably distributed similarly. Excretion occurs almost entirely through the biliary system. The whole body half-life of methyl mercury in humans ranges between 45 and 56 days.[49] Significant conversion of organic mercury to inorganic mercury occurs in the red cell.

Early symptoms of poisoning are metallic taste in the mouth, numbness and tingling of the digits and face, tremor, headache, fatigue, emotional lability and difficulty in thinking. Manifestations of more severe poisoning are incoordination, slurred speech, loss of position sense, hearing loss, constriction of visual fields, spasticity or rigidity of muscle movements and deterioration of mental capacity. Many poisonings caused by ingestion of organic mercurials have been fatal and a large percentage of survivors have suffered severe, permanent neurologic damage.[46,47,48]

Phenyl mercuric acetate is not as extremely toxic as the alkyl mercury compounds and is not as efficiently absorbed from the gut as methyl mercury.[50] Phenyl mercuric acetate is used to prevent fungal growth in latex paint. A case of acrodynia in a child led to latex paint as a possible source for mercury exposure. Symptoms of acrodynia include fever, erythema and desquamation of hands and feet, muscular weakness, leg cramps and personality changes.[51] Phenyl mercuric compounds have been banned from latex paint since 1990.[52]

## Confirmation of Poisoning

Mercury content of blood and tissues can be measured by atomic absorption spectrometry. Blood levels of 5 µg/dL or greater are considered elevated for acute exposure.[21] Special procedures are needed for extraction and measurement of specific organic mercury compounds.

## Treatment of Organomercury Toxicosis

Every possible precaution should be taken to avoid potentially life-threatening ingestions of organic mercury fungicides. Very little can be done to mitigate neurologic damage caused by organic mercurials.

The following are the basic steps in the management of organomercury poisoning:

Decontaminate skin and eyes, as discussed in **Chapter 3, *General Principles*.**

Remove persons experiencing symptoms (metallic taste in mouth) after inhalation of volatile organic mercury compounds (methyl mercury is the most volatile) promptly from the contaminated environment and observe closely for indications of neurologic impairment. Every possible precaution should be taken to avoid further exposure to organic mercury compounds.

Consider gastrointestinal decontamination as outlined in **Chapter 3**.

Administer chelation therapy. Chelation is an essential part of the management of acute mercury poisoning. For dosages of specific agents, see **Chapter 15, *Arsenicals***. Succimer (DMSA) appears to be the most effective agent available in the United States. Dimercaprol (BAL) is contraindicated in these poisonings because of its potential to increase brain levels of mercury.[52] EDTA is apparently of little value in poisonings by organic mercury. D-penicillamine is probably useful, is available in the United States and has proven effective in reducing the residence half-life of methyl mercury in poisoned humans.[52] DMPS (2,3-dimercaptopropane-1-sulfonate acid) and NAP (n-acetyl-D,L-penicillamine) are probably also useful but are not currently approved for use in the United States.

Consider extracorporeal hemodialysis and hemoperfusion, although experience to date has not been encouraging.

*Organomercury*
### HIGHLIGHTS

Extreme acute and chronic toxicity

Efficiently absorbed across gut and possibly skin

### SIGNS & SYMPTOMS

Metallic taste in mouth

Numbness, tingling of digits & face

Tremor, headache, fatigue

Emotional lability, difficulty in thinking

Incoordination, slurred speech, hearing loss in more severe cases

### TREATMENT

Skin, eye decontamination

Consider GI decontamination

Chelation with DMSA or other appropriate agent

### CONTRAINDICATED

Use of BAL

*Organotin Compound*
**COMMERCIAL PRODUCTS**

triphenyl tin

fentin hydroxide (Suzu-H, Super Tin, Tubotin)

fentin chloride (Tinmate)

fentin acetate (Batasan, Brestan, Phenostat-A, Phentinoacetate, Suzu, TPTA)

**HIGHLIGHTS**

Wettable & flowable powders

Eye, skin, respiratory irritant

Most have been discontinued in U.S.

**SIGNS & SYMPTOMS**

Headache, nausea, vomiting, dizziness

Sometimes convulsions, loss of consciousness

Photophobia, mental disturbances

**TREATMENT**

Skin, eye decontamination

ICU supportive care for CNS effects

Consider GI decontamination for large ingestions

## ORGANOTIN COMPOUNDS

Triphenyl tin, fentin hydroxide, fentin chloride and fentin acetate are formulated as wettable and flowable powders for use mainly as fungicides to control blights on field crops and orchard trees. Fentin chloride was also prepared as an emulsifiable concentrate for use as a molluscicide (Aquatin 20 EC, discontinued in 1995). Tributyltin salts were at one time used as fungicides and antifouling agents on ships, but this use has been banned by most countries. They are somewhat more toxic by the oral route than triphenyltin, but toxic actions are otherwise probably similar. Most organotin compounds have been discontinued in the United States.

### Toxicology

**Triphenyl tin, fentin hydroxide, fentin chloride** and **fentin acetate** are irritating to the eyes, respiratory tract and skin. They are probably absorbed to a limited extent by the skin and gastrointestinal tract. Manifestations of toxicity are due principally to effects on the brain: headache, nausea, vomiting, dizziness and sometimes convulsions and loss of consciousness. Photophobia and mental disturbances occur. Epigastric pain is reported, even in poisoning caused by inhalation. Elevation of blood sugar, sufficient to cause glycosuria, has occurred in some cases. The phenyl tin fungicides are less toxic than ethyl, dimethyl and trimethyl tin compounds that are used in the production of plastics. Signs and symptoms for poisoning from those compounds have included disorientation and other mental status changes, cerebral edema, neurologic damage and death in severely poisoned individuals.[34,53,54] No deaths and very few poisonings have been reported as a result of occupational exposures to phenyltin pesticides.

### Treatment of Organotin Toxicosis

Remove skin contamination by washing with soap and water. Flush eyes free of contaminating material with clean water or saline. If irritation persists, expert medical treatment should be obtained.

Provide supportive care in an intensive care unit if neurological effects are evident.

If large amounts of phenyltin compound have been ingested in the past hour, take measures to decontaminate the gastrointestinal tract as outlined in **Chapter 3, *General Principles***.

BAL, penicillamine, or other chelating agents have not been effective in lowering tissue stores of organotin compounds in experimental animals.

## CADMIUM COMPOUNDS

**Cadmium chloride, cadmium sulfate** and **cadmium succinate** have been used to treat fungal diseases affecting turf and the bark of orchard trees. They were formulated as solutions and emulsions. **Miller 531** and **Crag Turf Fungicide 531** were complexes of cadmium, calcium, copper, chromium and zinc oxides. **Kromad** is a mixture of cadmium sebacate, potassium chromate and thiram. **Cad-Trete** is a mixture of cadmium chloride and thiram. All cadmium fungicides in the United States have been discontinued. Cadmium exposure may also occur in the occupational setting from other sources and uses of the toxic metal.

## Toxicology

Cadmium salts and oxides are very irritating to the respiratory and gastrointestinal tracts. Inhaled cadmium dust or fumes can cause respiratory toxicity after a latency period of several hours, including a mild, self-limited illness of fever, cough, malaise, headache and abdominal pain, similar to metal fume fever. A more severe form of toxicity includes chemical pneumonitis and is associated with labored breathing, chest pain and a sometimes fatal hemorrhagic pulmonary edema.[55,56,57,58] Symptoms may persist for weeks.

Ingested cadmium causes nausea, vomiting, diarrhea, abdominal pain and tenesmus. Relatively small inhaled and ingested doses produce serious symptoms. Protracted absorption of cadmium has led to renal damage (proteinuria and azotemia), anemia, liver injury (jaundice) and defective bone structure (pathologic fractures) in chronically exposed persons. Prolonged inhalation of cadmium dust has contributed to chronic obstructive pulmonary disease.[59]

## Confirmation of Poisoning

Cadmium can be measured in body fluids by several methods, including electrothermal atomic absorption spectroscopy, graphite furnace atomic spectrophotometry, and potentiometric stripping analysis.[60,61,62] It is reported that blood cadmium concentrations tend to correlate with acute exposure and urine levels tend to reflect total body burden. Blood levels exceeding 5 µg/dL suggest excessive exposure.[55] Urinary excretion in excess of 100 µg per day suggests an unusually high body burden.

## Treatment of Cadmium Poisoning

Decontaminate skin and eyes as outlined in **Chapter 3, *General Principles*.**

For severe reactions such as pulomonary edema and pneumonitis, use aggressive measures in an intensive care setting, including positive end-expiratory pressure mechanical ventilation, monitoring of blood gases and administration of diuretics, steroid medications and antibiotics.[55,63] Codeine sulfate may be needed to control cough and chest pain. Respiratory irritation resulting from inhalation of small amounts of cadmium dust may resolve spontaneously, requiring no treatment.

Consider decontaminating the lower GI tract as outlined in **Chapter 3** if retention of some cadmium is suspected. The irritant action of ingested cadmium products on the gastrointestinal tract is so strong that spontaneous vomiting and diarrhea often eliminate nearly all unabsorbed cadmium from the gut.

Administer intravenous fluids to overcome dehydration caused by vomiting and diarrhea. Fluids also limit cadmium toxicity affecting the kidneys and liver. However, great care must be taken to monitor fluid balance and blood electrolyte concentrations so that failing renal function does not lead to fluid overload.

Consider chelation therapy with calcium disodium EDTA for acute poisoning, depending on measured cadmium in blood and urine and the status of renal function. Chelation therapy has been shown to increase urinary excretion of cadmium. Its therapeutic value in cadmium poisoning has not been established, and use of the agent carries the risk that unduly rapid transfer of cadmium to the kidney may precipitate renal failure. Monitor urine protein and blood urea nitrogen and creatinine carefully during therapy.

***Cadmium Compound***

### COMMERCIAL PRODUCTS

cadmium chloride (Caddy)

cadmium sulfate (generic, 14% solution)

cadmium succinate (Cadminate)

Miller 531 and Crag Turf Fungicide 531 (generic) were complexes of cadmium, calcium, copper, chromium, and zinc oxides

Kromad is a mixture of cadmium sebacate, potassium chromate, and thiram

Cad-Trete is a mixture of cadmium chloride and thiram

### HIGHLIGHTS

Discontinued in U.S.

### SIGNS & SYMPTOMS

Inhalation: Fever, cough, malaise, headache, abdominal pain

Ingestion: Nausea, vomiting, diarrhea, abdominal pain, tenesmus

### TREATMENT

Skin, eye decontamination

Consider lower GI decontamination if retained

Aggressive ICU measures for severe reactions

IV fluids for dehydration

*Miscellaneous Organic Fungicide*
**COMMERCIAL PRODUCTS**

anilazine (Dyrene)

benomyl (Benlate, Tersan 1991, Benex)

cycloheximide (naramycin)

dodine (Carpene, Curitan, Melprex, Venturol)

iprodione (Rovral, Glycophene)

metalaxyl (Ridomil, Subdue)

etridiazole (Terrazole, Aaterra, Ethazol, Koban, Pansoil, Truban)

thiabendazole (Apl-Luster, Arbotect, Mertect, Tecto, Thibenzole)

triforine (Funginex, Saprol, Denarin)

**Dosage of Calcium Disodium EDTA**

- ***75 mg/kg/day in three to six divided doses for 5 days. The total dose for the 5-day course should not exceed 500 mg/kg.[65]***

Succimer (DMSA) has also been used in this poisoning but has not been demonstrated to be efficacious.

6. Because of the risk of renal injury by mobilized cadmium, do not use dimercaprol (BAL) for treatment of cadmium poisoning.

7. Monitor urinary protein and cells regularly and measure hepatocellular enzymes and creatinine for indications of injury to these organ systems.

## MISCELLANEOUS ORGANIC FUNGICIDES

Some modern organic fungicides are widely used. Reports of adverse effects on humans are few. Some of the known properties of these agents follow.

**Anilazine** is supplied as wettable and flowable powders. It was used on vegetables, cereals, coffee, ornamentals and turf. No products are currently registered in the United States. This product has caused skin irritation in exposed workers. Acute oral and dermal toxicity in laboratory animals is low. Human systemic poisonings have not been reported in the published medical literature.

**Benomyl** is a synthetic organic fungistat having little or no acute toxic effect in mammals. There are no active products in the United States. No systemic poisonings have been reported in humans in the published literature. Although the molecule contains a carbamate grouping, benomyl is not a cholinesterase inhibitor. It is poorly absorbed across skin, and what is absorbed is promptly metabolized and excreted.

Skin injuries to exposed individuals have occurred, and dermal sensitization has been found among agricultural workers exposed to foliage residues.[3,66]

**Cycloheximide** is formulated as wettable powders, sometimes combined with other fungicides. There are no registered products in the United States. Cycloheximide is a product of fungal culture, effective against fungal diseases of ornamentals and grasses. It is selectively toxic to rats and much less toxic to dogs and monkeys. No human poisonings have been reported. Animals given toxic doses exhibit salivation, bloody diarrhea, tremors and excitement, leading to coma and death due to cardiovascular collapse. Hydrocortisone increases the rate of survival in deliberately poisoned rats. Atropine, epinephrine, methoxyphenamine and hexamethonium all relieved the symptoms of poisoning, but did not improve survival.[67]

**Dodine** is formulated as a wettable powder. It is commonly applied to berries, nuts, peaches, apples, pears and trees afflicted with leaf blight. Dodine is a cationic surfactant with antifungal activity. It is absorbed across the skin. In animal studies, it causes severe irritation to the eye, and also is a skin irritant. Acute oral and dermal toxicity in laboratory animals is moderate. Poisonings in humans have not been reported in the published medical literature. Based on animal studies, ingestion would probably cause nausea, vomiting and diarrhea.[68]

**Iprodione** is supplied as wettable powder and other formulations. It is used on berries, grapes, fruit, vegetables, grasses and ornamentals. It is also used as seed dressing. Iprodione exhibits low acute oral and dermal toxicity in laboratory animals.[69] No human poisonings have been reported in the published medical literature.

**Metalaxyl** is supplied as emulsifiable and flowable concentrates. It is a systemic fungicide used to control soil-borne fungal diseases on fruit trees, cotton, hops, soybeans, peanuts, ornamentals and grasses. It is also used as seed dressing. It exhibits low acute oral and dermal toxicity in laboratory animals.[70] No human poisonings have been reported in the published medical literature.

**Etridiazole** is supplied as wettable powder and granules for application to soil as a fungicide and nitrification inhibitor. There are no registered products in the United States. Human poisonings have not been reported in the published literature.

**Thiabendazole** is widely used as an agricultural fungicide, but most experience with its toxicology in humans has come from medicinal use against intestinal parasites. Oral doses administered for this purpose are far greater than those likely absorbed in the course of occupational exposure. Thiabendazole is rapidly metabolized and excreted in the urine, mostly as a conjugated hydroxy-metabolite. Symptoms and signs that sometimes follow ingestion are: dizziness, nausea, vomiting, diarrhea, epigastric distress, lethargy, headache and tinnitus.[71] Blood enzyme tests may indicate liver injury. Persons with liver and kidney disease may be unusually vulnerable to toxic effects. Adverse effects in humans from use of thiabendazole as a fungicide have not been reported in the published literature.

**Triforine** is supplied as emulsifiable concentrate and wettable powder. It is used on berries, fruit, vegetables and ornamentals. Mammals rapidly excrete it chiefly as a urinary metabolite. It exhibits low acute oral and dermal toxicity in laboratory animals.[72] No human poisonings have been reported in the published literature.

## Confirmation of Poisoning

Laboratory tests for these organic fungicides or their metabolites in body fluids are not generally available.

## Treatment of Organic Fungicide Toxicosis

See treatment for substituted benzenes, page 145.

## References

1. United States Environmental Protection Agency. *Reregistration Eligibility Decision (RED) for Chloroneb.* Sep 2005. EPA 738-R-04-012.

2. Lensen G, Jungbauer F, Goncalo M, Coenraads PJ. Airborne irritant contact dermatitis and conjunctivitis after occupational exposure to chlorothalonil in textiles. *Contact Dermatitis.* Sep 2007;57(3):181-186.

3. Penagos H, Ruepert C, Partanen T, Wesseling C. Pesticide patch test series for the assessment of allergic contact dermatitis among banana plantation workers in panama. *Dermatitis.* Sep 2004;15(3):137-145.

4. Penagos HG. Contact dermatitis caused by pesticides among banana plantation workers in Panama. *Int J Occup Environ Health.* Jan-Mar 2002;8(1):14-18.

5. Bruynzeel DP, van Ketel WG. Contact dermatitis due to chlorothalonil in floriculture. *Contact Dermatitis.* Jan 1986;14(1):67-68.

6. Dannaker CJ, Maibach HI, O'Malley M. Contact urticaria and anaphylaxis to the fungicide chlorothalonil. *Cutis.* Nov 1993;52(5):312-315.

7. Draper A, Cullinan P, Campbell C, Jones M, Newman Taylor A. Occupational asthma from fungicides fluazinam and chlorothalonil. *Occup Environ Med.* Jan 2003;60(1):76-77.

8. Peters HA, Gocmen A, Cripps DJ, Bryan GT, Dogramaci I. Epidemiology of Hexachlorobenzene-Induced Porphyria in Turkey: Clinical and Laboratory Follow-up After 25 Years. *Arch Neurol.* 1992;39(12):744-749.

9. Lilienthal H, Benthe C, Heinzow B, Winneke G. Impairment of schedule-controlled behavior by pre- and postnatal exposure to hexachlorobenzene in rats. *Arch Toxicol.* 1996;70(3-4):174-181.

10. Ribas-Fito N, Torrent M, Carrizo D, Julvez J, Grimalt JO, Sunyer J. Exposure to hexachlorobenzene during pregnancy and children's social behavior at 4 years of age. *Environ Health Perspect.* Mar 2007;115(3):447-450.

11. Larsen GL, Huwe JK, Bakke JE. Intermediary metabolism of pentachloronitrobenzene in the control and germ-free rat and rat with cannulated bile ducts. *Xenobiotica.* Oct 1998;28(10):973-984.

12. Renner G. Biotransformation of the fungicides hexachlorobenzene and pentachloronitrobenzene. *Xenobiotica.* Jul 1981;11(7):435-446.

13. Bartlett DW, Clough JM, Godwin JR, Hall AA, Hamer M, Parr-Dobrzanski B. The strobilurin fungicides. *Pest Manag Sci.* Jul 2002;58(7):649-662.

14. Center for Disease Control and Prevention. Acute Pesticide Poisoning Associated with Pyraclostrobin Fungicide—Iowa, 2007 *Morb Mortal Wkly Rep.* 2008;56(51):1343-1345.

15. van Ravenzwaay B, Akiyama M, Landsiedel R, et al. Toxicological overview of a novel strobilurin fungicide, oryastrobin. *J Pestic Sci.* 2007;32(3):270-277.

16. Dalvi RR. Toxicology of thiram (tetramethylthiuram disulfide): a review. *Vet Hum Toxicol.* Oct 1988;30(5):480-482.

17. Cole DC, Carpio F, Math JJ, Leon N. Dermatitis in Ecuadorean farmworkers. *Contact Dermatitis.* Jul 1997;37(1):1-8.

18. Nater JP, Terpstra H, Bleumink E. Allergic contact sensitization to the fungicide Maneb. *Contact Dermatitis.* Jan 1979;5(1):24-26.

19. Domico LM, Zeevalk GD, Bernard LP, Cooper KR. Acute neurotoxic effects of mancozeb and maneb in mesencephalic neuronal cultures are associated with mitochondrial dysfunction. *Neurotoxicology.* Sep 2006;27(5):816-825.

20. Pinkhas J, Djaldetii M, Joshua H, Resnick C, de Vries A. Sulfhemoglobinemia and Acute Hemolytic Anemia with Heinz Bodies Following Contact with a Fungicide— Zinc Ethylene Bisdithiocarbamate—in a Subject with Glucose-6-Phosphate Dehydrogenase Deficiency and Hypocatalasemia. *Blood.* 1963;21(4):484-494.

21. Koizumi A, Shiojima S, Omiya M, Nakano S, Sato N, Ikeda M. Acute renal failure and maneb (manganous ethylenebis [dithiocarbamate]) exposure. *JAMA.* Dec 7 1979;242(23):2583-2585.

22. Israeli R, Sculsky M, Tiberin P. Acute intoxication due to exposure to maneb and zineb. A case with behavioral and central nervous system changes. *Scand J Work Environ Health.* Feb 1983;9(1):47-51.

23. de Tollenaer SM, Buysse C, van den Anker JN, Touw DJ, de Hoog M. Life threatening central nervous system manifestations and hypothermia due to maneb intoxication in a child: a case report. *Ther Drug Monit.* Dec 2006;28(6):813-815.

24. Ferraz HB, Bertolucci PH, Pereira JS, Lima JG, Andrade LA. Chronic exposure to the fungicide maneb may produce symptoms and signs of CNS manganese intoxication. *Neurology.* Apr 1988;38(4):550-553.

25. Peluso AM, Tardio M, Adamo F, Venturo N. Multiple sensitization due to bis-dithiocarbamate and thiophthalimide pesticides. *Contact Dermatitis.* Nov 1991;25(5):327.

26. Vilaplana J, Romaguera C. Captan, a rare contact sensitizer in hairdressing. *Contact Dermatitis.* Aug 1993;29(2):107.

27. Royce S, Wald P, Sheppard D, Balmes J. Occupational asthma in a pesticides manufacturing worker. *Chest.* Jan 1993;103(1):295-296.

28. Chodorowski Z, Anand JS. Acute oral suicidal intoxication with Captan--a case report. *J Toxicol Clin Toxicol.* 2003;41(5):603.

29. Krieger RI, Thongsinthusak T. Captan metabolism in humans yields two biomarkers, tetrahydrophthalimide (THPI) and thiazolidine-2-thione-4-carboxylic acid (TTCA) in urine. *Drug Chem Toxicol.* 1993;16(2):207-225.

30. Goetz AK, Bao W, Ren H, et al. Gene expression profiling in the liver of CD-1 mice to characterize the hepatotoxicity of triazole fungicides. *Toxicol Appl Pharmacol.* Sep 15 2006;215(3):274-284.

31. Santana MB, Rodrigues KJ, Duran R, et al. Evaluation of the effects and mechanisms of action of flutriafol, a triazole fungicide, on striatal dopamine release by using *in vivo* microdialysis in freely moving rats. *Ecotoxicol Environ Saf.* Jul 2009;72(5):1565-1571.

32. Crofton KM, Boncek VM, Reiter LW. Hyperactivity induced by triadimefon, a triazole fungicide. *Fundam Appl Toxicol.* Apr 1988;10(3):459-465.

33. Reeves R, Thiruchelvam M, Richfield EK, Cory-Slechta DA. The effect of developmental exposure to the fungicide triadimefon on behavioral sensitization to triadimefon during adulthood. *Toxicol Appl Pharmacol.* Oct 1 2004;200(1):54-63.

34. United States Environmental Protection Agency. Chromated copper arsenate. 2008. http://www.epa.gov/oppad001/reregistration/cca/. Accessed December 18, 2012.

35. Yang CC, Wu ML, Deng JF. Prolonged hemolysis and methemoglobinemia following organic copper fungicide ingestion. *Vet Hum Toxicol.* Dec 2004;46(6):321-323.

36. Barceloux DG. Copper. *J Toxicol Clin Toxicol.* 1999;37(2):217-230.

37. Klein WJ, Jr., Metz EN, Price AR. Acute copper intoxication. A hazard of hemodialysis. *Arch Intern Med.* Apr 1972;129(4):578-582.

38. Agarwal SK, Tiwari SC, Dash SC. Spectrum of poisoning requiring haemodialysis in a tertiary care hospital in India. *Int J Artif Organs.* Jan 1993;16(1):20-22.

39. Lamont DL, Duflou JA. Copper sulfate. Not a harmless chemical. *Am J Forensic Med Pathol.* Sep 1988;9(3):226-227.

40. Chugh KS, Singhal PC, Sharma BK. Letter: Methemoglobinemia in acute copper sulfate poisoning. *Ann Intern Med.* Feb 1975;82(2):226-227.

41. Jantsch W, Kulig K, Rumack BH. Massive copper sulfate ingestion resulting in hepatotoxicity. *J Toxicol Clin Toxicol.* 1984;22(6):585-588.

42. Micromedex Poisondex. Copper poisoning. Englewood: Thomson Reuters; 1998.

43. Cartwright GE, Wintrobe MM. Copper Metabolism in Normal Subjects. *Am J Clin Nutr.* Apr 1964;14:224-232.

44. Friedman EM, Lovejoy FH, Jr. The emergency management of caustic ingestions. *Emerg Med Clin North Am.* Feb 1984;2(1):77-86.

45. Hantson P, Lievens M, Mahieu P. Accidental ingestion of a zinc and copper sulfate preparation. *J Toxicol Clin Toxicol.* 1996;34(6):725-730.

46. Bakir F, Rustam H, Tikriti S, Al-Damluji SF, Shihristani H. Clinical and epidemiological aspects of methylmercury poisoning. *Postgrad Med J.* Jan 1980;56(651):1-10.

47. Grandjean P, Weihe P, Nielsen JB. Methylmercury: significance of intrauterine and postnatal exposures. *Clin Chem.* Jul 1994;40(7 Pt 2):1395-1400.

48. Snyder RD. Congenital mercury poisoning. *N Engl J Med.* May 6 1971;284(18):1014-1016.

49. Smith JC, Farris FF. Methyl mercury pharmacokinetics in man: a reevaluation. *Toxicol Appl Pharmacol.* Apr 1996;137(2):245-252.

50. Mercury toxicity. Agency for Toxic Substance and Disease Registry. *Am Fam Physician.* Dec 1992;46(6):1731-1741.

51. Agocs MM, Etzel RA, Parrish RG, et al. Mercury exposure from interior latex paint. *N Engl J Med.* Oct 18 1990;323(16):1096-1101.

52. Clarkson TW. Mercury--an element of mystery. *N Engl J Med.* Oct 18 1990;323(16):1137-1139.

53. Yoo CI, Kim Y, Jeong KS, et al. A case of acute organotin poisoning. *J Occup Health.* Jul 2007;49(4):305-310.

54. Colosio C, Tomasini M, Cairoli S, et al. Occupational triphenyltin acetate poisoning: a case report. *Br J Ind Med.* Feb 1991;48(2):136-139.

55. Ando Y, Shibata E, Tsuchiyama F, Sakai S. Elevated urinary cadmium concentrations in a patient with acute cadmium pneumonitis. *Scand J Work Environ Health.* Apr 1996;22(2):150-153.

56. Barnhart S, Rosenstock L. Cadmium chemical pneumonitis. *Chest.* Nov 1984;86(5):789-791.

57. Okuda B, Iwamoto Y, Tachibana H, Sugita M. Parkinsonism after acute cadmium poisoning. *Clin Neurol Neurosurg.* Dec 1997;99(4):263-265.

58. Panchal L, Vaideeswar P. Acute lung injury due to cadmium inhalation--a case report. *Indian J Pathol Microbiol.* Apr 2006;49(2):265-266.

59. Hendrick DJ. Occupational and chronic obstructive pulmonary disease (COPD). *Thorax.* Sep 1996;51(9):947-955.

60. Christoffersson JO, Welinder H, Spang G, Mattsson S, Skerfving S. Cadmium concentration in the kidney cortex of occupationally exposed workers measured *in vivo* using X-ray fluorescence analysis. *Environ Res.* Apr 1987;42(2):489-499.

61. Mascagni P, Consonni D, Bregante G, Chiappino G, Toffoletto F. Olfactory function in workers exposed to moderate airborne cadmium levels. *Neurotoxicology.* Aug 2003;24(4-5):717-724.

62. Ostapczuk P. Direct determination of cadmium and lead in whole blood by potentiometric stripping analysis. *Clin Chem.* Oct 1992;38(10):1995-2001.

63. The Acute Respiratory Distress Syndrome Network. Ventilation with lower tidal volumes as compared with traditional tidal volumes for acute lung injury and the acute respiratory distress syndrome. *N Engl J Med.* May 4 2000;342(18):1301-1308.

64. Waters RS, Bryden NA, Patterson KY, Veillon C, Anderson RA. EDTA chelation effects on urinary losses of cadmium, calcium, chromium, cobalt, copper, lead, magnesium, and zinc. *Biol Trace Elem Res.* Dec 2001;83(3):207-221.

65. Klaassen CD. Heavy metals and heavy metal antagonists. In: Gilman AG, Rall TW, Niew AS, al. E, eds. *Goodman and Gilman's The Pharmacological Basis of Therapeutics.* 3rd ed. New York: Pergamon Press; 1990:1605-1606.

66. van Joost T, Naafs B, van Ketel WG. Sensitization to benomyl and related pesticides. *Contact Dermatitis.* Mar 1983;9(2):153-154.

67. Morgan DP. *Recognition and Management of Pesticide Poisonings.* 4th ed: United States EPA; 1989.

68. Reregistration Eligibility Decision (RED) for Dodine. United States EPA; 2005. http://www.epa.gov/oppsrrd1/REDs/dodine-red.pdf. Accessed January 3, 2011.

69. Reregistration Eligibility Decision (RED) Iprodione. United States EPA. 1998. http://www.epa.gov/oppsrrd1/REDs/2335.pdf. Accessed January 3, 2011.

70. Reregistration Eligibility Decision (RED) Metalaxyl. United States EPA. 1994. http://www.epa.gov/oppsrrd1/REDs/0081.pdf. Accessed January 3, 2011.

71. Tchao P, Templeton T. Thiabendazole-associated grand mal seizures in a patient with Down syndrome. *J Pediatr.* Feb 1983;102(2):317-318.

72. Reregistration Eligibility Decision (RED) for Triforine. United States EPA. 2008. http://www.epa.gov/oppsrrd1/REDs/triforine_red.pdf. Accessed January 3, 2011.

CHAPTER 17

# Fumigants

Packaging and formulation of fumigants are complex. Those that are gases at room temperature (methyl bromide, ethylene oxide, sulfur dioxide, sulfuryl fluoride) are provided in compressed gas cylinders. Liquids are marketed in cans or drums. Solids that sublime, such as naphthalene, must be packaged so as to prevent significant contact with air before they are used. Sodium cyanide is only available in an encapsulated form so that when wild canids attack livestock their bite releases the poison.

Mixtures of fumigants are sometimes used. For instance, chloropicrin, which has a strong odor and irritant effect, is often added as a "warning agent" to other liquid fumigants. It is important to be aware of the possibility of such mixtures.

Liquid halocarbons and carbon disulfide evaporate into the air while naphthalene sublimes. Paraformaldehyde slowly depolymerizes to formaldehyde. Aluminum phosphide slowly reacts with water vapor in the air to liberate phosphine, an extremely toxic gas.

Fumigants have remarkable capacities for diffusion (a property essential to their function). Some readily penetrate rubber and neoprene personal protective gear, as well as human skin. They are rapidly absorbed across the pulmonary membranes, gastrointestinal tract and skin. Special adsorbents are required in respirator canisters to protect exposed workers from airborne fumigant gases. Even these may not provide complete protection when air concentrations of fumigants are high.

## NAPHTHALENE

### Toxicology

Naphthalene is a solid white hydrocarbon long used in ball, flake or cake form as a moth repellent. It sublimes slowly. The vapor has a sharp, pungent odor that is irritating to the eyes and upper respiratory tract. Inhalation of high concentrations causes headache, dizziness, nausea and vomiting. Intensive, prolonged inhalation exposure, ingestion or dermal exposure (from contact with heavily treated fabric) may cause hemolysis, particularly in persons afflicted with glucose-6-phosphate dehydrogenase deficiency.[1] The metabolites of naphthalene actually are responsible for the hemolysis.[2] Secondary renal tubular damage may ensue from the naphthol and from the products of hemolysis. Convulsions and coma may occur, particularly in children. In infants, high levels of methemoglobin and bilirubin in the plasma may lead to encephalopathy. Kernicterus has been specifically described as a complication of exposure to naphthalene with severe hemolysis and resulting hyperbilirubinemia.[3] Some individuals exhibit dermal sensitivity to naphthalene.

## HALOCARBONS

### Toxicology

The halocarbons as a group are most commonly encountered as solvent agents. They have been associated with a wide variety of toxicities, including central nervous system, liver and renal toxicity, reproductive toxicity and carcinogenicity. However, not all are equipotent, nor do any of them routinely express this wide variety of effects.[4]

### HIGHLIGHTS

Easily absorbed in lung, gut, skin

### SIGNS & SYMPTOMS

Highly variable among agents

Many are irritants

Carbon disulfide, chloroform, ethylene dichloride, hydrogen cyanide, methyl bromide may have serious CNS effects

Methyl bromide, ethylene dibromide, ethylene oxide, aluminum phosphide (phosphine gas) can cause pulmonary edema

Chloroform, carbon tetrachloride, ethylene dichloride, ethylene dibromide, formaldehyde, carbon disulfide may have liver and/or kidney impacts

Hydrogen cyanide causes severe hypoxia without cyanosis in early stages

### TREATMENT

Skin, eye decontamination

Ensure breathing, pulse

Control seizures

Consider GI decontamination

Specific measures needed for various agents

### CONTRAINDICATED

Catecholamine-releasing agents in carbon disulfide poisoning

Ipecac in cyanide poisoning

## COMMERCIAL PRODUCTS

**Hydrocarbon**: naphthalene

**Halocarbons**: methylene chloride,* methyl bromide, methyl iodide, chloroform,* carbon tetrachloride,* chloropicrin, ethylene dichloride, ethylene dibromide,* 1,3-dichloropropene, 1,2-dichloropropane,* dibromochloropropane, paradichlorobenzene

**Oxides and Aldehydes**: ethylene oxide, propylene oxide,* formaldehyde and paraformaldehyde, acrolein

**Sulfur Compounds**: sulfur dioxide, sulfuryl fluoride, carbon disulfide*

**Phosphorus Compounds**: phosphine

**Nitrogen Compounds**: sodium/hydrogen cyanide, acrylonitrile*

**Methyl Isothiocyanate Generators:** Metam sodium, metam potassium, dazomet

*□□i□continued in t□e □□□□*

The individual characteristics of each registered or previously registered as pesticides will be discussed.

**Methylene chloride** is one of the less toxic halocarbons. It is absorbed by inhalation and to a limited extent across the skin. Exposure to high concentrations may cause central nervous system depression, manifesting as fatigue, weakness and drowsiness. A case has been described of severe optic atrophy after high level exposure to this agent.[5] Some absorbed methylene chloride is degraded to carbon monoxide in humans, yielding increased blood concentrations of carboxyhemoglobin.[6] However, concentrations are rarely high enough to cause symptoms of carbon monoxide poisoning. Ingestion has caused death from gastrointestinal hemorrhage, severe liver damage, coma, shock, metabolic acidosis and renal injury. In laboratory animals, extraordinary dosage has caused irritability, tremor and narcosis, leading to death. When heated to the point of decomposition, one of the products is the highly toxic phosgene gas that has caused significant, acute pneumonitis.[7]

**The methyl halides (methyl bromide** and **methyl iodide)** are similar in their toxicity and metabolic fate.[8] They are colorless and nearly odorless to moderately pungent (methyl iodide), but are severely irritating to the lower respiratory tract, sometimes inducing pulmonary edema, hemorrhage or a confluent pneumonia. The onset of respiratory distress may be delayed 4-12 hours after exposure. The methyl halides are central nervous system depressants but may also cause convulsions. Early symptoms of acute poisoning include headache, dizziness, nausea, vomiting, tremor, slurred speech and ataxia. The more severe cases of poisoning exhibit myoclonic and generalized tonic-clonic seizures, which are sometimes refractory to initial therapy. Residual neurological deficits including myoclonic seizures, ataxia, muscle weakness, tremors, behavioral disturbances and diminished reflexes may persist in more severely poisoned patients.[8,9,10] If liquid methyl halides contact the skin, severe burning, itching and blistering occurs. Skin necrosis may be deep and extensive.[11]

**Chloroform** has an agreeable, sweet odor and is only slightly irritating to the respiratory tract. It is well absorbed from the lungs and is also absorbed from the skin and gastrointestinal tract. It is a powerful central nervous system depressant (in fact, it has been used as an anesthetic).[12] Inhalation of toxic concentrations in air leads to dizziness, loss of sensation and motor power, and then unconsciousness. Inhalation of large amounts causes cardiac arrhythmias, sometimes progressing to ventricular fibrillation.[13] Large absorbed doses damage the functional cells of the liver and kidney. Ingestion is more likely to cause serious liver and kidney injury than is inhalation of the vapor.

**Carbon tetrachloride** is less toxic than chloroform as a central nervous system depressant but is much more severely hepatotoxic, particularly following ingestion. Liver cell damage is apparently due to free radicals generated in the process of initial dechlorination.[14] Sporadic arrhythmias, progressing to fibrillation, may follow inhalation of high concentrations of carbon tetrachloride or ingestion of the liquid. Kidney injury also occurs sometimes with minimal hepatic toxicity. The kidney injury may be manifested by acute tubular necrosis or by azotemia and general renal failure. Even topical exposure has resulted in acute renal toxicity.[15]

**Chloropicrin** is severely irritating to the upper respiratory tract, eyes and skin. Inhalation of an irritant concentration sometimes leads to vomiting. Ingestion could be expected to cause a corrosive gastroenteritis.[16,17]

**1,2-dichloroethane (ethylene dichloride)** is moderately irritating to the eyes and respiratory tract. Respiratory symptoms may have a delayed onset. It depresses the central nervous system, induces cardiac arrhythmias and damages the liver. Additional manifestations of poisoning include headache, nausea, vomiting, dizziness, diarrhea, hypotension, cyanosis and unconsciousness.[18]

**Ethylene dibromide** is a severe irritant to skin, eyes and respiratory tract. The liquid causes blistering and erosion of skin and is corrosive to the eyes. Once absorbed, it may cause pulmonary edema and central nervous system depression. Damage to testicular tissue has occurred in animals.[19] Its chemical similarity to DBCP (dibromochlorpropane) suggests this compound may have some damaging effect on testicular tissue with long-term exposure.[20] Persons poisoned by ingestion have suffered chemical gastroenteritis, liver necrosis and renal tubular damage. Death is usually due to respiratory or circulatory failure.[21] A powerful disagreeable odor is advantageous in warning occupationally exposed workers of the presence of this gas.

**Dichloropropene** and **dichloropropane** are strongly irritating to the skin, eyes and respiratory tract. Bronchospasm may result from inhalation of high concentrations. Liver, kidney and cardiac toxicity are seen in animals, but there are limited data for humans.[22] It appears that the risk of such toxicity is relatively low for humans except in large exposures, especially by ingestion.

**Paradichlorobenzene** is solid at room temperature. It is now widely used as a moth repellent, air freshener and deodorizer in homes and in public facilities. The vapor is only mildly irritating to the nose and eyes. Liver injury may occur following ingestion of large amounts. Although accidental ingestions, especially by children, have been fairly common, symptomatic human poisonings have been rare. The last report in the peer-reviewed literature of acute poisoning was in 1959.[23] Chronic intentional exposure has led to severe encephalopathy and serious withdrawal symptoms.[24]

## OXIDES AND ALDEHYDES

**Ethylene oxide** and **propylene oxide** are irritants to all tissues they contact. Aqueous solutions of ethylene oxide can cause blistering and erosion of the affected skin. The area of skin may thereafter be sensitized to the fumigant. Inhalation of high concentrations is likely to cause pulmonary edema and cardiac arrhythmias. Headache, nausea, vomiting, weakness and a persistent cough are common early manifestations of acute poisoning.[25] Coughing of bloody, frothy sputum is characteristic of pulmonary edema.

Airborne **formaldehyde** is irritating to the eyes and to membranes of the upper respiratory tract. In some individuals, it is a potent sensitizer, causing allergic dermatitis. In addition, it has been associated with asthma-like symptoms, though there remains some controversy as to whether these represent true allergic asthma caused by formaldehyde.[26,27,28] High air concentrations may cause laryngeal edema, asthma or tracheobronchitis, but apparently not pulmonary edema. Aqueous solutions in contact with the skin cause hardening and roughness due to superficial coagulation of the keratin layer. Ingested formaldehyde attacks the lining membrane of the stomach and intestine, causing necrosis and ulceration. Absorbed formaldehyde is rapidly converted to formic acid. The latter is partly responsible for the metabolic acidosis that is characteristic of formaldehyde poisoning. Circulatory collapse and renal failure may follow the devastating effects of ingested formaldehyde on the gut, leading to death.[29] **Paraformaldehyde** is a polymer that slowly releases formaldehyde into the air. Toxicity is somewhat less than that of formaldehyde because of the slow evolution of gas.

**Acrolein** (acrylaldehyde) is an extremely irritating gas used as a fumigant and an aquatic herbicide. The vapor causes lacrimation and upper respiratory tract irritation, which may lead to laryngeal edema, bronchospasm and delayed pulmonary edema. The consequences of ingestion are essentially the same as those that follow ingestion of formaldehyde. Contact with the skin may cause blistering.[30]

## SULFUR COMPOUNDS

**Sulfur dioxide** is a highly irritating gas, so disagreeable that persons inhaling it are usually prompted to seek uncontaminated air as soon as possible. However, laryngospasm and pulmonary edema have occurred, occasionally leading to severe respiratory distress and death. It is sometimes a cause of reactive airways disease in occupationally exposed persons.[31]

**Sulfuryl fluoride** has been used extensively for structural fumigation. Generally, use experience has been good, but some fatalities have occurred when fumigated buildings have been prematurely reentered by unprotected individuals.[32] Since this material is heavier than air, fatal hypoxia may follow early reentry. Manifestations of poisoning have been nose, eye and throat irritation, weakness, nausea, vomiting, dyspnea, cough, restlessness, muscle twitching and seizures.[33,34]

**Carbon disulfide** vapor is only moderately irritating to upper respiratory membranes. It has an offensive "rotten cabbage" odor. Acute toxicity is due chiefly to effects on the central nervous system. Inhalation of high concentrations for short periods has caused headache, dizziness, nausea, hallucinations, delirium, progressive paralysis and death from respiratory failure.[35] More prolonged exposure to lesser amounts has led to blindness, deafness, paresthesia, painful neuropathy and paralysis.[36] Carbon disulfide is a potent skin irritant, often causing severe burns. Long-term occupational exposures have been shown to accelerate atherosclerosis, leading to ischemic myocardiopathy, polyneuropathy and gastrointestinal dysfunction.[37] Toxic damage to the liver and kidneys may result in severe functional deficits of these organs.[38] Reproductive failure has been noted.

## PHOSPHORUS COMPOUNDS

**Phosphine gas** is extremely irritating to the respiratory tract. It also produces severe systemic toxicity. It is used as a fumigant by placing solid aluminum phosphide (phostoxin) near produce or in other storage spaces. By way of hydrolysis, phosphine gas is slowly released. Most severe acute exposures have involved ingestion of the solid aluminum phosphide, which is rapidly converted to phosphine by acid hydrolysis in the stomach. Poisoning due to ingestion carries a high mortality rate (50% to 90%).[39,40] The complex chemistry and toxic mechanisms of phosphine were recently reviewed. Three interdependent mechanisms contribute to phosphine toxicity: disruption of the sympathetic nervous system, suppressed energy metabolism and oxidative damage to the cells.[41] Extracellular magnesium levels have been found to be slightly elevated, suggesting a depletion of intracellular magnesium from myocardial damage.[42]

Poisonings had become quite frequent during the late 1980s and early 1990s in some parts of India.[39,40] The principal manifestations of poisoning are fatigue, nausea, headache, dizziness, thirst, cough, shortness of breath, tachycardia, chest tightness, paresthesia and jaundice. Cardiogenic shock is present in more severe cases. Pulmonary edema is a common cause of death. In other fatalities, ventricular arrhythmias, conduction disturbances and asystole developed.[39,43] The odor of phosphine is said to resemble that of decaying fish.

## NITROGEN COMPOUNDS

**Sodium cyanide/hydrogen cyanide** gas causes poisoning by inactivating cytochrome oxidase, the final enzyme essential to mammalian cellular respiration. The patient will have signs of severe hypoxia, but in some cases may not appear cyanotic. This is due to the failure of hemoglobin reduction in the face of loss of cellular respiration. This will result in a pink or red color to the skin and arteriolization of retinal veins. In addition to the suggestive physical findings, one may also find an unusually high $pO_2$ on a venous blood gas.[44] Cyanosis is a late sign and indicates circulatory collapse.

The cells of the brain appear to be the most vulnerable to cyanide action. Presenting signs are nonspecific and can be found with many poisonings. Unconsciousness and death may occur immediately following inhalation of a high cyanide concentration, respiratory failure being the principal mechanism. Metabolic acidosis is another common presenting sign. Low-dose exposures cause a constriction and numbness in the throat, stiffness of the jaw, salivation, nausea, vomiting, lightheadedness and apprehension. Worsening of the poisoning manifests as violent tonic or clonic convulsions. Fixed, dilated pupils, bradycardia and irregular gasping respiration (or apnea) are typical of profound poisoning. The heart often continues to beat after breathing has stopped.[44,45] A bitter almond odor to the breath or vomitus may be a clue to poisoning, but not all individuals are able to detect this odor.[44]

**Acrylonitrile** is biotransformed in the body to hydrogen cyanide. Toxicity and mechanisms of poisoning are essentially the same as have been described for cyanide, except that acrylonitrile is irritating to the eyes and the upper respiratory tract.

## METHYL ISOTHIOCYANATE GENERATORS

**Metam sodium, metam potassium** and **dazomet**, when used as fumigants, all rely on conversion to methyl isothiocyanate.[46] There is very limited literature on the effects of these agents when used as fumigants, but the toxicity appears to be related to exposure to methyl isothiocyanate. This is discussed in more detail in **Chapter 16, *Fungicides***, in the subsection, *Thiocarbamates*.

### Confirmation of Poisoning

**Naphthalene** is converted mainly to alpha naphthol in the body and promptly excreted in conjugated form in the urine. Alpha naphthol can be measured by gas chromatography. Many halocarbons can be measured in blood by gas chromatographic methods. Some can be measured in the expired air as well.

**Methylene chloride** is converted to carbon monoxide in the body, generating carboxyhemoglobin, which can be measured by clinical laboratories.

**Paradichlorobenzene** is metabolized mainly to 2,5-dichlorophenol, which is conjugated and excreted in the urine. This product can be measured chromatographically.

**Methyl bromide** yields inorganic bromide in the body. Methyl bromide itself has a short half-life and is usually not detectable after 24 hours. The bromide anion is slowly excreted in the urine (half-life about 10 days) and is the preferred method of serum measurement.[10] The serum from persons having no exceptional exposure to bromide usually contains less than 1 mg bromide ion per 100 mL. The possible contributions of medicinal bromides to elevated blood content and urinary excretion must be considered, but if methyl bromide is the exclusive source, serum bromide exceeding 6 mg per 100 mL probably means some absorption, and 15 mg per 100 mL is consistent with symptoms of acute poisoning. Inorganic bromide is considerably less toxic than methyl bromide; serum concentrations in excess of 150 mg per 100 mL occur commonly in persons taking inorganic bromide medications. In some European countries, blood bromide concentrations are monitored routinely in workers exposed to methyl bromide. Blood levels over 3 mg per 100 mL are considered a warning that personal protective measures must be improved. A bromide concentration over 5 mg per 100 mL requires that the worker be removed from the fumigant-contaminated environment until blood concentrations decline to less than 3 mg per 100 mL.[47]

**Carbon disulfide** can be measured in urine by gas chromatography, but the test is not generally available.

Cyanide ion from cyanide itself or **acrylonitrile** can be measured in whole blood and urine by an ion-specific electrode or by colorimetry. Symptoms of toxicity may appear at blood levels above 0.10 mg per liter.[45] Urine cyanide is usually less than 0.30 mg per liter in nonsmokers, but as much as 0.80 mg per liter in smokers. Thiocyanate, the metabolite of cyanide, can also be measured in blood and urine. It is considered elevated at blood levels exceeding 12 mg per liter.[45] Urine thiocyanate is usually less than 4 mg per liter in nonsmokers, but may be as high as 17 mg per liter in smokers.

Serum fluoride concentrations have been measured in fatalities from **sulfuryl fluoride** fumigation. Ante-mortem concentrations have ranged from as low as 0.5 mg/liter in one chronic exposure case to the range of ~20 mg/liter in acute poisoning deaths.[34]

There are no practical tests for absorbed alkyl oxides, aldehydes or phosphine that would be helpful in diagnosis of poisoning.

Large industrial plants sometimes monitor human absorption of halocarbons by analysis of expired air. Similar technology is available in some departments of anesthesiology. These analyses are rarely needed to identify the offending toxicant because this is known from the exposure history. In managing difficult cases of poisoning, however, it may be helpful to monitor breath concentrations of toxic gas to evaluate disposition of the fumigant. Protein and red cells levels in the urine may indicate renal injury. Free hemoglobin in urine most likely reflects hemolysis, as from naphthalene. Elevations of alkaline phosphatase, lactate dehydrogenase (LDH), serum GGT, ALT, AST and certain other enzymes are sensitive indices of insult to liver cells. More severe damage increases plasma concentrations of bilirubin. A chest X-ray may be used to confirm the occurrence of pulmonary edema. Electromyography may be useful in evaluating peripheral nerve injury. Sperm counts may be appropriate for workers exposed to **dibromochloropropane** and **ethylene dibromide**.

Some occupational health agencies now urge periodic neurologic and neuropsychological testing of workers heavily exposed to fumigants and solvents to detect injury to the nervous system as early as possible. This would be particularly desirable in the case of exposures to such agents as methyl bromide and carbon disulfide that have well documented chronic neurotoxic effects.

## Treatment of Fumigant Toxicosis

1. Flush contaminating fumigants from the skin and eyes with copious amounts of water or saline for at least 15 minutes. Some fumigants are corrosive to the cornea and may cause blindness. Specialized medical treatment should be obtained promptly following flushing. Skin contamination may cause blistering and deep chemical burns. Absorption of some fumigants across the skin may be sufficient to cause systemic poisoning in the absence of fumigant inhalation. For all these reasons, decontamination of eyes and skin must be immediate and thorough.

2. Remove victims of fumigant inhalation to fresh air immediately. Even though initial symptoms and signs are mild, keep the victim quiet, in a semi-reclining position. Minimal physical activity limits the likelihood of pulmonary edema.

3. If victim is not breathing, clear the airway of secretions and resuscitate with positive pressure oxygen apparatus. If this is not available, use chest compression to sustain respiration. If victim is pulseless, employ cardiac resuscitation.

4. Manage patients with signs and symptoms of severe poisoning, including pulmonary edema, respiratory failure, shock, renal failure and seizures in an intensive care unit.

5. Control convulsions. Seizures are most likely to occur in poisonings by methyl bromide, hydrogen cyanide, acrylonitrile, phosphine and carbon disulfide. See **Chapter 3, *General Principles*** for seizure management. In some cases of methyl bromide poisoning, seizures have been refractory to benzodiazepines and diphenylhydantoin, so consider resorting to anesthesia using thiopental.[10]

6. If a fumigant liquid or solid has been ingested less than an hour prior to treatment, consider gastric emptying, followed by activated charcoal, as suggested in **Chapter 3**.

7. Monitor fluid balance and check urine sediment regularly for indications of tubular injury. Measure serum alkaline phosphatase, LDH, ALT, AST and bilirubin to assess liver injury.

### Specific Treatment Measures for Particular Fumigants

Specific additional measures recommended in poisonings by particular fumigants follow.

## Naphthalene

1. If naphthalene toxicosis is caused by vapor inhalation, this can usually be managed simply by removing the individual to fresh air.

2. Decontaminate skin promptly by washing with soap and water. Remove eye contamination by flushing with copious amounts of clean water. Eye irritation may be severe, and if it persists, should receive ophthalmologic attention. See **Chapter 3, *General Principles*** for more information on decontamination.

3. Examine the plasma for evidence of hemolysis: a reddish-brown tinge. Examine the blood smear for "ghosts" and Heinz bodies. If present, monitor red blood cell count and hematocrit for anemia and urine for protein and cells. Measure direct- and indirect-reacting bilirubin in the plasma. Monitor fluid balance and blood electrolytes. If possible, monitor urinary excretion of naphthol to assess severity of poisoning and clinical progress.

4. If hemolysis is clinically significant, administer intravenous fluids to accelerate urinary excretion of the naphthol metabolite and protect the kidney from products of hemolysis. Use Ringer's lactate or sodium bicarbonate to keep urine pH above 7.5.

5. Consider the use of mannitol or furosemide to promote diuresis. If urine flow declines, intravenous infusions must be stopped to prevent fluid overload and hemodialysis should be considered.[2]

6. If anemia is severe, blood transfusions may be needed.

## Carbon Tetrachloride

For carbon tetrachloride poisoning, several treatment measures have been suggested to limit the severity of hepatic necrosis. The limited experience is outlined below.

1. Consider using hyperbaric oxygen, which has been used with some success.[14]

2. Administer n-acetyl cysteine (Mucomyst) orally as a means of reducing free radical injury.[48]

### Dosage of Mucomyst

- ***Dilute the proprietary 20% product 1:4 in a carbonated beverage, and give about 140 mg/kg body weight of the diluted solution as a loading dose. Then give 70 mg/kg every 4 hours after the loading dose for a total of 17 doses (this dosage schedule is used for acetaminophen poisonings).***

Administration via duodenal tube may be necessary in patients who cannot tolerate Mucomyst.[49] Intravenous administration of n-acetyl cysteine may be used; more information is available through the poison control centers.

## Carbon Disulfide

Mild poisonings by carbon disulfide inhalation may be managed best by no more than careful observation, even though sensory hallucinations, delirium and behavioral aberrations can be alarming. Severe poisonings may require specific measures.

1. If manic behavior threatens the safety of the victim, administer diazepam as a tranquilizer.

### Dosage of Diazepam

- ***Adults: 5-10 mg administered slowly, intravenously***
- ***Children: 0.2-0.4 mg/kg, administered slowly, intravenously***

Give as much as is necessary to achieve sedation.

2. Do not give catecholamine-releasing agents, such as reserpine or amphetamines.

## Phosphine Gas

Experience in India suggests that therapy with magnesium sulfate may decrease the likelihood of a fatal outcome.[39,43,50] The mechanism is unclear, but may possibly be due to the membrane stabilization properties of magnesium in protecting the heart from fatal arrhythmias. In one series of 90 patients, magnesium sulfate was found to decrease the mortality from 90% to 52%.[39] Two controlled studies have been done, one of which showed a reduction in mortality from 52% to 22%.[50] The other study found no effect on mortality.[51]

**Dosage for Magnesium Sulfate**

- ***3 grams during the first 3 hours as a continuous infusion, followed by 6 grams per 24 hours for the next 3 to 5 days.[39]***

## Hydrogen Cyanide and Acrylonitrile

Poisonings by hydrogen cyanide and acrylonitrile gases or liquids are treated essentially the same as poisoning by cyanide salts.

1. Because cyanide is so promptly absorbed following ingestion, commence treatment with prompt administration of oxygen and antidotes. The three antidotes – amyl nitrite, sodium nitrite and sodium thiosulfate – are available in cyanide antidote kits, available from various sources. Read and follow the package insert.[52] The nitrates are intended to produce methemoglobin, which binds cyanide, which is then released and metabolized by rhodanese with the help of thiosulfate.

2. Hydroxycobalamin has been known from animal studies to be an effective antidote for cyanide poisoning.[53,54] Hydroxycobalamin has a higher affinity for cyanide than do tissue cytochromes, thereby competitively binding and inactivating both free and cytochrome-bound cyanide. The cyanocobalamin formed is readily excreted by the kidney. The product became commercially available in 2007 in the United States (Cyanokit, Merck).[55]

3. Administer oxygen continuously. Hyperbaric oxygen has been evaluated as effective in this condition.[56] If respiration fails, maintain pulmonary ventilation mechanically.

4. Measure hemoglobin and methemoglobin in blood. If more than 50% of total hemoglobin has been converted to methemoglobin, consider blood transfusion or exchange transfusion, because conversion back to normal hemoglobin proceeds slowly.

Although various cobalt salts, chelators and organic combinations have shown some promise as antidotes to cyanide, they are not generally available in the United States. None has been shown to surpass the effectiveness of the nitrite-thiosulfate regimen.

## References

1. Shannon K, Buchanan GR. Severe hemolytic anemia in black children with glucose-6-phosphate dehydrogenase deficiency. *Pediatrics.* Sep 1982;70(3):364-369.

2. Gosselin RE, Smith HC, Hodge HC, eds. *Napthalene.* Clinical toxicology of commercial products. 5th ed. Baltimore: Williams & Wilkins; 1984.

3. Naiman JL, Kosoy MH. Red Cell Glucose-6-Phosphate Dehydrogenase Deficiency--a Newly Recognized Cause of Neonatal Jaundice and Kernicterus in Canada. *Can Med Assoc J.* Dec 12 1964;91:1243-1249.

4. Ruder AM. Potential health effects of occupational chlorinated solvent exposure. *Ann N Y Acad Sci.* Sep 2006;1076:207-227.

5. Kobayashi A, Ando A, Tagami N, et al. Severe optic neuropathy caused by dichloromethane inhalation. *J Ocul Pharmacol Ther.* Dec 2008;24(6):607-612.

6. Amsel J, Soden KJ, Sielken RL, Jr., Valdez-Flora C. Observed versus predicted carboxyhemoglobin levels in cellulose triacetate workers exposed to methylene chloride. *Am J Ind Med.* Aug 2001;40(2):180-191.

7. Snyder RW, Mishel HS, Christensen GC, 3rd. Pulmonary toxicity following exposure to methylene chloride and its combustion product, phosgene. *Chest.* Dec 1992;102(6):1921.

8. Schwartz MD, Obamwonyi AO, Thomas JD, Moorhead JF, Morgan BW. Acute methyl iodide exposure with delayed neuropsychiatric sequelae: report of a case. *Am J Ind Med.* Jun 2005;47(6):550-556.

9. Deschamps FJ, Turpin JC. Methyl bromide intoxication during grain store fumigation. *Occup Med (Lond).* Feb 1996;46(1):89-90.

10. Hustinx WN, van de Laar RT, van Huffelen AC, Verwey JC, Meulenbelt J, Savelkoul TJ. Systemic effects of inhalational methyl bromide poisoning: a study of nine cases occupationally exposed due to inadvertent spread during fumigation. *Br J Ind Med.* Feb 1993;50(2):155-159.

11. Hezemans-Boer M, Toonstra J, Meulenbelt J, Zwaveling JH, Sangster B, van Vloten WA. Skin lesions due to exposure to methyl bromide. *Arch Dermatol.* Jun 1988;124(6):917-921.

12. Dykes MH. Halogenated hydrocarbon ingestion. *Int Anesthesiol Clin.* Summer 1970;8(2):357-368.

13. Himmel HM. Mechanisms involved in cardiac sensitization by volatile anesthetics: general applicability to halogenated hydrocarbons? *Crit Rev Toxicol.* 2008;38(9):773-803.

14. Truss CD, Killenberg PG. Treatment of carbon tetrachloride poisoning with hyperbaric oxygen. *Gastroenterology.* Apr 1982;82(4):767-769.

15. Perez AJ, Courel M, Sobrado J, Gonzalez L. Acute renal failure after topical application of carbon tetrachloride. *Lancet.* 1987;1(8531):515-516.

16. Oriel M, Edmiston S, Beauvais S, Barry T, O'Malley M. Illnesses associated with chloropicrin use in California agriculture, 1992-2003. *Rev Environ Contam Toxicol.* 2009;200:1-31.

17. Barry T, Oriel M, Verder-Carlos M, Mehler L, Edmiston S, O'Malley M. Community exposure following a drip-application of chloropicrin. *J Agromedicine.* Jan 2010;15(1):24-37.

18. Liu JR, Fang S, Ding MP, et al. Toxic encephalopathy caused by occupational exposure to 1, 2-Dichloroethane. *J Neurol Sci.* May 15 2010;292(1-2):111-113.

19. Amir D. The spermicidal effect of ethylene dibromide in bulls and rams. *Mol Reprod Dev.* Jan 1991;28(1):99-109.

20. Slutsky M, Levin JL, Levy BS. Azoospermia and oligospermia among a large cohort of DBCP applicators in 12 countries. *Int J Occup Environ Health.* Apr-Jun 1999;5(2):116-122.

**21.** Singh N, Jatav OP, Gupta RK, Tailor MK, Jain R. Outcome of sixty four cases of ethylene dibromide ingestion treated in tertiary care hospital. *J Assoc Physicians India.* Dec 2007;55:842-845.

**22.** Stott WT, Gollapudi BB, Rao KS. Mammalian toxicity of 1,3-dichloropropene. *Rev Environ Contam Toxicol.* 2001;168:1-42.

**23.** Hallowell M. Acute haemolytic anaemia following the ingestion of paradichlorobenzene. *Arch Dis Child.* Feb 1959;34(173):74-77.

**24.** Cheong R, Wilson RK, Cortese IC, Newman-Toker DE. Mothball withdrawal encephalopathy: case report and review of paradichlorobenzene neurotoxicity. *Subst Abus.* Dec 2006;27(4):63-67.

**25.** Lin TJ, Ho CK, Chen CY, Tsai JL, Tsai MS. Two episodes of ethylene oxide poisoning--a case report. *Kaohsiung J Med Sci.* Jun 2001;17(7):372-376.

**26.** Harving H, Korsgaard J, Pedersen OF, Molhave L, Dahl R. Pulmonary function and bronchial reactivity in asthmatics during low-level formaldehyde exposure. *Lung.* 1990;168(1):15-21.

**27.** Krzyzanowski M, Quackenboss JJ, Lebowitz MD. Chronic respiratory effects of indoor formaldehyde exposure. *Environ Res.* Aug 1990;52(2):117-125.

**28.** Smedley J. Is formaldehyde an important cause of allergic respiratory disease? *Clin Exp Allergy.* Mar 1996;26(3):247-249.

**29.** Pandey CK, Agarwal A, Baronia A, Singh N. Toxicity of ingested formalin and its management. *Hum Exp Toxicol.* Jun 2000;19(6):360-366.

**30.** Beauchamp RO, Jr., Andjelkovich DA, Kligerman AD, Morgan KT, Heck HD. A critical review of the literature on acrolein toxicity. *Crit Rev Toxicol.* 1985;14(4):309-380.

**31.** Rabinovitch S, Greyson ND, Weiser W, Hoffstein V. Clinical and laboratory features of acute sulfur dioxide inhalation poisoning: two-year follow-up. *Am Rev Respir Dis.* Feb 1989;139(2):556-558.

**32.** Scheuerman EH. Suicide by exposure to sulfuryl fluoride. *J Forensic Sci.* Jul 1986;31(3):1154-1158.

**33.** Fatalities resulting from sulfuryl fluoride exposure after home fumigation--Virginia. *MMWR Morb Mortal Wkly Rep.* Sep 18 1987;36(36):602-604, 609-611.

**34.** Schneir A, Clark RF, Kene M, Betten D. Systemic fluoride poisoning and death from inhalational exposure to sulfuryl fluoride. *Clin Toxicol (Phila).* Nov 2008;46(9):850-854.

**35.** Spyker DA, Gallanosa AG, Suratt PM. Health effects of acute carbon disulfide exposure. *J Toxicol Clin Toxicol.* Mar 1982;19(1):87-93.

**36.** Huang CC, Chu CC, Wu TN, Shih TS, Chu NS. Clinical course in patients with chronic carbon disulfide polyneuropathy. *Clin Neurol Neurosurg.* May 2002;104(2):115-120.

**37.** Wilcosky TC, Tyroler HA. Mortality from heart disease among workers exposed to solvents. *J Occup Med.* Dec 1983;25(12):879-885.

**38.** Klemmer PJ, Harris AA. Carbon disulfide nephropathy. *Am J Kidney Dis.* Sep 2000;36(3):626-629.

**39.** Katira R, Elhence GP, Mehrotra ML, et al. A study of aluminum phosphide (AlP) poisoning with special reference to electrocardiographic changes. *J Assoc Physicians India.* Jul 1990;38(7):471-473.

**40.** Singh S, Singh D, Wig N, Jit I, Sharma BK. Aluminum phosphide ingestion--a clinicopathologic study. *J Toxicol Clin Toxicol.* 1996;34(6):703-706.

**41.** Nath NS, Bahattacharya I, Tuck AG, Schlipalius DI, Ebert PR. Mechanisms of phosphine toxicity. *Journal of Toxicology.* 2011;Article ID 494168.

**42.** Singh RB, Singh RG, Singh U. Hypermagnesemia following aluminum phosphide poisoning. *Int J Clin Pharmacol Ther Toxicol.* Feb 1991;29(2):82-85.

43. Gupta S, Ahlawat SK. Aluminum phosphide poisoning--a review. *J Toxicol Clin Toxicol.* 1995;33(1):19-24.

44. Johnson RP, Mellors JW. Arteriolization of venous blood gases: a clue to the diagnosis of cyanide poisoning. *J Emerg Med.* Sep-Oct 1988;6(5):401-404.

45. Yen D, Tsai J, Wang LM, et al. The clinical experience of acute cyanide poisoning. *Am J Emerg Med.* Sep 1995;13(5):524-528.

46. Dourson ML, Kohrman-Vincent MJ, Allen BC. Dose response assessment for effects of acute exposure to methyl isothiocyanate (MITC). *Regul Toxicol Pharmacol.* Nov 2010;58(2):181-188.

47. Zatuchni J, Hong K. Methyl bromide poisoning seen initially as psychosis. *Arch Neurol.* Aug 1981;38(8):529-530.

48. Ruprah M, Mant TG, Flanagan RJ. Acute carbon tetrachloride poisoning in 19 patients: implications for diagnosis and treatment. *Lancet.* May 4 1985;1(8436):1027-1029.

49. Anker AL, Smilkstein MJ. Acetaminophen. Concepts and controversies. *Emerg Med Clin North Am.* May 1994;12(2):335-349.

50. Chugh SN, Kamar P, Sharma A, Chugh K, Mittal A, Arora B. Magnesium status and parenteral magnesium sulphate therapy in acute aluminum phosphide intoxication. *Magnes Res.* Dec 1994;7(3-4):289-294.

51. Siwach SB, Singh P, Ahlawat S, Dua A, Sharma D. Serum & tissue magnesium content in patients of aluminium phosphide poisoning and critical evaluation of high dose magnesium sulphate therapy in reducing mortality. *J Assoc Physicians India.* Feb 1994;42(2):107-110.

52. Akorn. Cyanide Antidote Package For the Treatment of Cyanide Poisoning. In: Pharmaceuticals T, ed2009.

53. Ivankovich AD, Braverman B, Kanuru RP, Heyman HJ, Paulissian R. Cyanide antidotes and methods of their administration in dogs: a comparative study. *Anesthesiology.* Mar 1980;52(3):210-216.

54. Posner MA, Tobey RE, McElroy H. Hydroxocobalamin therapy of cyanide intoxication in guinea pigs. *Anesthesiology.* Feb 1976;44(2):157-160.

55. Rodgers GC, Jr., Condurache CT. Antidotes and treatments for chemical warfare/terrorism agents: an evidence-based review. *Clin Pharmacol Ther.* Sep 2010;88(3):318-327.

56. Myers RA, Schnitzer BM. Hyperbaric oxygen use. Update 1984. *Postgrad Med.* Oct 1984;76(5):83-86, 89-91, 94-85.

CHAPTER 18

# Rodenticides

Rodent poisons are usually added to baits (palatable grain or paste intended to encourage consumption). Safety for animals and humans depends on the toxicity of the agents, concentration of the active ingredient in the bait, the likelihood that a toxic dose will be consumed by non-target species, and bioaccumulation and persistence in body tissues. The first-generation anticoagulants, for example, are reasonably effective against pest rodents and are less toxic than second-generation anticoagulants (see discussion of first- and second-generation anticoagulants in subsection *Coumarins and Indandiones,* following). Rodents are more likely than domestic animals or humans to consume quantities of treated bait that will cause poisoning. However, accidental ingestion by young children or intentional ingestion by individuals with suicidal intent is possible with any poison.

Very small amounts of the extremely toxic rodenticides – **sodium fluoroacetate**, **fluoracetamide, strychnine, crimidine, yellow phosphorus, zinc phosphide** and **thallium sulfate** – can cause severe and even fatal poisoning. **Cholecalciferol** is also a highly toxic agent. The **anticoagulants, indandiones** and **red squill** are less hazardous to humans and domestic animals. Some of the newer anticoagulant compounds, termed "second-generation anticoagulants," may cause human toxicity at a much lower dose than conventional "first-generation anticoagulants"[1,2,3] and can bioaccumulate in the liver.[2]

Yellow phosphorus is not sold in the United States. Zinc phosphide is still registered in the United States and can be found in U.S. retail stores. Thallium sulfate is no longer registered for pesticidal use, but is used by government agencies and in medical diagnostic testing.

Strychnine and sodium fluoroacetate are still used for control of some mammal pests such as coyotes, as is cyanide (see **Chapter 17, *Fumigants*** for cyanide). Only specially trained personnel are allowed to use them.

Crimidine and fluoroacetamide are no longer registered in the United States for use as pesticides. TETS is banned worldwide.

## COUMARINS AND INDANDIONES

### Toxicology

**Anticoagulants** (**warfarin** and related compounds, **coumarins** and **indandiones**) are the most commonly used rodenticides in the United States. While there has been a modest decline in the number of exposures in 2008 compared to 1996, from 13,345 to 11,487, they still account for the largest number of reported rodenticide exposures.[4,5] Gastrointestinal absorption of these toxicants is efficient.

Certain agents in this category are referred to as "first-generation" or "second-generation" anticoagulants. "First-generation" anticoagulants, as the name implies, were developed earlier (during World War II), and include hydroxycoumarin derivatives such as warfarin, coumachlor and coumatetralyl. The "second-generation" anticoagulants, sometimes referred to as "superwarfarins," are generally more toxic. These include bromodiolone, brodifacoum and difenacoum.

Coumarins and indandiones depress the hepatic synthesis of vitamin K-dependent blood-clotting factors (II [prothrombin] and VII, IX and X). The antiprothrombin

*Coumarins & Indandiones*
**HIGHLIGHTS**

Most commonly used rodenticides in U.S.

Efficient GI absorption

Depress blood clotting and capillary permeability

**SIGNS & SYMPTOMS**

Bleeding nose/gums, hematuria, melena, eccymoses days after ingestion

For indandiones, headache, confusion, loss of consciousness, seizures

Increase in PT/INR

**TREATMENT**

Monitor PT/INR

Give Vitamin $K_1$ upon PT/INR evidence

**CONTRAINDICATED**

Vitamin $K_3$ or $K_4$

effect is best known and is the basis for detection and assessment of clinical poisoning. The agents also increase permeability of capillaries throughout the body, predisposing the animal to widespread internal hemorrhage. This generally occurs in the rodent after several days of warfarin ingestion because of the long half-lives of the vitamin K-dependent clotting factors,[1,2] although lethal hemorrhage may follow smaller doses of the modern, more toxic compounds.[2,3]

To identify potential toxic effects from the coumarins or indandiones, a prothrombin time (PT) is measured. Most laboratories report the PT as being adjusted to the International Normalized Ratio (INR) for patients on anticoagulant medication. Therefore, one may see PT or INR reported by a laboratory. The prolonged prothrombin time (PT/INR) from a toxic dose of coumarins or indandiones may be evident within 24 hours, but usually reaches a maximum in 36-72 hours.[2,6,7] Prolonged PT/INR occurs in response to doses much lower than that necessary to cause hemorrhage. There is concern that the more toxic modern compounds, such as brodifacoum and difenacoum, may cause serious poisoning of non-target mammals, including humans, at much lower dosage. **Brodifacoum**, one of the "second-generation anticoagulants," is much more toxic, partly due to a longer half-life; a dose as low as 1 mg in an adult or 0.014 mg/kg in a child is sufficient to produce toxicity.[2]

Symptomatic poisoning, with prolonged symptoms due to the long half-lives of second-generation anticoagulants, has been reported even with single exposures; however, these are usually intentional and are large, single dosages.[1] Because of their toxicity in relation to warfarin, patients may require higher dosages of vitamin K and will require longer monitoring of their PT. One patient required vitamin K for several months following discharge.[8] Another was released from the hospital with significant clinical improvement and only slightly elevated coagulation studies after brodifacoum ingestion. Two-and-a-half weeks later, this patient presented in a comatose state and was found to have massive intracranial hemorrhage.[9] In situations of purposeful ingestion, it is difficult to know if the patient is re-exposing himself or herself. Since 1999, individual case reports continue to appear in the medical literature. Nearly all are suicidal ingestions, although there are occasional reports of intentional subacute ingestion or Munchausen by proxy.[10,11,12,13,14,15]

In contrast to the intentional ingestions from suicide attempts, accidental single ingestions are more common, particularly when toddlers ingest a few pellets. The majority of these incidents did not result in significant bleeding, and most patients did not have a prolonged PT/INR.[7,16,17] It should also be noted that beginning June 2011, all rodenticide bait products available for sale on the residential consumer market in the United States must be in the form of blocks (pellets or loose bait no longer allowed) and be contained a tamper-resistant bait station. Also, since 2011, the second generation anticoagulants (brodifacoum, bromadiolone, difenacoum, difethialone) are not allowed in residential consumer products.[18]

Dermal exposure to the long-acting indandiones has also been reported to cause symptomatic bleeding. One 18-year-old patient presented with flank pain and gross hematuria following dermal exposure to 0.106% diphacinone.[19] Another patient had hematuria following exposure to 0.25% chlorophacinone on his torso.[20]

Clinical effects of these agents usually begin several days after ingestion, because of the long half-life of the factors. Primary manifestations include nosebleeds, bleeding gums, hematuria, melena and extensive ecchymoses.[1,2,8,9,21] Patients may also have symptoms of anemia including fatigue and dyspnea on exertion.[21] If the poisoning is severe, the patient may progress to shock and death.

Unlike the coumarin compounds, some indandiones cause symptoms and signs of neurologic and cardiopulmonary injury in laboratory rats. These lead to death before hemorrhage occurs, which may account for the greater toxicity of indandiones in rodents. In several cases of human poisonings, some of the presenting signs included headache, confusion, loss of consciousness and seizures. These CNS symptoms were found to be related to intracranial hemorrhage.[22,23] One other patient was reported to present in a comatose state. He had severe intra-abdominal bleeding without any intra-cranial bleeding and eventually recovered.[24]

## Confirmation of Poisoning

Coumarin or indandione poisoning results in an increase in PT/INR, the result of reduced plasma prothrombin concentration. This is a reliable test for absorption of physiologically significant doses. Detectable reduction in prothrombin occurs within 24-48 hours of ingestion and persists for 1-3 weeks.[2,6,7] Blood levels of the second-generation anticoagulants can be measured, however the test is not immediately available, nor does it aid in immediate treatment decisions as does the PT or INR.[21]

## Treatment of Anticoagulant Toxicosis

If the amount of agent ingested was assuredly no more than a few mouthfuls of coumarin- or indandione-treated bait, or a single mouthful of bait treated with the more toxic brodifacoum or bromadiolone compounds, medical treatment is probably unnecessary. Otherwise:

1. If there is an unknown amount or deliberate ingestion, assess PT/INR at baseline and then daily. While the anticoagulant effects of the coumarins might be noted within 12-24 hours of ingestion, some agents such as brodifacoum may not show an elevation until 48 hours after ingestion, if it does occur.[7] A normal PT/INR 48-72 hours after ingestion makes a significant bleeding event very unlikely.

2. Give phytonadione (vitamin $K_1$) orally to protect against the anticoagulant effect of these rodenticides, with essentially no risk to the patient. The indication for vitamin $K_1$ in these patients is laboratory evidence (elevated PT/INR) of excessive anticoagulation after ingestion. It is not recommended empirically after ingestion. On one hand, laboratory evidence may indicate it is not needed. However, most important, vitamin $K_1$ administration prior to PT/INR elevation may delay the lab abnormalities and the seriousness of the ingestion can be missed.

   CAUTION: ***Phytonadione, specifically, is required. Neither vitamin K[illegible] [illegible] en[illegible]dione, Hykinone) nor vitamin K[illegible] (menadiol) is an antidote for these anticoagulants. These need to be metabolized by the liver to active vitamin K, and with the potential of significant bleeding, liver function may be impaired. They were not effective as antidotes in prior poisonings.***[illegible]

3. If possible, administer vitamin $K_1$ by mouth. While vitamin $K_1$ can be given orally, subcutaneously (SC), intramuscularly (IM) and intravenously (IV), oral use is preferred because of its good adverse event profile. Anaphylactoid reactions resulting in death have been reported via the IV route; therefore, IVs should be restricted to those patients who are critically ill and cannot take it by any other route. IM or SC use can result in significant hematoma in anti-coagulated patients and, again, use of these routes should be reserved for those patients unable to take Vitamin $K_1$ orally.

***Coumarins & Indandiones***

**COMMERCIAL PRODUCTS**

***Anticoagulants***

brodifacoum (Havoc, Talon)

bromadiolone (Contrac, Maki)

coumachlor

coumatetralyl

difenacoum

difethialone

warfarin (Cov-R-Tox, Liqua-Tox)

***Indandiones***

chlorophacinone (Rozol)

diphacinone (diphacin, Ditrac, Ramik, Tomcat)

pivalyn (Pival)

radione

*Inorganic Rodenticides*
**COMMERCIAL PRODUCTS**

Yellow phosphorus
Zinc phosphide
Thallium sulfate

4. Begin dosing with vitamin $K_1$. Dosing can be variable and may depend on the level of anticoagulation and the agent ingested. The usual starting dose is:

**Dosage of Vitamin $K_1$**

- ***Adults: 10-50 mg orally, 2-4 times per day***
- ***Children: 5-10 mg (or 0.4 mg/kg/dose) orally, 2-4 times per day***

5. Monitor PT/INR for response to vitamin $K_1$ and, once declining, doses can be decreased accordingly. Patients who ingest large amounts, particularly of the superwarfarin compounds, will likely have a very prolonged period of decreased prothrombin activity. Patients may need to be treated for as long as 3 or 4 months.[8,9]

6. With ingestions of certain agents such as the second-generation anticoagulants, very large doses of vitamin $K_1$, between 100 mg up to 400 mg, have been needed initially to reverse the anticoagulation.[6,26] These large doses may be required from weeks to months, depending on the extent of the ingestion and anticoagulation. Monitor patients closely to assure that they are taking the vitamin $K_1$ and not deliberately re-exposing themselves.

7. Give patients who present with active bleeding fresh frozen plasma or whole blood, while also receiving vitamin $K_1$. This will temporize the bleeding until the vitamin $K_1$ has time to replenish the missing factors.

## INORGANIC RODENTICIDES

### Toxicology

**Yellow phosphorus** is a corrosive agent that damages all tissues it contacts, including skin and the gastrointestinal epithelium. A similar compound, **white phosphorus**, is used as an explosive agent in ammunition and fireworks, and some recent reports of toxicity have been from this source.[27,28] The skin is subject to severe burns from white phosporus, which can be third degree and require grafting.[27]

Initial symptoms of yellow phosphorus ingestion usually reflect mucosal injury and occur a few minutes to 24 hours following ingestion. The first symptoms include severe vomiting and burning pain in the throat, chest and abdomen. The emesis may be bloody (either red, brown or black)[29] and on occasion may have a garlic smell.[30,31,32] In some cases, central nervous system signs such as lethargy, restlessness and irritability are the earliest symptoms followed by symptoms of gastrointestinal injury. Shock and cardiopulmonary arrest leading to death may occur within hours in severe ingestions.[31]

If the patient survives the initial toxic effects, there may be a second stage, characterized by a period of apparent improvement that can last a few hours or days, although this is not always the case.[29] In severe poisoning, a third stage of toxicity then ensues with systemic signs indicating severe injury to the liver, myocardium and brain. Nausea and vomiting recur. There is a severe toxic hepatitis with elevated liver transaminases, and jaundice with elevation in both total and direct bilirubin levels.[29,32,33] Neutropenia has also been reported.[34] Hypovolemic shock and toxic myocarditis may develop. Brain injury is manifested by convulsions, delirium and coma. The coma may result from hyperammonemia due to severe hepatic failure.[32,33] Anuric renal failure commonly develops because of shock and the toxic effects of phosphorus products and accumulating bilirubin on renal tubules. The mortality rate of phosphorous poisonings may be as high as 50 percent.[29]

**Zinc phosphide** is much less corrosive to skin and mucous membranes than yellow phosphorus. Both zinc phosphide and aluminum phosphide (often used as a fumigant) can be very toxic following ingestion, although zinc phosphide ingestion is relatively less common than its aluminum counterpart.[35] In terms of toxicity following ingestion, phosphine is thought to be released from the metal phosphide following contact with fluids in the GT tract.[35] While the emetic effect of zinc released in the gut may provide a measure of protection for humans, fatal ingestion of zinc phosphide has been reported.[36,37] Nausea and vomiting, agitation, chills, chest tightness, dyspnea and cough may progress to pulmonary edema. These symptoms, as well as systemic toxicity, may be delayed or present after initial benign exam. Patients face many of the same systemic toxicities encountered with yellow phosphorous, including hepatic failure with jaundice and hemorrhage, delirium, convulsions and coma (from toxic encephalopathy); tetany from hypocalcemia and anuria from renal tubular damage. Ventricular arrhythmias from cardiomyopathy and shock also occur and are another common cause of death.[30,38] Severe hypoglycemia has also been reported, which is also thought to be of hepatic origin by inhibition of glycogenolysis and gluconeogenesis.[38,39] (Conversely, hyperglycemia has been reported following ingestion of the fumigant aluminum phosphide.)[40] Inhalation of phosphine gas from improper use of phosphide rodenticides has resulted in pulmonary edema, myocardial injury and multisystem involvement.[41] For more information about the effects of phosphine gas poisoning, see the section on aluminum phosphide in **Chapter 17, *Fu□ig□n□s***.

**Thallium sulfate** is well absorbed from the gut and across the skin. It exhibits a very large volume of distribution and is distributed chiefly to the kidney and liver, both of which participate in thallium excretion. Most blood-borne thallium is in the red cells. Thallium is thought to exert its toxic effects by competing with intracellular potassium and interfering with intracellular enzyme reactions.[42] Elimination half-life from blood in the adult human is about 1.9 days. Most authors report the $LD_{50}$ in humans to be between 10-15 mg/kg.[43]

Unlike other inorganic rodenticides such as yellow phosphorous and zinc phosphide, thallium poisoning tends to have a more insidious onset with a wide variety of toxic manifestations. The gastrointestinal, central nervous, cardiovascular, renal and integumentary systems are prominently affected by toxic intakes of thallium. Early symptoms include abdominal pain, nausea, vomiting, bloody diarrhea, stomatitis, salivation and ileus. Elevated liver enzymes may occur, indicating tissue damage. Proteinuria and hematuria may also occur. Other patients experience signs of central nervous system toxicity including headache, lethargy, muscle weakness or even paralysis, loss of deep tendon reflexes, paresthesias, tremor, ptosis and ataxia. These signs and symptoms usually occur several days to more than a week after exposure.[42,43,44] Extremely painful paresthesias, either in the presence or absence of gastrointestinal signs, may be the primary presenting complaint.[42,45,46,47] Myoclonic movements, convulsions, delirium and coma reflect more severe neurologic involvement.

Cardiovascular effects include hypotension, due at least in part to a toxic myocardiopathy, myocardial ischemia and ventricular arrhythmias.[48] Hypertension occurs later and is probably a result of peripheral arterial vasoconstriction. Patients may also develop alveolar edema and hyaline membrane formation in the lungs, consistent with a diagnosis of Acute Respiratory Distress Syndrome.[48] Death from thallium poisoning may be caused by respiratory failure or cardiovascular collapse.[47,48] Absorption of nonlethal doses of thallium has caused protracted, painful neuropathies and paresis, optic nerve atrophy, persistent ataxia, dementia, seizures and coma.[45]

***Phosphorus***

## HIGHLIGHTS

Phosphorus poisonings are often fatal

## SIGNS & SYMPTOMS

Yellow phosphorus usually causes mucosal injury, emesis, burning throat

## TREATMENT

Decontaminate skin using PPE

Supportive treatment

Monitor urine albuman, glucose, sediment

Monitor serum alkaline phosphatase, LDH, ALT, AST, prothrombin time, bilirubin

## CONTRAINDICATED

Emesis induction

Potassium permanganate lavage

*Zinc Phosphide*
**HIGHLIGHTS**

Less corrosive than yellow phosphorus

Can be very toxic following ingestion

**SIGNS & SYMPTOMS**

Nausea, vomiting, agitation, chills, cough

**TREATMENT**

Supportive treatment

Control airway

Consider GI decontamination

Alopecia is a fairly consistent feature of thallium poisoning that may be helpful in diagnosing a case of chronic poisoning. Since it occurs 2 weeks or more after the onset of acute symptoms, it is not diagnostically helpful early in the presentation.[42,43,45,49]

## Confirmation of Poisoning

Phosphorus and phosphides sometimes impart a foul rotten fish odor to vomitus, feces, and the breath. Luminescence of vomitus or feces is an occasional feature of phosphorus ingestion. Hyperphosphatemia and hypocalcemia occur in some cases but are not consistent findings.

Thallium can be measured in the serum, whole blood, urine and hair. The most reliable method for diagnosis is considered a 24-hour urine excretion. Values in non-exposed individuals have been reported to be less than 10 µg/liter per 24 hours.[43,46,47,48,49,50] Urinary excretion in the range of 10-20 mg/ liter indicates severe poisoning.[46,47,48,50] Hair analysis is likely to be useful only in establishing protracted prior absorption. Normal serum concentrations are less than 2 micrograms per liter.[47,50] Whole blood thallium levels greater than 100 micrograms/dL indicate poisoning.[51]

## Treatment of Yellow Phosphorus Toxicosis

1. Brush or scrape non-adherent phosphorus from the skin. Wash skin burns with copious amounts of water. Make sure all particles of phosphorus have been removed. If burned area is infected, cover with an antimicrobial cream.

2. Take special care to prevent your and other healthcare personnel's exposure when treating a patient poisoned by yellow phosphorus. While this is true of most pesticide poisonings, with a yellow phosphorus poisoning, personal protection, as outlined in **Chapter 3, *General Principles***, must be worn to avoid secondary contamination and burns from phosphorous particles in the patient's bodily fluids.

3. Provide supportive, symptomatic treatment. Poisonings by ingested yellow phosphorus are extremely difficult to manage. Control of airway must be established prior to considering gastrointestinal decontamination as described in **Chapter 3**.

4. Do not induce emesis. Lavage with potassium permanganate solution had historically been recommended in the management of phosphorus ingestion; however, there is not sufficient evidence for its efficacy. It is not recommend it because of the corrosive nature of yellow phosphorus.

5. Treat patients with shock in an intensive care unit.

6. Monitor urine albumin, glucose and sediment to detect early renal injury. Extracorporeal hemodialysis will be required if acute renal failure occurs, but it does not enhance excretion of phosphorus. Monitor ECG to detect myocardial impairment.

7. Monitor serum alkaline phosphatase, LDH, ALT, AST, prothrombin time and bilirubin to evaluate liver damage. Administer Aquamephyton[R] (vitamin $K_1$) if prothrombin level declines.

8. Administer parenteral pain medication for pain from burns.

## Treatment of Poisoning by Zinc Phosphide

1. Provide supportive, symptomatic treatment in an intensive care setting. There is no specific antidote for the treatment of phosphine exposure, and poisonings by ingestion are extremely difficult to manage.

2. Establish control of airway before considering gastrointestinal decontamination as described in **Chapter 3**, ***General Principles***.

   CAUTION: ***Highly toxic phosphine gas may evolve from emesis, lavage fluid and feces of victims of these poisons. The patient's room should be well ventilated. Persons attending the patient must wear gloves to avoid contact with the phosphorus. Air purifying or supplied-air respiratory equipment should be worn by healthcare providers if available.***

3. See the section on treatment of aluminum phosphide poisoning in **Chapter 17, *Fumigants*** for specific therapy for phosphine gas.

## Treatment of Thallium Sulfate Toxicosis

1. If thallium sulfate was swallowed less than an hour prior, consider gastrointestinal decontamination as outlined in **Chapter 3 *General Principles***. Multiple doses of activated charcoal may be helpful in increasing thallium elimination.[46]

2. Give electrolyte and glucose solutions by intravenous infusion to support urinary excretion of thallium by diuresis. Monitor fluid balance carefully to ensure that fluid overload does not occur. If shock develops, treat the patient in an intensive care unit.

3. Control seizures and myoclonic jerking as outlined in **Chapter 3**.

4. Consider hemodialysis. It has been used on victims of severe poisoning.[46,50]

5. Use potassium ferric ferrocyanide (Prussian blue) orally to enhance fecal excretion of thallium by exchange of potassium for thallium in the gut. It is available in the United States as an insoluble preparation, primarily for the purposes of radioactive thallium and radioactive cesium poisoning.[53] However, Prussian blue therapy has often been reported as a therapy in rodenticide poisonings.[42,46,50,43]

### Dosage of Prussian Blue

- ***Adults and adolescents over 13: 3 grams orally, 3 times a day.[53] Unfortunately, there is no established pediatric dosage by weight. The only available guidance for dosage by weight is from two adult cases that reported using a dosage of 250 mg/kg/day.[42,54]***

Several methods for chelating and/or accelerating disposition of thallium have been tested and found either relatively ineffective or hazardous. Chelating agents are not recommended in thallium poisoning. While potassium chloride has been recommended, it has been reported to increase toxicity to the brain[45,48] and has not been shown to increase elimination in some cases.[52]

***Thallium Sulfate***

### HIGHLIGHTS

Well absorbed from gut, across skin

Distributed chiefly to kidney, liver

### SIGNS & SYMPTOMS

Wide variety of symptoms

Alopecia in chronic cases

### TREATMENT

Consider GI decontamination

IV electrolytes, glucose

Administer Prussian Blue

### CONTRAINDICATED

Chelating agents

Potassium chloride

***Severe Multiorgan Metabolic Toxicant***

**COMMERCIAL PRODUCTS**

Sodium fluoroacetate, also known as Compound 1080

Fluoroacetamide (discontinued)

Strychnine

Crimidine (Castrix, discontinued)

Tetramethylenedisulfotetramine (TETS)

## SEVERE MULTIORGAN METABOLIC TOXICANTS

### Toxicology

**Sodium fluoroacetate** and **fluoroacetamide** are readily absorbed by the gut but only to a limited extent across skin. The toxic mechanism is distinct from that of fluoride salts. Three molecules of fluoroacetate or fluoroacetamide are combined in the liver to form a molecule of fluorocitrate, which poisons critical enzymes of the tricarboxylic acid (Krebs) cycle, blocking cellular respiration. The heart, brain and kidneys are the organs most prominently affected. The effect on the heart is to cause arrhythmias, progressing to ventricular fibrillation, which is a common cause of death. Metabolic acidosis, shock, electrolyte imbalance and respiratory distress are all signs of a poor prognostic. Neurotoxicity is expressed as violent tonic-clonic convulsions, spasms and rigor, sometimes not occurring until hours after ingestion.[55]

**Strychnine** is a natural toxin (nux vomica) that causes violent convulsions by direct excitatory action on the cells of the central nervous system, chiefly the spinal cord. Death is caused by convulsive-induced muscle spasms of the diaphragm and thoracic intercostals, resulting in impaired respiration.[56,57] The severe seizures can cause rhabdomyolysis, and the released myoglobin can result in renal failure. Other symptoms may include muscle stiffness, pain and hyperreflexia. Patients are generally conscious and oriented between seizures, which can be helpful in making the diagnosis in an unknown seizure event.[56,57] Since strychnine is rapidly absorbed and distributed, the onset of symptoms is usually within 15-20 minutes of ingestion. Strychnine is detoxified in the liver. Residence half-life is about 10 hours in humans. The lethal dose in adults is reported to be between 50 and 100 mg, although as little as 15 mg can kill a child.[58]

**Crimidine** is a synthetic chlorinated pyrimidine compound that inhibits pyridoxine (vitamin $B_6$). In adequate dosage, it causes violent convulsions similar to those produced by strychnine. Cases of human poisoning are very rare, though one is presented in the Belgian literature.[59]

**Tetramethylenedisulfotetramine (TETS)** is a tasteless and odorless convulsant rodenticide that is highly toxic, with an estimated human adult lethal dose of about 7-10 mg. It works by antagonizing g-amino butyricacid (GABA). It has been banned worldwide since 1984. Unfortunately, it is still available through illegal means and has been used in mass poisonings in the past. A recent case in the United States reported refractory seizures in a 15-month-old after playing with a white powder for rodent control that the parents had purchased in China and brought with them to New York City. Seizures began within 15 minutes of exposure and became refractory. She required intubation and mechanical ventilation for 3 days. Following recovery from the acute seizures, she was noted to have persistent neurological problems including cortical blindness, absence seizures and severe developmental delays. Eventually, through gas chromatography/mass spectrometry after testing negative for the other well known convulsant rodenticides and bromethalin, the powder was identified as TETS.[60]

### Confirmation of Poisoning

Strychnine and crimidine can both be detected in the blood using high performance thin layer chromatography.[61] For strychnine detection in small samples (as little as 0.1 mL), another method using liquid chromatography with photodiode-array detection has been described.[57,61] Strychnine levels that exceed 1 mg/L may be lethal, although one patient survived a blood concentration of 4.73 mg/L.[57,62,63,64] Because of the infrequency of crimidine poisoning, human blood levels are not well known. These tests are likely to be only available at specialized laboratories.

## Treatment of Sodium Fluoroacetate and Fluoroacetamide Toxicosis

Poisonings with these compounds have occurred almost entirely as a result of accidental and suicidal ingestions.

1. If the patient is seen within an hour of exposure and is not convulsing, consider gastrointestinal decontamination as outlined in **Chapter 3,** ***General Principles***. If the victim is already convulsing, control the seizures before undertaking gastric lavage and catharsis.

2. Control seizures as outlined in **Chapter 3**. Seizure activity from these compounds may be so severe that doses necessary for seizure control may paralyze respiration. For this reason, as well as severe cardiac toxicity, these patients should be managed in a critical care environment, with pulmonary ventilation supported mechanically. This has the added advantage of protecting the airway from aspiration of regurgitated gastric contents.

3. Treat patients in an intensive care unit, with special attention to fluids, electrolytes and cardiovascular monitoring.

4. Give calcium gluconate (10% solution) slowly, intravenously to relieve hypocalcemia. Take care to avoid extravasation.

**Dosage of Calcium Gluconate**

***supplied as 100 mg/mL (10% solution)***

- ***Adults and children over 12 years: 10 mL of 10% solution, given slowly, intravenously. Repeat as necessary.***
- ***Children under 12 years: 200-500 mg/kg/24 hr divided Q6 hour. For cardiac arrest, 100/kg/dose. Repeat dosage as needed.***

Antidotal efficacy of glycerol monacetate and ethanol, observed in animals, has not been substantiated in humans. These therapies are not recommended in humans.

## Treatment of Strychnine, Crimidine or TETS Toxicosis

Strychnine and crimidine cause violent convulsions shortly following ingestion of toxic doses. Both poisons are probably well adsorbed by charcoal.

1. Control seizures as outlined in **Chapter 3**, ***General Principles***. Because of the severity of toxicity from this poison, patients should be treated in an intensive care unit setting if available. If seen within an hour of ingestion and the patient is conscious and not convulsing, consider gastrointestinal decontamination with charcoal. If seizures have already begun, they should be controlled prior to attempting decontamination.

2. Administer intravenous fluids to support excretion of absorbed toxicants. Inclusion of sodium bicarbonate in the infusion fluid counteracts metabolic acidosis generated by convulsions caused by strychnine.[64,65,66,67,68]

3. Avoid hemodialysis and hemoperfusion. Their effectiveness has not been tested.

***Severe Multiorgan Metabolic Toxicant***

### HIGHLIGHTS

Sodium fluoroacetate/ fluoroacetamide readily absorbed by gut

Strychnine acts on CNS

Crimidine inhibits Vitamin $B_6$

TETS antagonizes GABA

TETS banned worldwide

### SIGNS & SYMPTOMS

***Sodium Fluoroacetate/ Fluoroacetamide***

Blocks cellular respiration

Affects heart, brain, kidneys

Heart arrhythmia

***Strychnine/Crimidine***

Violent convulsions

***TETS***

Seizures

### TREATMENT

***Sodium Fluoroacetate/ Fluoroacetamide***

Control seizures

Decontaminate GI tract as appropriate

Monitor fluids, electrolytes, cardiovascular

IV calcium gluconate

***Strychnine/Crimidine***

Control seizures

Decontaminate GI tract as appropriate

IV fluids with sodium bicarbonate

Consider pyridoxine

***TETS***

Supportive treatment

*Miscellaneous Rodenticides*
**COMMERCIAL PRODUCTS**

Red squill (Dethdiet, Rodine; red squill is no longer registered for rodenticidal use in the U.S.)

Cholecalciferol (Quintox, Rampage)

Bromethalin

4. Consider administering pyridoxine. Because of the mechanism of crimidine toxicity, pyridoxine (vitamin $B_6$) has been suggested for seizures due to this convulsant, though there are no humans reports of its use.[69]

5. Provide supportive treatment of TETS poisoning. The patient will likely require intensive care in order to survive.

## MISCELLANEOUS RODENTICIDES

Red squill and cholecalciferol are no longer registered as rodenticides. Bromethalin is widely used and is available in the form of bait stations and loose pellets.

### Toxicology

**Red squill** is an ancient rodenticide, consisting of the inner portions of a small cabbage plant grown in eastern Mediterranean countries. The toxic properties are probably due to cardioactive glycosides. For several reasons, mammals other than rodents are unlikely to be poisoned: (1) red squill is an intense nauseant, so animals that vomit (rodents do not) are unlikely to retain the poison; (2) the glycoside is not efficiently absorbed from the gut; and (3) absorbed glycoside is rapidly excreted. Injection of the glycosides leads to effects typical of digitalis: alterations in cardiac impulse conduction and arrhythmias.

**Cholecalciferol** is the activated form of vitamin D (vitamin $D_3$). Although it is registered for use as a rodenticide, most toxic exposures from vitamin D result from over supplementation or ingestion of multivitamins. Cholecalciferol's toxic effect is probably a combination of actions on the liver, kidneys and possibly the myocardium, the last two toxicities being the result of hypercalcemia. Symptoms and signs of vitamin D-induced hypercalcemia in humans are fatigue, weakness, headache and nausea.[70] Polyuria, polydipsia, nephrocalcinosis, hypertension and proteinuria may result from acute renal tubular injury by hypercalcemia.[71,72] The rodenticide form of cholecalciferol has been implicated in numerous poisonings of domestic animals, however, there are no reports in the published medical literature of toxicity.

**Bromethalin** is a neurotoxin that works by uncoupling oxidative phosphorylation, thus depleting adenosine triphosphate (ATP). The resultant loss of ATP results in disruption of the sodium/potassium channels causing cerebral edema and eventually increased intracranial pressure.[73] Human poisonings are rare, although there is one reported fatality in the medical literature. Symptoms mirror what may be found in animal poisonings (dog and cat), which include mental status changes, stupor, coma, and cerebral edema.[74,75,76,77] Although not reported in these case examples, seizures may also occur. Treatment for bromethalin poisoning is supportive as outlined in the **Chapter 3, *General Principles***. There are no known antidotes.

### Confirmation of Poisoning

Cholecalciferol intoxication is indicated by an elevated concentration of calcium (chiefly the unbound fraction) in the serum. There are no generally available tests for the other rodenticides or their biotransformation products.

### Treatment of Red Squill Toxicosis

Red squill is unlikely to cause poisoning unless ingested at substantial dosage. The problem is usually self-correcting because of its intense emetic effect. If, for some reason, the squill is retained, consider gastrointestinal decontamination as outlined in **Chapter 3, *General Principles***. Monitor cardiac status electrocardiographically.

## Treatment of Cholecalciferol Toxicosis

Cholecalciferol at high dosage may cause severe poisoning and death. Human poisonings from its use as a rodenticide have not been reported, but vitamin D overdosage has occurred under clinical circumstances. Treatment is directed at limiting gastrointestinal absorption, accelerating excretion and counteracting the hypercalcemic effect.

1. If cholecalciferol has been ingested within an hour prior to treatment consider gastric decontamination as outlined in **Chapter 3, *General Principles***. Repeated administration of charcoal at half or more the initial dosage every 2-4 hours may be beneficial.

2. Administer intravenous fluids (normal saline or 5% glucose) at moderate rates to support excretory mechanisms and excretion. Monitor fluid balance to avoid overload, and measure serum electrolytes periodically. Measure total and ionized calcium levels in the blood 24 hours after cholecalciferol ingestion to determine severity of toxic effect. Monitor urine for protein, red and white cells to assess renal injury. Patients should be managed in an intensive care unit setting with nephrological consultation if possible.

3. Consider using prednisone and similar glucocorticoids to reduce elevated blood calcium levels. Although prednisone and glucocorticoids have not been tested in cholecalciferol overdosage, it is possible that they would be beneficial.

### Dosage of Prednisone and Glucocorticoids

- ***Approximately 1 mg per kilogram per day, to a maximum of 20 mg per day***

4. Consider administering calcitonin (salmon calcitonin, Calcimar). It is a logical antidote for cholecalciferol actions, but has only very limited use in human poisoning.[78] Calcium gluconate for intravenous injection should be immediately available if indications of hypocalcemia (carpopedal spasm, cardiac arrhythmias) appear.

### Dosage of Calcitonin

- ***In other conditions, the usual dosage is 4 International Units per kg body weight every 12 hours, by intramuscular or subcutaneous injection, continued for 2-5 days. The dose may be doubled if calcium-lowering effect is not sufficient. Consult package insert for additional directions and warnings.***

5. Consider administering cholestryamine. It appears to be effective in the treatment of Vitamin D toxicity in animals[79] but has seen very limited use in humans.[80,81]

***Miscellaneous Rodenticides***

### HIGHLIGHTS

Red squill toxicity probably due to cardioactive glycosideds

Cholecalciferol is activated form of vitamin $D_3$

Bromethalin depletes ATP, disrupting sodium/potassium channels

### SIGNS & SYMPTOMS

***Cholecalciferol***
- Fatigue, weakness, headache, nausea

***Bromethalin***
- Mental changes, stupor, coma, cerebral edema

### TREATMENT

***Red Squill***
- Consider GI decontamination if no emesis
- ECG monitoring

***Cholecalciferol***
- Consider GI decontamination as appropriate
- IV fluids
- Consider prednisone
- Consider calcitonin
- Consider cholestryamine

***Bromethalin***
- Supportive; no known antidote

## References

1. Katona B, Wason S. Superwarfarin poisoning. *J Emerg Med.* Nov-Dec 1989;7(6):627-631.
2. Huckle, KR, Hutson, DH and Warburton, PA. Elimination and accumulation of the rodenticide flocoumafen in rats following repeated oral administration. *Xenobiotica*, 1988:18(12): 1465-1479.
3. Mack RB. Not all rats have four legs. Superwarfarin poisoning. *N C Med J.* Nov 1994;55(11):554-556.
4. Bronstein AC, Spyker DA, Cantilena LR, Jr., Green JL, Rumack BH, Giffin SL. 2008 Annual Report of the American Association of Poison Control Centers' National Poison Data System (NPDS): 26th Annual Report. *Clin Toxicol (Phila).* Dec 2009;47(10):911-1084.
5. Litovitz TL, Smilkstein M, Felberg L, Klein-Schwartz W, Berlin R, Morgan JL. 1996 annual report of the American Association of Poison Control Centers Toxic Exposure Surveillance System. *Am J Emerg Med.* Sep 1997;15(5):447-500.
6. Burucoa C, Mura P, Robert R, Boinot C, Bouquet S, Piriou A. Chlorophacinone intoxication. A biological and toxicological study. *J Toxicol Clin Toxicol.* 1989;27(1-2):79-89.
7. Smolinske SC, Scherger DL, Kearns PS, Wruk KM, Kulig KW, Rumack BH. Superwarfarin poisoning in children: a prospective study. *Pediatrics.* Sep 1989;84(3):490-494.
8. Lipton RA, Klass EM. Human ingestion of a 'superwarfarin' rodenticide resulting in a prolonged anticoagulant effect. *JAMA.* Dec 7 1984;252(21):3004-3005.
9. Helmuth R, McCloskey D, Doedens D, Hawley D. Fatal ingestion of a brodifacoum-containing rodenticide. *Lab Med.* 1989;20:25-27.
10. Olmos V, Lopez CM. Brodifacoum poisoning with toxicokinetic data. *Clin Toxicol (Phila).* Jun-Aug 2007;45(5):487-489.
11. Palmer RB, Alakija P, de Baca JE, Nolte KB. Fatal brodifacoum rodenticide poisoning: autopsy and toxicologic findings. *J Forensic Sci.* Jul 1999;44(4):851-855.
12. Soubiron L, Hantson P, Michaux I, Lambert M, Mahieu P, Pringot J. Spontaneous haemoperitoneum from surreptitious ingestion of a rodenticide. *Eur J Emerg Med.* Dec 2000;7(4):305-307.
13. Spahr JE, Maul JS, Rodgers GM. Superwarfarin poisoning: a report of two cases and review of the literature. *Am J Hematol.* Jul 2007;82(7):656-660.
14. Tahir M, Khan MF, Tourbaf K. Impending compartment syndrome and hemothorax after brodifacoum ingestion. *South Med J.* Dec 2008;101(12):1277.
15. Terneu S, Verhelst D, Thys F, Ketelslegers E, Hantson P, Wittebole X. An unusual cause of abdominal pain. *Acta Clin Belg.* Jul-Aug 2003;58(4):241-244.
16. Mullins ME, Brands CL, Daya MR. Unintentional pediatric superwarfarin exposures: do we really need a prothrombin time? *Pediatrics.* Feb 2000;105(2):402-404.
17. Shepherd G, Klein-Schwartz W, Anderson BD. Acute, unintentional pediatric brodifacoum ingestions. *Pediatr Emerg Care.* Jun 2002;18(3):174-178.
18. EPA. Cancellation Process for Certain Rodenticide Products. US Environmental Protection Agency website. http://www.epa.gov/pesticides/mice-and-rats/cancellation-process.html. Accessed 11-21-12.
19. Spiller HA, Gallenstein GL, Murphy MJ. Dermal absorption of a liquid diphacinone rodenticide causing coagulaopathy. *Vet Hum Toxicol.* Dec 2003;45(6):313-314.
20. Binks S, Davies P. Case of the month: "Oh! Drat!--A case of transcutaneous superwarfarin poisoning and its recurrent presentation". *Emerg Med J.* Apr 2007;24(4):307-308.
21. Norcross WA, Ganiats TG, Ralph LP, Seidel RG, Ikeda TS. Accidental poisoning by warfarin-contaminated herbal tea. *West J Med.* Jul 1993;159(1):80-82.

22. Kruse JA, Carlson RW. Fatal rodenticide poisoning with brodifacoum. *Ann Emerg Med.* Mar 1992;21(3):331-336.

23. Ornstein DL, Lord KE, Yanofsky NN, Cornell CJ, Zacharski LR. Successful donation and transplantation of multiple organs after fatal poisoning with brodifacoum, a long-acting anticoagulant rodenticide: case report. *Transplantation.* Feb 15 1999;67(3):475-478.

24. Corke PJ. Superwarfarin (brodifacoum) poisoning. *Anaesth Intensive Care.* Dec 1997;25(6):707-709.

25. Murdoch DA. Prolonged anticoagulation in chlorphacinone poisoning. *Lancet.* Feb 12 1983;1(8320):355-356.

26. Hoffman RS, Smilkstein MJ, Goldfrank LR. Evaluation of coagulation factor abnormalities in long-acting anticoagulant overdose. *J Toxicol Clin Toxicol.* 1988;26(3-4):233-248.

27. Frank M, Schmucker U, Nowotny T, Ekkernkamp A, Hinz P. Not all that glistens is gold: civilian white phosphorus burn injuries. *Am J Emerg Med.* Oct 2008;26(8):974 e973-975.

28. Santos O, Restrepo JC, Velasquez L, et al. Acute liver failure due to white phosphorus ingestion. *Ann Hepatol.* Apr-Jun 2009;8(2):162-165.

29. McCarron MM, Gaddis GP, Trotter AT. Acute yellow phosphorus poisoning from pesticide pastes. *Clin Toxicol.* Jun 1981;18(6):693-711.

30. Dipalma J. Human toxicity from rat poisons. *Amer Fam Physician.* 1981;24:186-189.

31. Simon FA, Pickering LK. Acute yellow phosphorus poisoning. "Smoking stool syndrome". *JAMA.* Mar 29 1976;235(13):1343-1344.

32. Elizabeth J, Kelkar PN, Weishali G. Yellow phosphorus poisoning - an unusual presentation. *J Assoc Physicians India.* May 1995;43(5):371-372.

33. Karanth S, Nayyar V. Rodenticide-induced hepatotoxicity. *J Assoc Physicians India.* Aug 2003;51:816-817.

34. Tafur AJ, Zapatier JA, Idrovo LA, Oliveros JW, Garces JC. Bone marrow toxicity after yellow phosphorus ingestion. *Emerg Med J.* Mar 2004;21(2):259-260.

35. Proudfoot AT. Aluminium and zinc phosphide poisoning. *Clin Toxicol (Phila).* Feb 2009;47(2):89-100.

36. Azoury M, Levin N. Identification of zinc phosphide in a falsely labeled rodenticide bait. *J Forensic Sci.* May 1998;43(3):693-695.

37. Orak M, Ustundag M, Sayhan MB. Severe metabolic acidosis secondary to zinc phosphide poisoning. *J Pak Med Assoc.* May 2008;58(5):289-290.

38. Patial RK, Bansal SK, Kashyap S, Sharma AK, Sharma B. Hypoglycaemia following zinc phosphide poisoning. *J Assoc Physicians India.* Apr 1990;38(4):306-307.

39. Frangides CY, Pneumatikos IA. Persistent severe hypoglycemia in acute zinc phosphide poisoning. *Intensive Care Med.* Feb 2002;28(2):223.

40. Mehrpour O, Alfred S, Shadnia S, et al. Hyperglycemia in acute aluminum phosphide poisoning as a potential prognostic factor. *Hum Exp Toxicol.* Jul 2008;27(7):591-595.

41. Schoonbroodt D, Guffens P, Jousten P, Ingels J, Grodos J. Acute phosphine poisoning? A case report and review. *Acta Clin Belg.* 1992;47(4):280-284.

42. Atsmon J, Taliansky E, Landau M, Neufeld MY. Thallium poisoning in Israel. *Am J Med Sci.* Nov 2000;320(5):327-330.

43. Mayfield S, Morgan D, Roberts R. Acute thallium poisoning in a 3-year old child. A case report. *Clin Ped.* 1983;23(8):461-462.

44. Fred HL, Accad MF. Abdominal pain, leg weakness, and alopecia in a teenage boy. *Hosp Pract (Minneap).* Apr 15 1997;32(4):69-70.

45. Bank WJ, Pleasure DE, Suzuki K, Nigro M, Katz R. Thallium poisoning. *Arch Neurol.* May 1972;26(5):456-464.

46. Meggs WJ, Hoffman RS, Shih RD, Weisman RS, Goldfrank LR. Thallium poisoning from maliciously contaminated food. *J Toxicol Clin Toxicol.* 1994;32(6):723-730.

47. Sharma AN, Nelson LS, Hoffman RS. Cerebrospinal fluid analysis in fatal thallium poisoning: evidence for delayed distribution into the central nervous system. *Am J Forensic Med Pathol.* Jun 2004;25(2):156-158.

48. Roby DS, Fein AM, Bennett RH, Morgan LS, Zatuchni J, Lippmann ML. Cardiopulmonary effects of acute thallium poisoning. *Chest.* Feb 1984;85(2):236-240.

49. Feldman J, Levisohn DR. Acute alopecia: clue to thallium toxicity. *Pediatr Dermatol.* Mar 1993;10(1):29-31.

50. Malbrain ML, Lambrecht GL, Zandijk E, et al. Treatment of severe thallium intoxication. *J Toxicol Clin Toxicol.* 1997;35(1):97-100.

51. Thallium. Micromedex 2.0. Thomson Reuters; 2011. http://www.thomsonhc.com/ Accessed January 3, 2011.

52. Koshy KM, Lovejoy FH, Jr. Thallium ingestion with survival: ineffectiveness of peritoneal dialysis and potassium chloride diuresis. *Clin Toxicol.* May 1981;18(5):521-525.

53. Thompson DF, Callen ED. Soluble or insoluble prussian blue for radiocesium and thallium poisoning? *Ann Pharmacother.* Sep 2004;38(9):1509-1514.

54. DeBacker W, Zachee P, Verpooten GA, Majelyne W, Vanheule A, DeBroe ME. Thallium intoxication treated with combined hemoperfusion–hemodialysis. *J Toxicol Clin Toxicol* 1982;19:259-264.

55. Chi CH, Chen KW, Chan SH, Wu MH, Huang JJ. Clinical presentation and prognostic factors in sodium monofluoroacetate intoxication. *J Toxicol Clin Toxicol.* 1996;34(6):707-712.

56. Santhosh GJ, Joseph W, Thomas M. Strychnine poisoning. *J Assoc Physicians India.* Jul 2003;51:739-740.

57. Duverneuil C, de la Grandmaison GL, de Mazancourt P, Alvarez JC. Liquid chromatography/photodiode array detection for determination of strychnine in blood: a fatal case report. *Forensic Sci Int.* Apr 20 2004;141(1):17-21.

58. Benomran FA, Henry JD. Homicide by strychnine poisoning. *Med Sci Law.* Jul 1996;36(3):271-273.

59. Besnard T, Sadeg N, Ricart N, et al. Serial determination of crimidine by HPLC/SE/SM in a patient ingesting a "mouse trap". *Acta Clin Belg Suppl.* 2002(1):8-11.

60. Barrueto F, Jr., Furdyna PM, Hoffman RS, Hoffman RJ, Nelson LS. Status epilepticus from an illegally imported Chinese rodenticide: "tetramine". *J Toxicol Clin Toxicol.* 2003;41(7):991-994.

61. De Saqui-Sannes P, Nups P, Le Bars P, Burgat V. Evaluation of an HPTLC method for the determination of strychnine and crimidine in biological samples. *J Anal Toxicol.* May-Jun 1996;20(3):185-188.

62. Marques EP, Gil F, Proença P, et al. Analytical method for the determination of strychnine in tissues by gas chromatography/mass spectrometry: two case reports. *Forensic Sci Int.* May 15 2000;110(2):145-152.

63. Perper JA. Fatal strychnine poisoning--a case report and review of the literature. *J Forensic Sci.* Oct 1985;30(4):1248-1255.

64. Wood D, Webster E, Martinez D, Dargan P, Jones A. Case report: Survival after deliberate strychnine self-poisoning, with toxicokinetic data. *Crit Care.* Oct 2002;6(5):456-459.

65. Shadnia S, Moiensadat M, Abdollahi M. A case of acute strychnine poisoning. *Vet Hum Toxicol.* Apr 2004;46(2):76-79.

66. Dittrich K, Bayer MJ, Wanke LA. A case of fatal strychnine poisoning. *J Emerg Med.* 1984;1(4):327-330.

67. Edmunds M, Sheehan TM, van't Hoff W. Strychnine poisoning: clinical and toxicological observations on a non-fatal case. *J Toxicol Clin Toxicol.* 1986;24(3):245-255.

68. Boyd RE, Brennan PT, Deng JF, Rochester DF, Spyker DA. Strychnine poisoning. Recovery from profound lactic acidosis, hyperthermia, and rhabdomyolysis. *Am J Med.* Mar 1983;74(3):507-512.

69. Lheureux P, Penaloza A, Gris M. Pyridoxine in clinical toxicology: a review. *Eur J Emerg Med.* Apr 2005;12(2):78-85.

70. Vieth R, Pinto TR, Reen BS, Wong MM. Vitamin D poisoning by table sugar. *Lancet.* Feb 23 2002;359(9307):672.

71. Jehle DR, Keller F, Schwarz A, Jehle PM. Hypercalcemia-induced renal insufficiency during therapy with dihydrotachysterol. *J Med.* 1999;30(1-2):39-50.

72. Titan SM, Callas SH, Uip DE, Kalil-Filho R, PC AG. Acute renal failure and hypercalcemia in an athletic young man. *Clin Nephrol.* Apr 2009;71(4):445-447.

73. van Lier RB, Cherry LD. The toxicity and mechanism of action of bromethalin: a new single-feeding rodenticide. *Fundam Appl Toxicol.* Nov 1988;11(4):664-672.

74. Pasquale-Styles MA, Sochaski MA, Dorman DC, Krell WS, Shah AK, Schmidt CJ. Fatal bromethalin poisoning. *J Forensic Sci.* Sep 2006;51(5):1154-1157.

75. Dorman DC, Simon J, Harlin KA, Buck WB. Diagnosis of bromethalin toxicosis in the dog. *J Vet Diagn Invest.* Apr 1990;2(2):123-128.

76. Martin T, Johnson B. A suspected case of bromethalin toxicity in a domestic cat. *Vet Hum Toxicol.* Jun 1989;31(3):239-240.

77. Dorman DC, Parker AJ, Dye JA, Buck WB. Bromethalin neurotoxicosis in the cat. *Prog Vet Neurol.* 1990;1:189-196.

78. Buckle RM, Gamlen TR, Pullen IM. Vitamin D intoxication treated with porcine calcitonin. *Br Med J.* Jul 22 1972;3(5820):205-207.

79. Queener SF, Bell NH. Treatment of experimental vitamin d3 intoxication in the rat with cholestyramine. *Clin Res.* 1976;24:583A.

80. Jibani M, Hodges NH. Prolonged hypercalcaemia after industrial exposure to vitamin D3. *Br Med J (Clin Res Ed).* Mar 9 1985;290(6470):748-749.

81. Thomson RB, Johnson JK. Another family with acute vitamin D intoxication: another cause of familial hypercalcaemia. *Postgrad Med J.* 1986;62:1025-1028.

4-Aminopyridine
## COMMERCIAL PRODUCTS

Avitrol

## HIGHLIGHTS

Toxic to all vertebrates

Used as bird repellent in grain baits or powdered sugar formulation

Rapidly absorbed across gut; less so on skin

Enhances cholinergic transmission in nervous system

## SIGNS & SYMPTOMS

Severe muscle spasms

Reported abdominal discomfort, nausea, vomiting, weakness, dizziness, diaphoresis

## TREATMENT

Decontaminate skin, eyes

Consider GI decontamination

Control seizures

Treat severe spasms in ICU

IV fluids if needed

CHAPTER 19

# Miscellaneous Pesticides, Synergists, Solvents and Adjuvants

## MISCELLANEOUS PESTICIDES

### 4-Aminopyridine

#### *Toxicology*

4-Aminopyridine is a highly toxic white powder used as a bird repellent. It works by making one or two birds acutely ill, and they warn off the remaining birds with their cries of distress.[1] It is toxic to all vertebrates.[2] It is usually added to grain baits in 0.5%-3.0% concentration, but 25% and 50% concentrates in powdered sugar are available. The avian $LD_{50}$ is between 1mg/kg and 8 mg/kg.[1] Information on human exposure has come from its use as an investigational drug in the treatment of multiple sclerosis.[3,4] It is rapidly absorbed by the gut and less effectively across skin. The chief mechanism of toxicity is enhancement of cholinergic transmission in the nervous system through the release of acetylcholine both centrally and peripherally. Because of enhanced transmission at neuromuscular junctions, severe muscle spasms may be a prominent manifestation of toxicity.[3] 4-aminopyridine is rapidly metabolized and excreted.

No human poisonings have occurred as a result of ordinary use, but toxic effects following intentional ingestion have been reported. In a report of ingestion of about 60 mg, patients experienced immediate abdominal discomfort, nausea and vomiting, weakness, dizziness and profuse diaphoresis. One went on to develop a tonic-clonic seizure and required ventilatory support. Acidosis was also present in those cases.[2] Dizziness and gait disturbances are additional reported findings. As seen above, seizures may also occur and be severe, although recovery with supportive therapy and ventilatory support has been the usual outcome.[2,3,4]

#### *Treatment of 4-Aminopyridine Toxicosis*

1. Decontaminate the skin with soap and water, as outlined in the **Chapter 3, *General Principles***. Treat eye contamination by irrigating the exposed eyes with copious amounts of clean water or saline for at least 15 minutes. Remove contact lenses, if present, prior to irrigation. If irritation persists after irrigation, specialized medical treatment in a healthcare facility should be obtained.

2. If ingested, consider gastrointestinal decontamination as outlined in **Chapter 3**.

3. Control seizures with benzodiazepines. See **Chapter 3** for specific medications and dosages.

4. Treat severe muscular spasms that may occur with this agent with neuromuscular blockade in an intensive care unit setting.[2]

5. Treat dehydration with intravenous fluids if oral fluids cannot be retained.

## Calcium Cyanamide

This synthetic compound is marketed as granules containing 44% calcium cyanamide, yielding 19.5% nitrogen. It is incorporated into soil to serve as fertilizer, fungicide and herbicide. In contact with water, especially under acidic conditions, hydrogen cyanamide is released. Hydrogen cyanamide is a solid with considerable vapor pressure. It has toxic properties totally different from those of cyanide, and it does not degrade to cyanide.

### Toxicology

While the initial ingredient, **calcium cyanamide**, is only moderately irritating to skin, the byproduct hydrogen cyanamide is severely irritating and caustic to skin. Dermal and mucosal lesions in the mouth, tongue and upper esophagus have occurred after exposure. Contact dermatitis has been reported manifested by exfoliation. Additional symptoms include fever, pruritus, anorexia, insomnia and malaise.[5] Lichen planus has also been reported.[6] If hydrogen cyanamide is inhaled, it can be strongly irritating to mucous membranes.[7] Systemic poisonings have followed inhalation of hydrogen cyanamide and ingestion of the salt. Manifestations of systemic poisoning include flushing, headache, vertigo, dyspnea, tachycardia and hypotension, sometimes progressing to shock.[7]

Cyanamide is an inhibitor of acetaldehyde dehydrogenase; therefore, as with disulfiram (Antabuse), ingestion of alcohol may significantly worsen the toxic effects.[8] A pharmaceutical form of calcium cyanamide has been used in alcohol aversion therapy, and one patient treated with this experienced peripheral neuropathy.[9] Long-term use of cyanamide has been reported to cause hepatocellular damage.[10]

### Treatment of Cyanamide Toxicosis

1. Decontaminate the skin with soap and water, as outlined in the **Chapter 3, *General Principles***. Treat eye contamination by irrigating the exposed eyes with copious amounts of clean water or saline for at least 15 minutes. Remove contact lenses, if present, prior to irrigation. If irritation persists after irrigation, specialized medical treatment in a healthcare facility should be obtained.

2. If ingested, consider gastrointestinal decontamination as outlined in **Chapter 3**.

3. If hypotension occurs, provide supportive care including intravenous fluids. If severe, the patient may need vasopressors and intensive care management.[11]

## Creosote

Creosote is a wood preservative that is registered as a restricted use pesticide, which limits its use to non-residential sites and requires strict worker protection standards. Creosote was first registered in 1948. It was used in the past as an animal dip and disinfectant, but is currently only registered for use on heavy-duty, pressure-treated industrial products, such as railroad ties and utility poles.[12]

### Toxicology

**Creosote** is obtained by distillation of the tar formed by heating wood or coal in the absence of oxygen. It is purified by extraction into oils. Creosote from wood consists of caustic phenol compounds, mainly guaiacol (methoxy phenol) and cresol (methyl

***Calcium Cyanamide***

**COMMERCIAL PRODUCTS**

cyanamide

nitrolime (has been discontinued)

**HIGHLIGHTS**

Marketed as multi-function granules

Byproduct hydrogen cyanamide is severe skin irritant

Alcohol may worsen toxic effects

**SIGNS & SYMPTOMS**

Mouth, tongue, upper esophageal lesions

Possible contact dermatitis with exfoliation

Fever, pruritus, anorexia, insomnia, malaise

Systemic: flushing, headache, vertigo, dyspnea, tachycardia, hypotension

**TREATMENT**

Decontaminate skin, eyes, GI as appropriate

Supportive care, IV fluids for hypotension

### *Creosote*
## HIGHLIGHTS

Restricted use pesticide

Used only on heavy-duty, pressure-treated industrial wood

Skin, eye, mucous membrane irritant

Absorbed by gut, lung

## SIGNS & SYMPTOMS

Skin irritation, possible eruptions, pigmentation

GI lesions, pain

Systemic: salivation, vomiting, dyspnea, headache, dizziness, loss of pupil reflex, cyanosis, hypothermia, convulsions, coma

Dark, "smoky" urine

## TREATMENT

Decontaminate skin, eyes

Consider GI decontamination with caution

ICU support may be indicated

phenol). Coal-derived creosote contains, in addition, some phenol, pyridine and pyridinol. Much of human exposure is in the form of various phenol compounds. Some phenolic compounds such as cresol are also used as disinfectants; more information on toxicity from these compounds is found in **Chapter 20, *Disinfectants***. Creosote is carcinogenic in animals and epidemiological evidence has suggested an association with some human cancers. This is discussed in the **Chapter 21, *Chronic Effects***.

Creosote is irritating to skin, eyes and mucous membranes. Workers in contact with technical creosote or with treated timbers sometimes develop skin irritation, vesicular or papular eruptions, dermal pigmentation and occasionally gangrene and skin cancer.[13] Photosensitization has been reported. Eye contamination has resulted in conjunctivitis and keratitis, sometimes resulting in corneal scarring.[14] The constituents of creosote are efficiently absorbed across the skin, but acute systemic poisonings following dermal absorption have occurred very rarely.

Absorption of ingested phenolic compounds from the gut occurs promptly, and there may be significant absorption of vapor by the lung. Conjugates of absorbed phenolic constituents are excreted mainly in the urine. Acute toxic effects are similar to those of Lysol, but the corrosive nature of creosote is somewhat less because of greater dilution of phenol in the creosote.[15] Irritation of the gastrointestinal tract including mucosal lesions, esophageal ulcers, abdominal pain, toxic encephalopathy and renal tubular injury are all principal effects following ingestion or inhalation exposure from phenolic compounds.[16,17,18,19,20]

Manifestations of acute systemic poisoning are salivation, vomiting, dyspnea, headache, dizziness, loss of pupillary reflexes, cyanosis, hypothermia, convulsions and coma. Death is due to multiorgan system failure as patients develop shock, acidosis, respiratory depression and anuric renal failure. Acute respiratory distress syndrome (ARDS) has also been reported and is potentially fatal.[17,21,22] Some reports of poisoning have been from related phenolic compounds as opposed to the wood preservative itself. A chronic toxicosis from continuing gastrointestinal absorption (creosote used medicinally) has been described, consisting of gastroenteritis and visual disturbances.[19]

### *Confirmation of Poisoning*

The presence of phenolic oxidation products imparts a dark, smoky color to the urine.[15,20] If creosote poisoning is suspected, addition of a few drops of ferric chloride solution to the urine yields a violet or blue color indicating the presence of phenolic compounds. Methods to determine urinary levels of phenolic compounds using capillary gas chromatography have been described.[23] Data are limited in determining a "normal range"; however, in separate fatal cases of phenol poisoning, peak blood and urine levels were 58-60 µg/mL (blood) and 20-208 µg/mL (urine) using GC/MS.[24,25]

### *Treatment of Creosote Toxicosis*

1. Decontaminate the skin with soap and water, as outlined in **Chapter 3, *General Principles***. Treat eye contamination by irrigating the exposed eyes with copious amounts of clean water or saline for at least 15 minutes. Remove contact lenses, if present, prior to irrigation. If irritation persists after irrigation, specialized medical treatment in a healthcare facility should be obtained.

2. If ingested, consider gastrointestinal decontamination as outlined in **Chapter 3**. Given the corrosive nature of phenolic compounds such as creosote, efforts to use an adsorbent such as charcoal (or repeated use of charcoal) depend on whether there has been corrosive injury to the esophagus. If pharyngeal redness

and swelling are evident, emesis, whether induced or exacerbated by activated charcoal, is not advisable because of potential re-exposure of the esophagus to the creosote. Risk of perforation from a gastric tube precludes the use of gastric lavage.

3. Treat severe systemic creosote poisoning in an intensive care unit setting with appropriate supportive care including respiratory support, intravenous fluids, cardiac monitoring and renal function support as necessary.

4. Draw a blood sample to test for methemoglobinemia, to measure BUN and blood electrolytes and to check for signs of liver injury (bilirubin, GGT, LDH, ALT, AST and alkaline phosphatase). Examine the urine for protein and cells, and for "smoky" phenolic excretion products.

5. Control seizures with benzodiazepines. See **Chapter 3** for specific medications and dosages. Hemoperfusion over charcoal has been reported to be successful.[26]

6. Methemoglobinemia is rarely severe, but consicer administration of methylene blue if 25%-30% of hemoglobin is converted.

**Dosage of Methylene Blue**

- ***1-2 mg/kg body weight, IV, over 5 minutes, repeated as needed every 4 hours.***

Methylene blue is contraindicated in patients with G6PD deficiency.

7. Refer patients for endoscopic evaluation following a deliberate ingestion. If there are any signs or symptoms suggestive of mucosal pharyngeal or esophageal injury (visible burns in the oral mucosa, stridor, drooling, dysphagia, refusal to swallow or abdominal pain) following inadvertent or unintentional ingestion, those patients should also have an endoscopy.[27,28,29]

## Endothall

As the free acid or as sodium, potassium or amine salts, endothall is used as a contact herbicide, defoliant, aquatic herbicide and algaecide. It is formulated in aqueous solutions and granules at various strengths.

### *Toxicology*

**Endothall** is irritating to skin, eyes and mucous membranes. It is well absorbed across abraded skin and from the gastrointestinal tract. Recognized systemic toxic mechanisms in mammals include a corrosive effect on the gastrointestinal tract (particularly from high concentrations of the free acid), cardiomyopathy and vascular injury leading to shock, and central nervous system injury, causing convulsions and respiratory depression. A single case has been reported of a lethal poisoning in a previously healthy 21-year-old man who died after ingestion of 7-8 grams of endothall. In this patient, hemorrhage and edema were noted in the gastrointestinal tract and lungs.[30]

***Endothall***

**COMMERCIAL PRODUCTS**

Accelerate
Aquathol
Des-i-cate
Endothall Turf Herbicide
Herbicide 273
Hydrothol

**HIGHLIGHTS**

Well absorbed by gut and abraded skin
GI corrosive
Cardiac, respiratory, CNS impacts

**SIGNS & SYMPTOMS**

Convulsions
Respiratory depression
"Smoky" urine

**TREATMENT**

Decontaminate skin, eyes
Consider GI decontamination with caution
ICU support may be indicated
Control seizures

**CONTRAINDICATED**

Gastric lavage in most cases

*Metaldehyde*
## COMMERCIAL PRODUCTS

Antimilace

Biodehido Snailkill

Cekumeta

Halizan

META

Metason

Namekil

and others

## HIGHLIGHTS

Slug baits and other uses

Inhalation and ingestion hazard

## SIGNS & SYMPTOMS

Nausea, vomiting, dizziness after ingestion

Severe toxicity: pyrexia, seizures, metabolic acidosis, mental changes

Possible headache, hypersalivation, facial flushing, tachypnea

## TREATMENT

Consider GI decontamination

ICU treatment may be indicated

### *Confirmation of Poisoning*

There are no standards for endothall levels, and they are not considered useful in the management of acute poisoning.

### *Treatment of Endothall Toxicosis*

1. Decontaminate the skin with soap and water as outlined in the **Chapter 3, *General Principles***. Irrigate exposed eyes with copious amounts of clean water or saline for at least 15 minutes. Remove contact lenses, if present, prior to irrigation. If irritation persists after irrigation, obtain specialized medical treatment in a health-care facility.

2. If ingested, consider gastrointestinal decontamination as outlined in **Chapter 3**, but use treatment procedures appropriate for ingestions of corrosives (strong acids and alkalis). Due to the corrosive nature of this agent, gastric lavage is usually contraindicated, because of the risk of esophageal perforation. Refer patient to a surgeon or gastroenterologist for consideration of endoscopy.

3. Treat severe systemic endothall poisoning in an intensive care unit setting with appropriate supportive care, including respiratory support, intravenous fluids, cardiac monitoring and renal function support as necessary.

4. Draw a blood sample to test for methemoglobinemia, measure BUN and blood electrolytes and check for signs of liver injury (bilirubin, GGT, LDH, ALT, AST and alkaline phosphatase). Examine the urine for protein and cells and for "smoky" phenolic excretion products.

5. Control seizures with benzodiazepines. See **Chapter 3** for specific medications and dosages.

## Metaldehyde

### *Toxicology*

Metaldehyde is a 4-unit cyclic polymer of acetaldehyde long used to kill slugs and snails, which are attracted to it without the use of bait. Occasional poisonings of animals and children have resulted from ingestion of pellets intended as molluscicide, and tablets designed as a combustible fuel ("meta-fuel") have also been responsible for human poisonings.[31] Another form of exposure is "snow storm tablets," which the user places at the end of a lighted cigarette to create snow. Toxicity occurs through inhalation of metaldehyde fumes.[32] The biochemical mechanism of poisoning is not known. Both acetaldehyde and metaldehyde produced similar effects in dogs; however, acetaldehyde was not detected in the plasma or urine of the metaldehyde-poisoned dogs.[33] Poisoned animals show muscle tremors, ataxia, hyperesthesia, salivation, tachycardia and seizures.[33,34]

Ingestion of a toxic dose is often followed shortly by nausea, vomiting and dizziness, if symptoms are even present.[35] Primary features of severe toxicity include pyrexia, generalized seizures, metabolic acidosis and mental status changes such as irritability, sometimes progressing to coma.[31,35,36,37] Other signs and symptoms that may occur include headache, hypersalivation, facial flushing and tachypnea.[31,32] Pneumonitis has followed inhalational exposure to metaldehyde.[32] Most cases are either self limiting or with significant but controllable seizures, and fatal events are infrequent.[31,35,36,37] One patient survived what has been considered a lethal dose of 600 mg/kg.[36] Autopsy findings in fatal human poisonings indicate severe damage to liver cells and renal tubular epithelium.

### Confirmation of Poisoning

Metaldehyde can be measured in the serum, although there are very few reports of levels among poisoned humans. Saito described a method to measure metaldehyde in serum using headspace solid-phase microextraction and gas chromatography-mass spectrometry.[38] One patient who had severe tonic-clonic seizures and was comatose had a metaldehyde level in the serum of 125 mg/l, with a half-life of 27 hours. This patient did not have detectable acetaldehyde in the serum.[37]

### Treatment of Metaldehyde Toxicosis

There is no specific antidote for metaldehyde poisoning.

1. If ingested, consider gastrointestinal decontamination as outlined in **Chapter 3, *General Principles*.**

2. Treat severe systemic metaldehyde poisoning in an intensive care unit setting with appropriate supportive care, including respiratory support, intravenous fluids, cardiac monitoring and renal function support as necessary. Early and aggressive treatment of all of the above may be life saving following a massive ingestion.[36]

3. Consider sodium bicarbonate in the event of severe metabolic acidosis.[36] Monitor fluid balance and electrolytes carefully to avoid fluid overload if renal failure supervenes.

4. Order liver function tests and urine sediment examination to assess liver and kidney injury in poisoned patients.

## Sodium Chlorate

Sodium chlorate is used in agriculture as a defoliant, nonselective contact herbicide and semi-permanent soil sterilant. Because of its explosive nature, it must be formulated with a water-soluble, fire-retardant material such as sodium metaborate, soda ash, magnesium chloride or urea. It is usually applied in water solution.

### Toxicology

**Sodium chlorate** is irritating to skin, eyes and mucous membranes of the upper respiratory tract.[39] Dermal absorption is slight. Even though gastrointestinal absorption is also inefficient, severe poisoning, sometimes fatal, follows ingestion of a toxic dose, said to be about 20 grams in the adult human. Excretion is chiefly in the urine.

Sodium chlorate poisoning can manifest in many ways. The principal mechanisms of toxicity are hemolysis, coagulation disturbances methemoglobin formation, cardiac arrhythmia (partly secondary to hyperkalemia) and renal tubular injury.[39,40,41] The most common cause of death in early stages of toxicity is anoxia from methemoglobinemia and hemolysis, resulting in disseminated intravascular coagulation. Renal failure may occur afterwards and figure prominently as a cause of death.[39,40,41]

Following ingestion, sodium chlorate may have an irritant action on the gut, causing nausea, vomiting and abdominal pain. Once absorbed, hemoglobin is rapidly

*Sodium Chlorate*
**COMMERCIAL PRODUCTS**

Bladvel
Defol
De-Fol-Ate
Dervan
Drop-Leaf
Fall
KM
Kusatol
Leafex

*Sodium Chlorate*
**HIGHLIGHTS**

Forms methemoglobin in unique fashion

Destroys erythrocytes and enzymatic systems

Irritating to skin, eyes, mucous membranes

Many poisoning manifestations

**SIGNS & SYMPTOMS**

Nausea, vomiting, abdominal pain

Cyanosis

Dark brown plasma, urine

**TREATMENT**

Decontaminate skin, eyes

Consider GI decontamination

ICU may be appropriate

Consider oral or IV sodium thiosulfate

IV fluids and sodium bicarbonate

**CONTRAINDICATED**

Methylene blue unless very early

oxidized to methemoglobin, and hemolysis and intravascular hemolysis subsequently occur.[39,40,41] Cyanosis is prominent if methemoglobinemia is severe and may be the only presenting sign.[41] Acute tubular necrosis and hemoglobinuria may result from the hemolysis or direct toxic injury. Plasma and urine are dark brown from presence of free hemoglobin and methemoglobin.[39,41,42,43] One fatal case presented with 30% methemoglobinemia.[42] Release of potassium from red cell destruction results in hyperkalemia that may be severe enough to cause life-threatening arrythmias.[43] The liver and spleen are often enlarged due to uptake of hemolyzed erythrocytes.[41] Hypoxemia may lead to convulsions. Death may be the result of shock, tissue hypoxia, renal failure, hyperkalemia or disseminated intravascular coagulation (DIC).[39,40,41,43]

Although other toxicants will induce methemoglobinemia formation, the mechanism associated with chlorates is unique. Chlorate not only forms methemoglobin, it also destroys erythrocytes and the enzymatic systems in the process.[40] Ordinarily, methylene blue is used as an antidote to reduce methemoglobin. This process depends on NADPH formation by the oxidation of glucose-6-phosphate. However, chlorate will denature the glucose-6-phosphate dehydrogenase, which will in turn render methylene blue ineffective.[40]

### *Confirmation of Poisoning*

Sodium chlorate poisoning can be detected by ion chromatography, although this test may not be widely available.[42] Chlorate poisoning should be considered when patients present with methemoglobinemia. Dark brown-to-black staining of the plasma and urine indicates the action of a strong oxidizing agent on hemoglobin.[41,44]

### *Treatment of Chlorate Toxicosis*

1. Decontaminate the skin with soap and water as outlined in **Chapter 3, *General Principles***. Irrigate exposed eyes with copious amounts of clean water or saline for at least 15 minutes. Remove contact lenses, if present, prior to irrigation. If irritation persists after irrigation, obtain specialized medical treatment in a healthcare facility.

2. If ingested, consider gastrointestinal decontamination as outlined in **Chapter 3**.

3. Treat severe systemic sodium chlorate poisoning in an intensive care unit setting, with appropriate supportive care including respiratory support, intravenous fluids, cardiac monitoring and renal function support as necessary.[44] In addition, monitor for methemoglobinemia.

4. Sodium thiosulfate has been used as an antidote against absorbed sodium chlorate.[39,44] Thiosulfate is thought to inactivate the chlorate ion to form the less toxic chloride ion.

**Dosage of Sodium Thiosulfate**

- ***2-5 gm dissolved in 200 mL of 5% sodium bicarbonate, given orally or as an IV infusion over 60-90 minutes[39]***

5. Administer intravenous fluids and sodium bicarbonate.[41,44] Monitor urine production closely, so that intravenous fluids can be slowed or discontinued if renal failure occurs.

6. Transfuse blood if hemolysis and methemoglobinemia are severe. Exchange transfusion has been recommended to enhance clearance and treat DIC.[43]

7. Consider hemodialysis, which has been used in several cases of chlorate poisoning.[42,44]

8. Consider administration of methylene blue, which has been used to reverse methemoglobinemia if as much as 25%-30% of hemoglobin is converted.

**Dosage of Methylene Blue**

- ***1-2 mg/kg body weight, IV, over 5 minutes, q 4 hours prn***

Methylene blue is generally not helpful with sodium chlorate poisoning unless given very early because of the unique characteristics described above in the *Toxicology* subsection.[40,44]

## SYNERGISTS

### Piperonyl Butoxide

Synergists are chemical agents included in pesticide products to enhance the killing power of the active ingredients. The widely used insecticide synergist piperonyl butoxide acts by inhibiting the enzymatic degradation of pyrethrins, rotenone, N-methyl carbamates, organophosphates and possibly some other insecticides. There is limited dermal absorption on contact. Inherent toxicity in mammals is low, with an oral $LD_{50}$ in rats of >4,500 mg/kg.[45] Large absorbed doses may theoretically enhance the toxic hazard of the rapidly metabolized insecticides used today, although inhibition of human drug metabolizing enzymes by these agents has not actually been demonstrated. Their presence in pesticide products to which humans are exposed does not change the basic approach to management of poisoning with the focus of treatment based on the active ingredient involved. The notable exception is that care providers should be aware of some possibility of enhanced toxicity of the active insecticidal ingredients.

## SOLVENTS AND ADJUVANTS

Liquid materials in which pesticides are dissolved or the solids on which they are adsorbed (sometimes called carriers or vehicles) are chosen by producers to achieve stability of the active ingredient, convenience in handling and application, and maximum killing power following application. The solvents and adjuvants pesticide manufacturers choose can give their commercial products a competitive edge. For this reason, their inclusion in marketed products is usually proprietary information, not available to the general public except under emergency circumstances. In a poisoning emergency, pesticide companies will usually cooperate in supplying physicians with information needed to provide treatment. The physician should seek this information to assist in evaluating all possible exposures. A direct request to the producer is necessary to secure this information. Some companies put the inert ingredients on the Material Safety Data Sheet (MSDS).

***Solvents & Adjuvants***

**HIGHLIGHTS**

Petroleum distillates are commonly used for lipophilic pesticides

Hydrocarbon pneumonitis: rapid respiration, cyanosis, tachycardia, fever

Some solvents may enhance dermal absorption of some pesticides

Some adjuvants, penetrants, safeners irritate skin, eyes, mucous membranes

Dust formulations can result in systemic poisoning via gut absorption

**TREATMENT**

Consider GI decontamination with caution, especially with petroleum distillates

Watch for and treat hydrocarbon pneumonitis

Monitor urine, heart

**Petroleum distillates** are the most commonly used solvents for lipophilic pesticides. (Most insecticides are lipophilic.) The distillates are mixtures of aliphatic and aromatic hydrocarbons with low boiling points.

Sometimes specific hydrocarbons, such as toluene or xylene (strongly odiferous), are added to stabilize the solution of insecticide or make it more emulsifiable. Hydrocarbon-dissolved pesticides are usually diluted for application by adding measured amounts of water to form emulsions. Some chlorinated hydrocarbons may be present in particular technical mixtures. A strong odor lingering after application of a structural pest control spray is often due to the solvent rather than the active ingredient. Rapid respiration, cyanosis, tachycardia and low-grade fever are the usual indications of frank hydrocarbon pneumonitis.[46]

Less lipophilic active ingredients are sometimes dissolved in mixtures of **alcohols, glycols, ethers** or **various chlorinated solvents**. It is possible that these enhance the dermal absorbability of some pesticides. A well-described example is increased dermal absorption of the insect repellent DEET when dissolved in 70% ethyl alcohol compared to the solvent polyethylene glycol.[47] Also, some solvents (*e.g.,* methanol and isopropanol) may represent a significant toxic hazard if swallowed in sufficient dosage. Symptoms may include central nervous system depression ranging from disorientation to lethargy and coma with severe overdose, as well as respiratory depression and ketosis.[48] The presence of chlorinated solvents in some formulations may add significantly to the toxic hazard, particularly if the product is ingested. Certain adjuvants are irritants to skin, eyes and mucous membranes and may account for the irritant properties of products with active ingredient(s) lacking this effect. With these exceptions, however, the presence of adjuvants in most finished pesticide products probably does not enhance or reduce systemic mammalian toxicity to any great extent.

**Granular formulations** utilize various clay materials that adsorb pesticide, retain it in more or less stable form until application and then desorb the material slowly into treated soil. There is some significant desorption when granules are in contact with human skin and very substantial desorption into gastrointestinal secretions if granules are swallowed. The clay materials themselves are not a toxic hazard.

**Dusts** are infrequently used today. Various forms of talc (magnesium silicate particles) have been used in the past to adsorb pesticides for application to foliage. Particle sizes are such that these dusts are usually trapped in the upper respiratory mucous when inhaled. When the mucous is swallowed, the particles desorb pesticide into gastrointestinal secretions. Dust formulations may, therefore, release enough of some pesticides to cause systemic poisonings.

**Stickers and spreaders** (film extenders) are organic substances added to formulations to disperse pesticide over treated foliage surfaces and enhance adhesion. The availability and persistence of residue on the leaf surfaces is thereby increased. Substances used include proteinaceous materials (milk products, wheat flour, blood albumin, gelatin), oils, gums, resins, clays, polyoxyethylene glycols, terpenes and other viscid organics. Some also include sulfated alcohols, fatty acid esters, alkyl and petroleum sulfonates. For persons exposed in the course of formulation or application of pesticides, these adjuvants probably add little or no toxic hazard to that inherent in the active pesticidal ingredients.

**Emulsifiers** serve to stabilize water-oil emulsions formed when water is added to technical hydrocarbon concentrates. Chemically, they are detergent-like (one part of the molecule lipophilic, the other hydrophilic). Long-chain alkyl sulfonate esters of polyethylene glycol and polyoxyethylene oleate are exemplary emulsifiers. They have low inherent mammalian toxicity, and their presence probably has little effect on the overall toxicity of formulated products that include them.

**Penetrants** facilitate the transfer of herbicide from foliage surface to the interior tissues. Some are lipids while others are detergent (surfactant) in nature. Substances used include heavy petroleum oils and distillates, polyol fatty acid esters, polyethoxylated fatty acid esters, aryl alkyl polyoxyethylene glycols, alkyl amine acetate, alkyl aryl sulfonates, polyhydric alcohols and alkyl phosphates. Some of these are eye and skin irritants and may account for the irritant effects of particular herbicide formulations whose active ingredients do not have this property.

**Safeners** are substances added to mixtures of fertilizers with pesticides (commonly herbicides) to limit the formation of undesirable reaction products. Some substances used are alcohol sulfates, sodium alkyl butane diamate, polyesters of sodium thiobutane dioate and benzene acetonitrile derivatives. Some are moderately irritating to the skin and eyes. Systemic toxicities are generally low.

**Anticaking agents** are added to granular and dust formulations to facilitate application by preventing cakes and clumps. Among several products used are the sodium salt of mono- and di-methyl naphthalene sulfonate, and diatomaceous earth. Diatomaceous earth has little adverse effect except a drying action on the skin. Methyl naphthalenes are said to be skin irritants and photosensitizers; whether their derivatlves have this effect is not known.

### *Treatment of Solvent and Adjuvant Toxicosis*

Petroleum distillates are mineral hydrocarbons that undergo limited absorption across the gut. In general, clinical toxicologists do not recommend induced emesis or gastric lavage in treating ingestions of these materials, because of the serious risk of hydrocarbon pneumonitis even if tiny amounts of the liquid are aspirated into the lungs. However, this injunction against emptying the stomach may be set aside when the petroleum distillate is a vehicle for toxic pesticides in significant concentration.

1. In such cases, if the patient is seen within 1 hour of exposure, consider gastrointestinal decontamination, as outlined in **Chapter 3, *General Principles*.**

2. Hospitalize patients with presumed hydrocarbon pneumonitis who are symptomatic. If the patient has pulmonary symptoms, order a chest X-ray to detect or confirm signs of pneumonitis. Mechanically assisted pulmonary ventilation with pure oxygen may be required. Hydrocarbon pneumonitis is sometimes fatal, and survivors may require several weeks for full recovery. In milder cases, clinical improvement usually occurs within several days, although radiographic findings will remain abnormal for longer periods.[49]

3. Examine the urine for protein, sugar, acetone, casts and cells; and examine an ECG for arrhythmias and conduction defects.

## References

1. Bischoff K, Morgan S, Chelsvig J, Spencer D. 4-aminopyridine poisoning of crows in the Chicago area. *Vet Hum Toxicol.* Dec 2001;43(6):350-352.

2. Spyker DA, Lynch C, Shabanowitz J, Sinn JA. Poisoning with 4-aminopyridine: report of three cases. *Clin Toxicol.* Jun 1980;16(4):487-497.

3. Pickett TA, Enns R. Atypical presentation of 4-aminopyridine overdose. *Ann Emerg Med.* Mar 1996;27(3):382-385.

4. Stork CM, Hoffman RS. Characterization of 4-aminopyridine in overdose. *J Toxicol Clin Toxicol.* 1994;32(5):583-587.

5. Kawana S. Drug eruption induced by cyanamide (carbimide): a clinical and histopathologic study of 7 patients. *Dermatology.* 1997;195(1):30-34.

6. Torrelo A, Soria C, Rocamora A, Moreno R, Ledo A. Lichen planus-like eruption with esophageal involvement as a result of cyanamide. *J Am Acad Dermatol.* Dec 1990;23(6 Pt 1):1168-1169.

7. Sittig M. *Handbook of Toxic and Hazardous Chemicals and Carcinogens.* 3rd ed. Park Ridge: Noyes Publications; 1991.

8. Kojima T, Nagasawa N, Yashiki M, Iwasaki Y, Kubo H, Kimura N. A fatal case of drinking and cyanamide intake. *Nihon Hoigaku Zasshi.* Apr 1997;51(2):111-115.

9. Reilly TM. Letter: Peripheral neuropathy associated with citrated calcium carbimide. *Lancet.* Apr 24 1976;1(7965):911-912.

10. Yokoyama A, Sato S, Maruyama K, et al. Cyanamide-associated alcoholic liver disease: a sequential histological evaluation. *Alcohol Clin Exp Res.* Oct 1995;19(5):1307-1311.

11. Calcium cyanamide. Thomson Reuters; 2011.

12. Agency USEP. *Reregistration Eligibility Decision (RED) for Creosote (Case 0139).* 2008.

13. Sittig M. *Handbook of Toxic and Hazardous Chemicals and Carcinogens.* 3rd ed. Park Ridge: Noyes Publications; 1991.

14. Lin TM, Lee SS, Lai CS, Lin SD. Phenol burn. *Burns.* Jun 2006;32(4):517-521.

15. Bowman CE, Muhleman MF, Walters E. A fatal case of creosote poisoning. *Postgrad Med J.* Jul 1984;60(705):499-500.

16. Bulut M, Turkmen N, Fedakar R, Aydin SA. A case report of fatal oral ingestion of resorcinol. *Mt Sinai J Med.* Nov 2006;73(7):1049-1051.

17. Cheng SL, Wang HC, Yang PC. Acute respiratory distress syndrome and lung fibrosis after ingestion of a high dose of ortho-phenylphenol. *J Formos Med Assoc.* Aug 2005;104(8):585-587.

18. Haddad LM, Dimond KA, Schweistris JE. Phenol poisoning. *JACEP.* Jul 1979;8(7):267-269.

19. Koruk ST, Ozyilkan E, Kaya P, Colak D, Donderici O, Cesaretli Y. Juniper tar poisoning. *Clin Toxicol (Phila).* 2005;43(1):47-49.

20. Sakai Y, Abo W, Yagita K, Tanaka T, Doi T, Fuke C. Chemical burn with systemic cresol intoxication. *Pediatr Int.* Apr 1999;41(2):174-176.

21. Fu HS, Chu YK, Liao SQ, Liu RS. Scintigraphic appearance in Lysol-induced lung toxicity. *Clin Nucl Med.* Jul 2001;26(7):655-656.

22. Gupta S, Ashrith G, Chandra D, Gupta AK, Finkel KW, Guntupalli JS. Acute phenol poisoning: a life-threatening hazard of chronic pain relief. *Clin Toxicol (Phila).* Mar 2008;46(3):250-253.

23. Bieniek G. Concentrations of phenol, o-cresol, and 2,5-xylenol in the urine of workers employed in the distillation of the phenolic fraction of tar. *Occup Environ Med.* May 1994;51(5):354-356.

24. Boatto G, Nieddu M, Carta A, et al. Determination of phenol and o-cresol by GC/MS in a fatal poisoning case. *Forensic Sci Int.* Jan 28 2004;139(2-3):191-194.

25. Tanaka T, Kasai K, Kita T, Tanaka N. Distribution of phenol in a fatal poisoning case determined by gas chromatography/mass spectrometry. *J Forensic Sci.* Sep 1998;43(5):1086-1088.

26. Christiansen RG, Klaman JS. Successful treatment of phenol poisoning with charcoal hemoperfusion. *Vet Hum Toxicol.* Feb 1996;38(1):27-28.

27. Crain EF, Gershel JC, Mezey AP. Caustic ingestions. Symptoms as predictors of esophageal injury. *Am J Dis Child.* Sep 1984;138(9):863-865.

28. Gaudreault P, Parent M, McGuigan MA, Chicoine L, Lovejoy FH, Jr. Predictability of esophageal injury from signs and symptoms: a study of caustic ingestion in 378 children. *Pediatrics.* May 1983;71(5):767-770.

29. Nuutinen M, Uhari M, Karvali T, Kouvalainen K. Consequences of caustic ingestions in children. *Acta Paediatr.* Nov 1994;83(11):1200-1205.

30. Allender WJ. Suicidal poisoning by endothall. *J Anal Toxicol.* Mar-Apr 1983;7(2):79-82.

31. Longstreth WT, Jr., Pierson DJ. Metaldehyde poisoning from slug bait ingestion. *West J Med.* Aug 1982;137(2):134-137.

32. Jay MS, Kearns GL, Stone V, Moss M. Toxic pneumonitis in an adolescent following exposure to Snow Storm tablets. *J Adolesc Health Care.* Sep 1988;9(5):431-433.

33. Booze TF, Oehme FW. An investigation of metaldehyde and acetaldehyde toxicities in dogs. *Fundam Appl Toxicol.* Apr 1986;6(3):440-446.

34. Yas-Natan E, Segev G, Aroch I. Clinical, neurological and clinicopathological signs, treatment and outcome of metaldehyde intoxication in 18 dogs. *J Small Anim Pract.* Aug 2007;48(8):438-443.

35. Shih CC, Chang SS, Chan YL, et al. Acute metaldehyde poisoning in Taiwan. *Vet Hum Toxicol.* Jun 2004;46(3):140-143.

36. Bleakley C, Ferrie E, Collum N, Burke L. Self-poisoning with metaldehyde. *Emerg Med J.* Jun 2008;25(6):381-382.

37. Moody JP, Inglis FG. Persistence of metaldehyde during acute molluscicide poisoning. *Hum Exp Toxicol.* Sep 1992;11(5):361-362.

38. Saito T, Morita S, Motojyuku M, et al. Determination of metaldehyde in human serum by headspace solid-phase microextraction and gas chromatography-mass spectrometry. *J Chromatogr B Analyt Technol Biomed Life Sci.* Nov 15 2008;875(2):573-576.

39. Helliwell M, Nunn J. Mortality in sodium chlorate poisoning. *Br Med J.* Apr 28 1979;1(6171):1119.

40. Steffen C, Wetzel E. Chlorate poisoning: mechanism of toxicity. *Toxicology.* Nov 12 1993;84(1-3):217-231.

41. Steffen C, Seitz R. Severe chlorate poisoning: report of a case. *Arch Toxicol.* Nov 1981;48(4):281-288.

42. Eysseric H, Vincent F, Peoc'h M, Marka C, Aitken Y, Barret L. A fatal case of chlorate poisoning: confirmation by ion chromatography of body fluids. *J Forensic Sci.* Mar 2000;45(2):474-477.

43. Smith EA, Oehme FW. A review of selected herbicides and their toxicities. *Vet Hum Toxicol.* Dec 1991;33(6):596-608.

44. Ranghino A, Costantini L, Deprado A, et al. A case of acute sodium chlorate self-poisoning successfully treated without conventional therapy. *Nephrol Dial Transplant.* Oct 2006;21(10):2971-2974.

45. Agency USEP. *Reregistration Eligibility Decision (RED) for piperonyl butoxide (PBO).* 2006.

46. Mastropietro CW, Valentine K. Early administration of intratracheal surfactant (calfactant) after hydrocarbon aspiration. *Pediatrics.* Jun;127(6):e1600-1604.

47. Stinecipher J, Shah J. Percutaneous permeation of N,N-diethyl-m-toluamide (DEET) from commercial mosquito repellents and the effect of solvent. *J Toxicol Environ Health.* Oct 10 1997;52(2):119-135.

48. Vujasinovic M, Kocar M, Kramer K, Bunc M, Brvar M. Poisoning with 1-propanol and 2-propanol. *Hum Exp Toxicol.* Dec 2007;26(12):975-978.

49. Anas N, Namasonthi V, Ginsburg CM. Criteria for hospitalizing children who have ingested products containing hydrocarbons. *JAMA.* Aug 21 1981;246(8):840-843.

*Alcohols*
**HIGHLIGHTS**

Often mixtures, usually ethanol & isopropanol

Most common household product 70% isopropyl alcohol

Well absorbed by GI, skin, inhalation

High concentrations can depress CNS leading to coma, death

**SIGNS & SYMPTOMS**

GI irrigation: gastritis, vomiting

Can be measured in blood and urine

**TREATMENT**

Support for hypotension, respiratory depression; best in ICU

Glucose if hypoglycemia occurs

Consider hemodialysis

**CONTRAINDICATED**

Induced emesis

CHAPTER 20

# Disinfectants

A wide variety of disinfectant agents are used to destroy microorganisms, and they differ greatly in their toxic effects. However, most can conveniently be grouped into a few categories, some of which are represented in other classes of pesticides. Many of these materials are not registered as pesticides but are registered for medical or medicinal use. This chapter reviews a few of the more common or more toxic disinfectants.

## ALCOHOLS

Alcohols have a long history of use as disinfectants. Often disinfectants are mixtures, usually of ethanol and isopropyl alcohol (isopropanol). The alcohol most commonly used in households as a disinfectant is isopropyl alcohol, commonly marketed as a 70% solution. It is a clear, colorless liquid with an odor similar to ethanol.

### Toxicology

**Isopropyl alcohol** is well and rapidly absorbed from the gastrointestinal tract. It is also well absorbed by skin and by inhalation. It is considered to be more toxic to the central nervous system than **ethanol**, with similar effects. Both ingestion and inhalation at high concentrations can result in the rapid onset of CNS depression with subsequent coma and death. Apnea commonly accompanies this CNS depression.[1,2] Similar neurological toxicity has been reported with excessive topical exposure to the umbilicus of a neonate.[3] Irritation of the gastrointestinal tract results in gastritis and severe vomiting. Isopropyl alcohol may also produce mild hepatic injury following acute exposure. Acute tubular necrosis has been reported with this agent,[1] but the renal toxicity is not as great as with methanol poisonings. Ketosis without metabolic acidosis can occur, but prominent hypoglycemia is common.[2,3] This ketosis is the result of direct metabolism of this compound to acetone.[1,3] Monitoring of isopropyl levels is useful, when available. In addition, blood levels of acetone and glucose should be determined to aid in management.

### Confirmation of Poisoning

Isopropyl alcohol can be measured in the blood and urine. Serum acetone can also be measured. Blood isopropyl alcohol levels of 128-200 mg/dL have been associated with death.

### Treatment of Isopropyl Alcohol Toxicosis

1. Do not induce emesis, since the onset of coma is often rapid with this poisoning. Spontaneous vomiting, however, often occurs.

2. Provide supportive care for hypotension and respiratory depression. This is critical to survival and should be administered whenever possible in an intensive care setting.

3. If hypoglycemia occurs, administer glucose.

4. Consider hemodialysis, which has been reported to be beneficial in patients with severe poisoning who are unresponsive to standard supportive therapy.[1,4]

## ALDEHYDES

The two aldehydes most commonly used as disinfectants are **formaldehyde** and **glutaraldehyde**. Formaldehyde is discussed in **Chapter 17, *Fumigants***. Glutaraldehyde is very similar to formaldehyde in its toxicity and treatment, although it is slightly less toxic. Glutaraldehyde is commonly prepared as an aqueous solution at a 2% concentration and is slightly alkaline in this solution. It has been reported to cause respiratory irritation, resulting in rhinitis[5,6] and occupational asthma.[5,7,8] It has also resulted rarely in palpitations and tachycardia in human subjects. At high dosage, given orally, it results in gastrointestinal irritation with diarrhea, which may be hemorrhagic.[9,10,11] Because of the irritant effects of glutaraldehyde, Occupational Safety and Health Administration (OSHA) standards may apply for wearing personal protective equipment to protect the skin (29 CFR 1910.132) and eyes (29 CFR 1910.133). OSHA standards may also require the use of appropriate respirators by employees who may be exposed to glutaraldehyde during routine or emergency work procedures (29 CRF 1910.134).[12]

*Aldehydes*
### HIGHLIGHTS

Formaldehyde also discussed in Ch. 17

Glutaraldehyde similar but less toxic

### SIGNS & SYMPTOMS

Respiratory irritation

GI irritation, diarrhea, possible hemorrhage if oral

### TREATMENT

Move to fresh air, administer oxygen as needed

Decontaminate skin if irritated

### Treatment of Aldehyde Toxicosis

1. If patient has been in an area with a strong odor of glutaraldehyde due to vaporization, move to fresh air and administer oxygen as needed.

2. If skin irritation is noted, decontaminate. Systemic toxicity from skin exposure is unlikely.

## CATIONIC DETERGENTS

Several cationic detergents are used as disinfectants. All share the capacity, in sufficient concentration, to cause rather severe caustic burns. Concentrations greater than approximately 7.5% appear necessary to produce significant caustic injuries. However, experience with human exposures to these compounds is very limited. The three agents most commonly used as detergent disinfectants are benzalkonium chloride, cetrimide and cetylpyridium chloride.

No cetrimide preparations are available in the United States; several are available in European Union countries. Concentrated solutions are usually only available in industrial settings, such as production of consumer products, or for use in hospitals for disinfectant purposes. Therefore, acute poisonings are uncommon.

### Toxicology

In low concentration solutions, **cationic detergents** have been reported to cause eye discomfort, as well as skin rashes and irritation. A severe contact dermatitis has been reported with a bath oil containing **benzalkonium chloride** and **triclosan**.[13]

*Cationic Detergent*
## HIGHLIGHTS

Caustic agents capable of causing burns

Most common: benzalkonium chloride, cetrimide, cetylpyridium chloride (no cetrimide in U.S.)

Acute poisonings are uncommon

## SIGNS & SYMPTOMS

Eye, skin, GI irritant

Severe exposures: CNS, liver, pulmonary impacts

## TREATMENT

Wash eyes, examine/treat corneas for burns

Endoscopy within 24 hours if indicated

Treat CNS, pulmonary, other systemic effects

## CONTRAINDICATED

GI decontamination

In stronger concentrations, they can cause severe corneal and skin burns. Likewise, strong concentrations will result in caustic burns to lips, oral mucosa, esophagus and stomach.[14,15] Vomiting, diarrhea and abdominal pain have been reported.[16] Necrosis of the gut, with peritonitis, has also been reported.[17] In severe exposures, there are also reports of CNS depression, liver injury and pulmonary edema.[14,16]

### Treatment of Cationic Detergent Toxicosis

1. If a high concentration solution is in contact with the eyes, wash the eyes profusely and then carefully examine the corneas. If burns have occurred, obtain ophthalmologic care.

2. Do not use any method of gastrointestinal decontamination, including gastric emptying. They are contraindicated in these poisonings. Some experts recommend cautious dilution with small amounts of milk or water.[14,18] Acidic solutions, such as juices, should never be offered for dilution.

3. Conduct an endoscopy if a highly concentrated solution was ingested or oral burns are noted. The patient needs urgent endoscopy for grading of the caustic injury. The endoscopy should be performed within 24 hours to minimize the risk of perforation.[17] A competent surgeon or gastroenterologist should provide subsequent care.

4. Treat CNS, pulmonary and other systemic effects symptomatically, consistent with sound medical practice.

Although corticosteroids are commonly used to treat these burns, their use remains controversial. Use of other agents, such as $H_2$ antagonists and sulcralfate, has been reported, but also remains controversial at this time.

## CHLORHEXIDINE

Chlorhexidine is a cationic biguanide, available in concentrations up to 4% as a topical agent used as a skin cleanser and mouthwash. Skin preparations of 0.5%-4% are marketed under the trade names Hibiclens and Hibistat. It is also marketed as a mouthwash in a 0.12% solution under the trade name Peridex. There is very little human experience with poisonings, as these concentrations do not appear to be significantly toxic.

### Toxicology

**Chlorhexidine** is poorly absorbed from skin or the gastrointestinal tract. Therefore, most effects noted have been primarily local. Low concentration solution ingested or applied to the skin can cause mild local irritation. Contact dermatitis, urticaria and anaphylaxis have followed repeated skin exposures to this agent.[19,20,21] Corneal injuries have been described in several cases after inadvertent exposure of the eyes to the 4% concentration. These injuries have resulted in permanent corneal scarring.[22] Esophageal burns have been reported in a single case after ingestion of a large quantity of a 20% solution of this agent.[23] Ulcerative colitis has been described after an enema of the 4% solution mixed with tap water (10 mL in 2 liters water).[24] Liver toxicity can occur with large exposures.[23]

## Treatment of Chlorhexidine Toxicosis

1. If a highly concentrated solution is ingested, manage as a caustic ingestion as described in the preceding *Treatment of Cationic Detergent Toxicosis* subsection, without gastrointestinal decontamination.

2. Perform liver injury panel with large ingestions.

3. If a high concentration solution is in contact with the eyes, wash eyes profusely and examine the corneas carefully. If burns have occurred, obtain ophthalmologic care.

## HYPOCHLORITES

Hypochlorites are implicated in a large proportion of the disinfectant exposures reported to poison control centers in the United States, with more than 30,000 reports in 2009.[25] Most are solutions of **sodium** or **calcium hypochlorite**. **Chloramine**, a disinfectant used in many municipal water supplies, is an infrequent cause of acute poisonings. Sodium and calcium hypochlorite solutions are of relatively low toxicity. They are mildly corrosive to eyes,[26] and mucous membrane burns have been reported.[27] Despite the large number of reports to poison control, significant poisonings are very infrequent with these agents in solution.[25,28]

When hypochlorite solutions are mixed with acids or ammonia solutions, chlorine or chloramine gas is produced, resulting in an irritant with pulmonary toxicity. Many brief exposures have led to transient symptoms requiring limited emergency department management.[29] Prolonged exposure or exposure to high concentrations carries the potential of severe toxic pneumonitis.[30] Great efforts should be made to discourage mixing of these materials with acid or ammonia.

## Treatment of Hypochlorite Toxicosis

1. After oral exposures, do not use gastric emptying. If a granular material is ingested and the patient has symptomatic mucosal burns, refer patient to a surgeon or gastroenterologist for consideration of endoscopy and management.

2. If vomiting has not occurred, give patient water or milk for dilution, not to exceed approximately 15 mL/kg in a child or 120-240 mL in an adult. Administration of acids is contraindicated, because of the risk or increasing generation of chlorine gas.

3. If a high concentration solution is in contact with the eyes, wash eyes profusely and examine corneas carefully. If burns have occurred, obtain ophthalmologic care.

4. Manage skin exposure with copious water dilutions.

5. If exposure to vapors or chlorine or chloramine gas has occurred, move patient immediately to fresh air. If symptoms occur or persist, oxygenation should be assessed and oxygen administered as needed. If persistent symptoms occur, obtain a chest film and consider hospitalization. Intensive care may be appropriate in severe inhalations.

***Chlorhexidine***

### HIGHLIGHTS

Used as skin cleanser, mouthwash

Poorly skin, gut absorption

### SIGNS & SYMPTOMS

Mild skin irritant, worse if repeat exposures

Corneal, esophageal injuries possible

### TREATMENT

For highly concentrated/ large doses

- Ingested: manage as caustic ingestion, perform liver panel
- Eye contact: wash, examine, treat corneas

***Hypochlorite***

### HIGHLIGHTS

Chloramine, sodium/calcium hypochlorite

Many exposures reported; significant poisonings few

### SIGNS & SYMPTOMS

Pulmonary irritation, toxicity when mixed with acids or ammonia solutions

GI, eye, skin, pulmonary impacts

### TREATMENT

If ingestions result in mucosal burns, refer to surgeon/ gastroenterologist

Decontaminate eyes, skin

Examine/treat corneas

If vapor exposure, move to fresh air and consider oxygen administration, chest film, hospitalization

### CONTRAINDICATED

Administration of acids

Gastric emptying

*Iodine*
### HIGHLIGHTS

Most common: 7.5%-10% povidone-iodine solution

Betadine is an example

At standard dilutions, poorly absorbed from GI, skin

Symptomatic poisonings possible on burned skin, wounds

### SIGNS & SYMPTOMS

Initial: headache, dizziness, delirium, hallucinations, seizures

Severe: hypotension, arrhythmias, cyanosis, metabolic acidosis, shock, renal failure

### TREATMENT

Decontaminate skin

Osmotic agents or diuretics if indicated

Treat seizures

Monitor thyroid

## IODINE

The most common iodine-containing disinfectant is povidone-iodine. A trade name often associated with this agent is Betadine (7.5%-10% solution). Povidine-iodine is described as an iodophor, which is a complex of iodine and polyvinylpyrrolidone, a solubilizing agent. It is intended to liberate free iodine in solution for its effect. Although reported concentrations of iodine in these solutions is only 80-120 µg/dL, the total available iodine is approximately 10% of the povidone-iodine. Therefore, a 10% solution will have in the range of 1% total available iodine.

### Toxicology

This compound is very poorly absorbed from the gastrointestinal tract, because of the rapid conversion of free iodine to iodide in the stomach. Though highly concentrated iodine solutions or iodine salts are corrosive to the gastrointestinal tract,[31] solutions of **povidone-iodine** have little caustic potential. It is likewise poorly absorbed from intact skin. All symptomatic poisonings reported have occurred either after repeated exposure to burned skin or following irrigation of wounds, joints or serosal surfaces, such as the mediastinum.[32,33,34,35] The one exception was an infant who received an enema of povidone-iodine in a polyethylene glycol solution, followed by whole bowel irrigation with polyethylene glycol mixed with povidone-iodine. This child died with severe hyperglycemia and very high iodine levels.[31]

In povidone-iodine exposures by these routes, the primary symptoms initially appear to be neurological, with headache, dizziness, delirium, hallucinations and seizures.[35] Hypotension, arrhythmias, cyanosis, metabolic acidosis, shock and acute renal failure occur in severe cases.[32,33,34] Hepatic injury, manifested by elevated serum transaminase levels, has also been reported with very high level exposures.[34] Hyperkalemia has occurred, and the serum chloride may be falsely elevated due to the presence of a second halide.[33]

### Treatment of Iodine Toxicosis

1. Remove skin contamination by vigorous washing with soap and water.
2. Use osmotic agents or diuretics in symptomatic poisonings, since iodine clearance is apparently enhanced by procedures that enhance chloride excretion.
3. Treat seizures with anticonvulsants, as outlined in **Chapter 3, *General Principles*.**
4. Monitor thyroid function following recovery to confirm euthyroid state.

## MERCURIALS

A wide variety of organic mercurials have been used as disinfectants and as preservatives. These included **phenylmercuric acetate, phenylmercuric nitrate, nitromersol, thimerosol, mercurochrome** and **mercurobutol**. None is currently registered with the U.S. Environmental Protection Agency. The toxicity and treatment of exposure to these compounds is described in detail in **Chapter 16, *Fungicides*** under the subsection *Organomercury Compounds*.

## PHENOLS

Several phenols are used as disinfectants, including cresol, phenol, thymol, hexachlorophene, o-phenylphenol, 4-tert-amylphenol, 2-benzyl-4-chlorophenol and triclosan. Cresol and thymol are alkyl derivatives of phenol, while hexachlorophene and triclosan are chlorinated phenols. Common trade names for commercial products are provided in the margin. One survey found that **triclosan** or a similar agent, **triclocarban**, was found in 45% of liquid and bar soaps available in consumer outlets.[36] However, no episodes of acute toxicity from triclosan have been reported, so the concerns with this agent relate to chronic effects, the development of triclosan resistance in microbial organisms, and reports of contact dermatitis caused by exposure to triclosan.[13,37,38] Cresols and hexachlorophene will be discussed individually; these compounds are familiar and some human data are available.

### Toxicology of Cresols

**Cresols**, in common with phenol and other phenolic compounds, are highly corrosive. Ingestion of concentrated forms causes severe corrosive injury to the mouth and upper gastrointestinal tract. Likewise, severe eye and skin caustic injuries can occur with cresol exposure.[39] Symptoms usually include nausea, vomiting and diarrhea. Hypotension, myocardial failure, pulmonary edema, neurological changes may also occur.[40] Liver and renal toxicity, methemoglobinemia and hemolysis have all been reported.[40,41] After long-term, repeated exposure, contact dermatitis may complicate these exposures. These compounds are well absorbed from the gastrointestinal tract and are also significantly absorbed from the skin and by inhalation.

### Treatment of Cresol Toxicosis

1. Do not attempt gastrointestinal decontamination because of the corrosive nature of these compounds. Consider dilution with milk or water if vomiting has not occurred.

2. If a corrosive injury has occurred with burns to the mouth, or if there is a clear history of gastrointestinal exposure, consider endoscopy and consult a gastroenterologist or surgeon for diagnosis and management.

3. If a high concentration solution is in contact with the eyes, wash eyes with profuse amounts of water and follow with a careful exam of the corneas. If burns have occurred, provide ophthalmologic care. Given the corrosive nature of the substance, referral to an ophthalmologist should be considered.

4. Provide respiratory and circulatory support in accordance with sound medical management. If severe systemic symptoms persist, the patient should be treated in an intensive care unit, if possible.

### Toxicology of Hexachlorophene

**Hexachlorophene** is well absorbed via the oral and dermal routes. Dermal exposures have led to severe toxicity and death in neonates, due to application to damaged skin or repeated or high-concentration skin exposures.[42] It should never be used as a disinfectant on open wounds or abraded or inflamed skin surfaces. It is not significantly caustic, however, and exposure does not result in the severe caustic injuries seen with other phenolic chemicals.

*Phenols*

**COMMERCIAL PRODUCTS**

Triclosan: many consumer soap products

Mixed cresols in soap: Lysol

Hexachlorophene: Phisohex, Bilevon, Dermaadex, Exofene, Gamophen, Texosan, Surgi-Cen, Surofene, various soap bars and cosmetics

**HIGHLIGHTS**

***Triclosan***
- Very common, no acute toxicity reports
- Contact dermatitis

***Cresols***
- Highly corrosive
- Can cause severe mouth, GI, eye, skin injury
- Well absorbed from GI, skin, inhalation

***Hexachlorophene***
- Well absorbed via skin, gut
- Not as caustic as other phenolic compounds
- Potent neurotoxicant

*continued next page*

*Phenols, cont.*

## SIGNS & SYMPTOMS

***Cresols***

Nausea, vomiting, diarrhea

Caustic skin, eye injuries

***Hexachlorophene***

Complex CNS effects

Lethargy, muscle weakness/fasciculation, irritability, cerebral edema, paralysis

Vomiting, diarrhea, anorexia

Skin rash

## TREATMENT

***Cresols***

Consider GI dilution with water/milk

Consider endoscopy with gastroenterologist consult

Decontaminate eyes, examine/treat corneas

Respiratory, circulatory support

***Hexachlorophene***

Consider activated charcoal

Decontaminate skin with soap and water

Control seizures

Support cardiovascular, respiratory systems

## CONTRAINDICATED

***Cresols***

GI decontamination

Hexachlorophene is a potent neurotoxicant. It causes brain edema and spongy degeneration of white matter.[43] This neurotoxicity can be seen after acute or chronic exposures, either by skin absorption or ingestion. The nervous system symptoms are complex. Lethargy is an early manifestation, followed by muscular weakness, muscular fasciculation, irritability, cerebral edema and paralysis, leading to coma and death. Seizures commonly occur in more severe cases.[42,44] Blindness and optic atrophy have also been seen following exposure to hexachlorophene.[45]

In addition to the neurological effects, common early symptoms of poisoning are vomiting, diarrhea and anorexia.[44] These findings have been accompanied in animals by significant hepatotoxicity.[46] With skin exposure, an erythematous, desquamative rash is often noted at the site of exposure.[44] With chronic exposure, contact dermatitis may be noted. In severe poisonings, cardiovascular symptoms, including hypotension and bradycardia have been noted.[47] In a single case, repeated exposure to this compound led to asthma in a pediatric nurse.[48]

## Treatment of Hexachlorophene Toxicosis

1. Although this compound is quite toxic systemically and enhanced clearance methods would appear beneficial, there is no evidence to support efficacy of hemodialysis, peritoneal dialysis, hemoperfusion or exchange transfusion.[47]

2. Consider using activated charcoal. Since hexachlorophene is thought to have an enterohepatic recirculation, it is possible that repeated dosing of activated charcoal, as outlined in the **Chapter 3,** ***General Principles***, will enhance clearance of this compound although hexachlorophene does not bind well to charcoal and there are no clinical trials of this therapy for this agent.

3. If exposure has occurred through the skin, wash skin aggressively with soap or detergent and water to remove any residues still on the skin. Since hexachlorophene is not soluble in water, washing with water alone will not provide significant benefit.

4. Perform neurological support and seizure control, as these are critical to survival. When possible, perform in an intensive care setting. Seizure control should be in accordance with recommendations in **Chapter 3.**

5. Provide cardiovascular and respiratory support, which are also very important to success in treating severe poisonings with this agent. This care should be provided in an intensive care unit in accordance with accepted medical practice.

## PINE OIL

### Toxicology

Exposures to **pine oil** detergent and disinfectant solutions are commonly reported to poison control centers in the United States.[49] Pine oil is an agent commonly contained in a variety of household and commercial cleaners and disinfectants. It is a mixture of monoterpenes derived from the distillation of wood from various pine species, with approximately 57% being alpha-pinene.[50] Its most common side effects in smaller dosage are irritation of mucous membranes, gastrointestinal irritation, mild respiratory and CNS depression and renal toxicity. Larger ingestions can result in severe respiratory distress, cardiovascular collapse, and severe CNS effects. Renal failure and myoglobinuria have also been reported in severe poisonings.[51] Since even small ingestions can result in severe aspiration pneumonia, all ingestions should be considered potentially hazardous.

While many of the reported effects of poisoning with this agent are related to direct irritant effect on mucous membranes, gastrointestinal tract and lungs (by aspiration), some reports suggest significant absorption from oral and rectal exposures. Other reports suggest a lesser rate of absorption.[50] While alpha terpineol can be measured in blood, there are no data relating terpineol levels to degree of toxicity; this measure, therefore, is not considered useful in guiding diagnosis and management.

### Treatment of Pine Oil Toxicosis

1. Do not induce emesis. Since there is a high risk of aspiration pneumonia, induced emesis is usually considered contraindicated in these poisonings. However, spontaneous emesis may occur because of direct irritation of the gastric mucosa.

2. If a high concentration solution is in contact with the eyes, flush eyes profusely and carefully examine corneas. If burns have occurred, obtain ophthalmologic care.

3. Observe the patient for at least 6 hours with any significant ingestion in order to observe the onset of any symptoms, particularly pulmonary symptoms.

4. Order chest films and measure oxygenation if any pulmonary symptoms are observed. If pulmonary symptoms occur, hospitalization is appropriate. With severe pulmonary symptoms transfer to an intensive care unit is usually appropriate. With severe aspiration, manage as with any severe aspiration pneumonia, in accordance with accepted medical practice.

5. Treat other severe systemic effects in accordance with accepted medical practice.

There is no evidence that activated charcoal is helpful in these poisonings. Likewise, although a variety of enhanced elimination methods have been proposed and tried, there is no evidence to support their efficacy.

*Pine Oil*

#### HIGHLIGHTS

Common household/ commercial cleaning ingredient

Monoterpene wood derivative

Primarily inhalation and ingestion route to poisoning

All ingestions have potential for severe aspiration pneumonia

Oral, rectal exposure routes also possible

#### SIGNS & SYMPTOMS

Mucous membrane, GI irritation

Mild respiratory, CNS depression

Severe respiratory, cardiovascular, CNS impacts from larger ingestions

#### TREATMENT

Flush eyes and examine/ treat corneas

Observe for 6 hours post-exposure, esp. for pulmonary symptoms

Pulmonary support

Order chest films and measure oxygenation

Hospitalize, consider ICU

In severe cases, manage as aspiration pneumonia

#### CONTRAINDICATED

Induced emesis

## References

1. Lacouture PG, Wason S, Abrams A, Lovejoy FH, Jr. Acute isopropyl alcohol intoxication. Diagnosis and management. *Am J Med.* Oct 1983;75(4):680-686.

2. Rich J, Scheife RT, Katz N, Caplan LR. Isopropyl alcohol intoxication. *Arch Neurol.* Mar 1990;47(3):322-324.

3. Vivier PM, Lewander WJ, Martin HF, Linakis JG. Isopropyl alcohol intoxication in a neonate through chronic dermal exposure: a complication of a culturally-based umbilical care practice. *Pediatr Emerg Care.* Apr 1994;10(2):91-93.

4. Manring E, Meggs W, Pape G, Ford M. Toxicity of an intravenous infusion of isopropyl alcohol. *J Toxicol Clin Toxicol.* 1997;35:503.

5. Corrado OJ, Osman J, Davies RJ. Asthma and rhinitis after exposure to glutaraldehyde in endoscopy units. *Hum Toxicol.* Sep 1986;5(5):325-328.

6. Norback D. Skin and respiratory symptoms from exposure to alkaline glutaraldehyde in medical services. *Scand J Work Environ Health.* Dec 1988;14(6):366-371.

7. Chan-Yeung M, McMurren T, Catonio-Begley F, Lam S. Occupational asthma in a technologist exposed to glutaraldehyde. *J Allergy Clin Immunol.* May 1993;91(5):974-978.

8. Stenton SC, Beach JR, Dennis JH, Keaney NP, Hendrick DJ. Glutaraldehyde, asthma and work--a cautionary tale. *Occup Med (Lond).* May 1994;44(2):95-98.

9. Symptoms of irritation associated with exposure to glutaraldehyde--Colorado. *MMWR Morb Mortal Wkly Rep.* Apr 3 1987;36(12):190-191.

10. Fukunaga K, Khatibi A. Glutaraldehyde colitis: a complication of screening flexible sigmoidoscopy in the primary care setting. *Ann Intern Med.* Aug 15 2000;133(4):315.

11. Stonehill AA, Krop S, Borick PM. Buffered glutaraldehyde -- a new chemical sterilization solution. *Am J Hosp Pharm.* 1963;20:458-465.

12. US Department of Labor OSaHA. *Best practices for the safe use of glutaraldehyde in health care.* 2006.

13. Storer E, Koh KJ, Warren L. Severe contact dermatitis as a result of an antiseptic bath oil. *Australas J Dermatol.* Feb 2004;45(1):73-75.

14. Mucklow ES. Accidental feeding of a dilute antiseptic solution (chlorhexidine 0.05% with cetrimide 1%) to five babies. *Hum Toxicol.* Nov 1988;7(6):567-569.

15. Wilson JT, Burr IM. Benzalkonium chloride poisoning in infant twins. *Am J Dis Child.* Oct 1975;129(10):1208-1209.

16. Chan TY. Poisoning due to Savlon (cetrimide) liquid. *Hum Exp Toxicol.* Oct 1994;13(10):681-682.

17. Zargar SA, Kochhar R, Mehta S, Mehta SK. The role of fiberoptic endoscopy in the management of corrosive ingestion and modified endoscopic classification of burns. *Gastrointest Endosc.* Mar-Apr 1991;37(2):165-169.

18. Consensus: POISONDEX Editorial Board consensus opinion poll, irritants/caustics specialty board. 1988.

19. Perrenoud D, Bircher A, Hunziker T, et al. Frequency of sensitization to 13 common preservatives in Switzerland. Swiss Contact Dermatitis Research Group. *Contact Dermatitis.* May 1994;30(5):276-279.

20. Okano M, Nomura M, Hata S, et al. Anaphylactic symptoms due to chlorhexidine gluconate. *Arch Dermatol.* Jan 1989;125(1):50-52.

21. Wong WK, Goh CL, Chan KW. Contact urticaria from chlorhexidine. *Contact Dermatitis.* Jan 1990;22(1):52.

22. Tabor E, Bostwick DC, Evans CC. Corneal damage due to eye contact with chlorhexidine gluconate. *JAMA.* Jan 27 1989;261(4):557-558.

23. Massano G, Ciocatto E, Rosabianca C, Vercelli D, Actis GC, Verme G. Striking aminotransferase rise after chlorhexidine self-poisoning. *Lancet.* Jan 30 1982;1(8266):289.

24. Hardin RD, Tedesco FJ. Colitis after Hibiclens enema. *J Clin Gastroenterol.* Oct 1986;8(5):572-575.

25. Bronstein AC, Spyker DA, Cantilena LR, Jr., Green JL, Rumack BH, Giffin SL. 2009 Annual Report of the American Association of Poison Control Centers' National Poison Data System (NPDS): 27th Annual Report. *Clin Toxicol (Phila).* Dec 2010;48(10):979-1178.

26. Ingram TA, 3rd. Response of the human eye to accidental exposure to sodium hypochlorite. *J Endod.* May 1990;16(5):235-238.

27. French RJ, Tabb HG, Rutledge LJ. Esophageal stenosis produced by ingestion of bleach: report of two cases. *South Med J.* Oct 1970;63(10):1140-1144.

28. Landau GD, Saunders WH. The Effect of Chlorine Bleach on the Esophagus. *Arch Otolaryngol.* Aug 1964;80:174-176.

29. Mrvos R, Dean BS, Krenzelok EP. Home exposures to chlorine/chloramine gas: review of 216 cases. *South Med J.* Jun 1993;86(6):654-657.

30. Reisz GR, Gammon RS. Toxic pneumonitis from mixing household cleaners. *Chest.* Jan 1986;89(1):49-52.

31. Kurt TL, Morgan ML, Hnilica V, Bost R, Petty CS. Fatal iatrogenic iodine toxicity in a nine-week old infant. *J Toxicol Clin Toxicol.* 1996;34(2):231-234.

32. Campistol JM, Abad C, Nogue S, Bertran A. Acute renal failure in a patient treated by continuous povidone-iodine mediastinal irrigation. *J Cardiovasc Surg (Torino).* Jul-Aug 1988;29(4):410-412.

33. Means LJ, Rescorla FJ, Grosfeld JL. Iodine toxicity: an unusual cause of cardiovascular collapse during anesthesia in an infant with Hirschsprung's disease. *J Pediatr Surg.* Dec 1990;25(12):1278-1279.

34. Pietsch J, Meakins JL. Complications of povidone-iodine absorption in topically treated burn patients. *Lancet.* Feb 7 1976;1(7954):280-282.

35. Ponn RB. Continuous povidone-iodine irrigation. *Ann Thorac Surg.* Feb 1987;43(2):239.

36. Perencevich EN, Wong MT, Harris AD. National and regional assessment of the antibacterial soap market: a step toward determining the impact of prevalent antibacterial soaps. *Am J Infect Control.* Oct 2001;29(5):281-283.

37. Robertshaw H, Leppard B. Contact dermatitis to triclosan in toothpaste. *Contact Dermatitis.* Dec 2007;57(6):383-384.

38. Wong CS, Beck MH. Allergic contact dermatitis from triclosan in antibacterial handwashes. *Contact Dermatitis.* Nov 2001;45(5):307.

39. Pegg SP, Campbell DC. Children's burns due to cresol. *Burns Incl Therm Inj.* Apr 1985;11(4):294-296.

40. Arthurs GJ, Wise CC, Coles GA. Poisoning by cresol. *Anaesthesia.* Jul-Aug 1977;32(7):642-643.

41. Chan TK, Mak LW, Ng RP. Methemoglobinemia, Heinz bodies, and acute massive intravascular hemolysis in lysol poisoning. *Blood.* Dec 1971;38(6):739-744.

42. Mullick FG. Hexachlorophene toxicity. Human experience at the Armed Forces Institute of Pathology. *Pediatrics.* Feb 1973;51(2):395-399.

43. Anderson JM, Cockburn F, Forfar JO, Harkness RA, Kelly RW, Kilshaw B. Neonatal spongioform myelinopathy after restricted application of hexachlorophane skin disinfectant. *J Clin Pathol.* Jan 1981;34(1):25-29.

44. Martin-Bouyer G, Lebreton R, Toga M, Stolley PD, Lockhart J. Outbreak of accidental hexachlorophene poisoning in France. *Lancet.* Jan 9 1982;1(8263):91-95.

45. Slamovits TL, Burde RM, Klingele TG. Bilateral optic atrophy caused by chronic oral ingestion and topical application of hexachlorophene. *Am J Ophthalmol.* May 1980;89(5):676-679.

46. Prasad GV, Rajendra W, Indira K. Brain ammonia metabolism in hexachlorophene-induced encephalopathy. *Bull Environ Contam Toxicol.* Apr 1987;38(4):561-564.

47. Boehm RM, Jr., Czajka PA. Hexachlorophene poisoning and the ineffectiveness of peritoneal dialysis. *Clin Toxicol.* Mar 1979;14(3):257-262.

48. Nagy L, Orosz M. Occupational asthma due to hexachlorophene. *Thorax.* Aug 1984;39(8):630-631.

49. Bronstein AC, Spyker DA, Cantilena LR, Jr., Green JL, Rumack BH, Giffin SL. 2008 Annual Report of the American Association of Poison Control Centers' National Poison Data System (NPDS): 26th Annual Report. *Clin Toxicol (Phila).* Dec 2009;47(10):911-1084.

50. Koppel C, Tenczer J, Tonnesmann U, Schirop T, Ibe K. Acute poisoning with pine oil - metabolism of monoterpenes. *Arch Toxicol.* Nov 1981;49(1):73-78.

51. Litovitz TL, Schmitz BF, Matyunas N, Martin TG. 1987 annual report of the American Association of Poison Control Centers National Data Collection System. *Am J Emerg Med.* Sep 1988;6(5):479-515.

*Section V*

# CHRONIC EFFECTS

CHAPTER 21

# Chronic Effects

This chapter is a departure from the format and content of the other chapters in this manual. Rather than discussing signs and symptoms of acute poisoning, this chapter addresses chronic (also known as persistent) effects that have been associated with pesticide exposure. The information in this chapter is designed to provide the practitioner with evidence for the better-established inferences for chronic effects of pesticides. This will offer some facility in the basic knowledge of chronic effects, allowing an approach to such effects, aiding the practitioner in answering questions from patients and the public, and providing a basis for further inquiry into areas of interest. Knowledge of chronic effects of pesticide exposure is evolving rapidly and providers will need to be alert to new findings as they become available. The chapter is not intended to be a comprehensive review; such reviews are referenced when they are available.

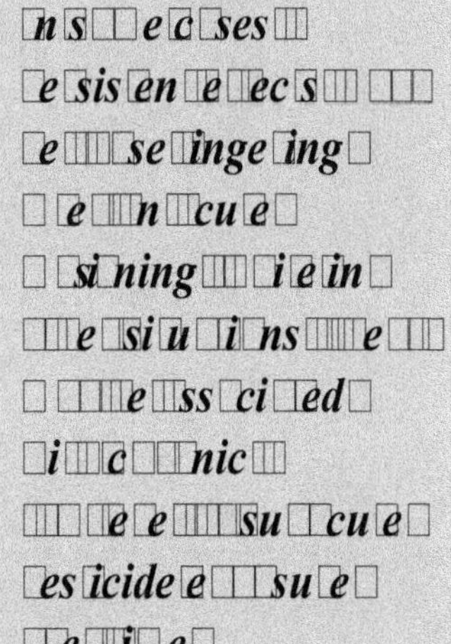

In some cases, persistent effects may be those lingering after an acute poisoning, while in other situations, persistent symptoms or demonstrable physiological alteration may be associated with chronic, low-level or subacute pesticide exposure over time. Evidence linking pesticide exposure to chronic health conditions relies on observational epidemiological studies and/or standard chronic toxicity testing using animal models. For obvious ethical reasons, experimental studies with purposeful dosing of pesticides are not conducted in humans. Therefore, while cause and effect is not proven with any one epidemiology study, several well designed studies in different populations, alone or combined with inferential evidence from animal exposures, can strongly support the likelihood that a given association is in fact causal in nature.

This chapter covers chronic health conditions that may have an association with pesticide exposure. Neurological effects, particularly neurodevelopmental abnormalities in children, have been implicated with exposure to insecticides that have toxicological activity on the central nervous system. Numerous studies have examined the effects of pesticides on the development of cancer in children and adults. Several classes of pesticides have properties that mimic endocrine hormones and may affect multiple organ systems and functions including reproductive health and cancer risk. Recently, data have emerged indicating a potential relationship between certain pesticides and asthma. Chronic, low-level arsenic exposure is associated with multiple chronic disease endpoints including skin disease, neuropathy and cancer.

## Evaluating Epidemiological Findings

The conditions that are traditionally used to consider an established statistical association as causal in nature were clearly articulated by Sir Bradford Hill in relation to epidemiological and other research on smoking and lung cancer. While only rarely will all conditions be met, the more that are met, the more confident one can be in the truth of a causal connection. The most important condition that must be met is a temporal relationship (*i.e.,* exposure precedes outcome). Other supportive conditions include the strength of effect (described as the size of the effect – *e.g.*, high relative risk or odds ratio), dose-response relationship (more exposure = more effect), consistency (*i.e.*, multiple studies with similar outcomes), biological plausibility (*i.e.*, the outcome can be explained biologically), experimental support (usually done on animal models), and analogy (similar exposures produce similar outcomes).[1]

*Supportive conditions:*

- *temporal relationship with exposure preceding outcome*
- *strength of effects*
- *dose-response relationship*
- *consistency*
- *biological plausibility*
- *experimental support*
- *analogy*

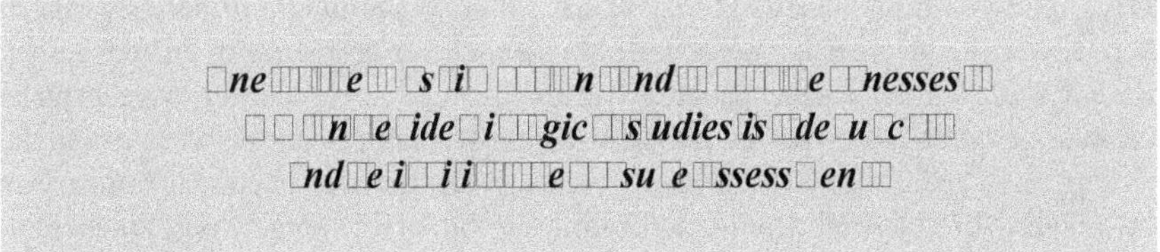

Disease processes with low incidence represent a particular challenge to evaluate using epidemiological methods. Although adult cancers are relatively common, cancer in childhood is rare. Consequently, to adequately study a disease with a low incidence, case-control studies rather than cohort designs provide adequate power, but they are subject to greater recall and classification bias. One of the most important and major weaknesses of many epidemiological studies is adequacy and reliability of exposure assessment.[2,3] Studies that incorporate pesticide-specific exposure assessment, markers of biological mechanisms, objective assessment of outcomes and consideration of the influence of timing of exposure across the lifespan are needed to better define the relationships.

## Differences between Children and Adults

When evaluating the effect of chronic, low-level exposures in humans, important differences in exposure sources and patterns between children and adults stemming from differences in physiology and behavior must be considered. From a developmental standpoint, children in the first few years of life spend a considerable amount of time on the floor, where residues following indoor application of pesticides (or outdoor application that may be tracked inside) accumulate.[4,5,6] Children have more frequent hand-to-mouth activity, which can be an added source of oral exposure.[7,8] Children ingest a larger amount of food and water per body weight than adults. For example, in the first year of life, infants may take in 100-150 cc/kg/day of liquids. For a 70-kg adult to ingest an equivalent amount of fluid, he/she would need to drink six 2-liter bottles of fluids a day. Dietary composition for children differs from that for adults. For example, U.S. children are much more likely to routinely ingest a variety of apple-based products on a daily basis, thus ingesting a greater amount of pesticide residue from apples than would the typical adult.[9]

## NEUROLOGICAL AND NEURODEVELOPMENTAL EFFECTS

Many registered pesticides are specifically toxic to the central nervous systems of target pests including insects and mammals such as rodents. Neurotoxicity to animals has been a useful attribute for the development of pesticides for use as insecticides and rodenticides. It is not surprising that these agents also have neurotoxic effects on large mammals including humans. However, many other pesticides, including herbicides, fumigants and fungicides, have human neurotoxicant properties. This section summarizes those effects that may persist following acute exposure, as well as describes subacute and chronic effects following long-term exposure.

### Chronic Effects Following Acute Exposure

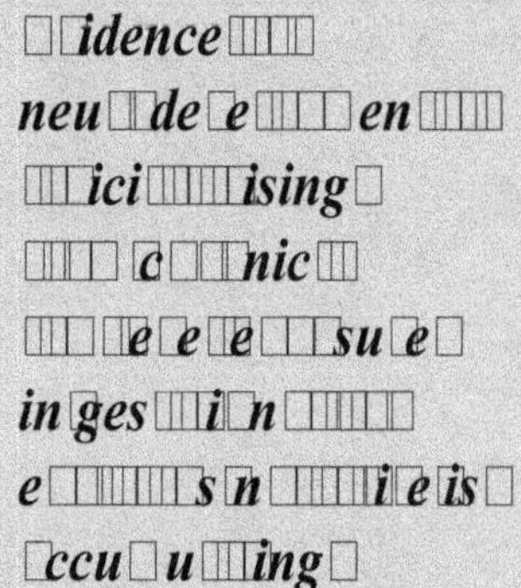

Acute pesticide intoxications may leave recovered individuals with residual neurologic impairment, particularly if they result in multiorgan failure or nervous system hypoxia. Such outcomes are noted for individual agents elsewhere in this document. Several studies document that patients with a history of a single acute organophosphate or other pesticide poisoning are at risk of neuropsychiatric sequelae when examined as long as 10 years after the episode. These show significantly impaired performance on a battery of validated neuro-behavioral tests and, and in some cases, compound-specific peripheral neuropathy. The findings are subtle and, in some cases, identified only through formal neuropsychologic testing rather than as frank abnormalities on clinical neurologic exam.[10,11,12]

Certain organophosphates have caused damage to the afferent fibers of peripheral and central nerves. The mechanism of this type of toxicity is the inhibition of "neuropathy target esterase" (NTE). This delayed syndrome has been termed organophosphate-induced delayed neuropathy (OPIDN) and is manifested chiefly by weakness or paralysis and paresthesias of the extremities. In addition to acute poisoning episodes and OPIDN, an intermediate syndrome has been described. This syndrome occurs after resolution of the acute cholinergic crisis, generally 24-96 hours after the acute exposure, with signs and symptoms lasting from several days to several weeks.[13] It is characterized by acute respiratory paresis and muscular weakness, primarily in the facial, neck and proximal limb muscles. In addition, it is often accompanied by cranial nerve palsies and depressed tendon reflexes. Both this syndrome and OPIDN lack muscarinic symptoms. The intermediate syndrome appears to result from a combined pre- and post-synaptic dysfunction of neuromuscular transmission. These syndromes are described in greater detail in **Chapter 5,** [illegible].

### Effects Following Low-Level, Chronic Exposure

The effects of chronic, low-level exposures to pesticides on the nervous system are less well understood, but consistent evidence of neurodevelopmental toxicity arising from chronic, low-level exposure in gestational or early postnatal life is accumulating. One well established example of such effects is arsenic exposure. Neurologic symptoms are also common with chronic exposure. Peripheral neuropathy, manifested by paresthesia, pain, anesthesia, paresis and ataxia, may be a prominent feature. These effects may begin with the sensory symptoms in the lower extremities and progress to muscular weakness and eventual paralysis and muscle wasting.[14,15,16] Central nervous system effects may also occur, including mood changes such as depression, irritability, anxiety and difficulty concentrating. Additional symptoms include insomnia, headaches and neurobehavioral impairment.[16]

## Low-Level Insecticide Exposure

Research on insecticide toxicity to the developing brain and neurodevelopmental outcomes has been reviewed.[17,18] Most studies focus on exposure to organophosphates and organochlorines. Since these pesticides have historically been and/or currently are in wide usage for household or agricultural pest control, exposures to the child and pregnant mother have been common. Since these exposures have been common and widespread over many years, it is not surprising that they would be studied and that association of effects from these agents would be among the first documented in epidemiological research. Little or no research has been done on the neurodevelopmental effects of other common agents, such as pyrethroids commonly used in households and agriculture or exposures to herbicides and fungicides used extensively in agriculture. One published longitudinal cohort study assessed prenatal exposure to household permethrin and piperonyl butoxide by maternal air monitoring and examination of maternal and cord blood plasma. When assessing neurodevelopment at 36 months, significant adverse impacts were observed for exposure to piperonyl butoxide (PBO), the most common synergist used in household pyrethroid products. No adverse associations were observed with exposure to the active ingredient permethrin. The authors note the more challenging task of measuring permethrin in biological and environmental samples compared to assessment of PBO and the need for confirmatory studies to clarify the roles of pyrethroids and PBO.[19]

The following sections review some of the data available for neurological and neurodevelopmental effects by age group of the studied population.

### *Longitudinal Studies in Preschool Children*

Two longitudinal birth cohorts have observed *in utero* organophosphate exposure associated with abnormal behavioral effects at birth. Using the Brazelton Neonatal Behavior Assessment Scale (BNBAS) in young infants born to mothers living and working in the Salinas Valley of California, a rich agricultural setting, increases in abnormal reflex functioning among the infants were associated with increases in maternal organophosphate urinary metabolite concentrations in pregnancy. The effects were not associated with early postnatal measurement of maternal urinary metabolite concentrations.[20] Using a similar design in a cohort of urban women, abnormal BNBAS responses in newborns were also related to maternal organophosphate metabolite levels during pregnancy. In this study, polychlorinated biphenyls (PCBs) and DDE (the primary metabolite of the organochlorine insecticide DDT) were also measured in maternal blood in the third trimester. There was no observed association between DDE and BNBAS scores and a very weak beneficial association between PCBs in one area of the BNBAS: "range of state."[21]

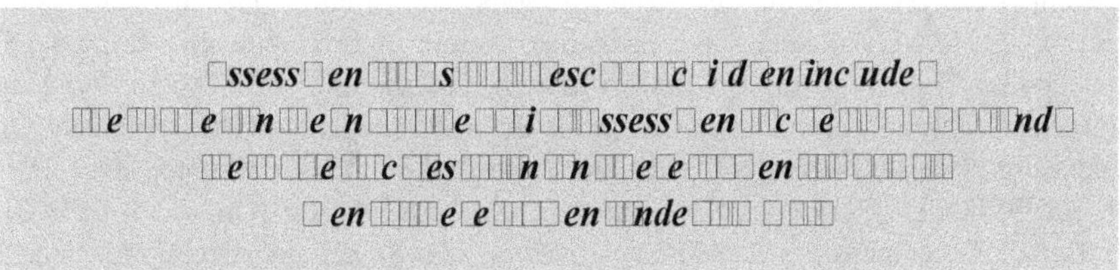

Another birth cohort residing adjacent to a PCB-contaminated harbor in Massachusetts evaluated the relationship between cord blood, PCBs and DDE and performance on the BNBAS. This study observed consistent inverse relationships between BNBAS measures of poor attention and levels of PCBs and DDE in the newborn infants.[22] In contrast, a similar study in an agricultural community failed to show a relationship between prenatal DDT/DDE exposure and BNBAS.[23] Prospective follow-up of these birth cohorts suggests that prenatal exposure to pesticides has long-term effects on children's neurodevelopment.[24,25]

In a study of children exposed to hexachlorobenzene (HCB) during gestation, a relationship at 4 years of age between HCB exposure and the California Preschool Social Competence Scale and an ADHD scoring scale was demonstrated.[26] The greater the exposure, the worse the performance on the Social Competence Scale and the higher the ADHD scores. In the Salinas Valley agricultural cohort, effects of organophosphate insecticide exposure assessed based on maternal urinary metabolite monitoring during pregnancy and postnatal urinary levels assessed in young children has been investigated. Higher rates of symptoms associated with pervasive developmental disorder have been observed in association with both prenatal and postnatal exposure. Prenatal exposure was also associated with lower performance scores on the Bayley Scales of Infant Development (BSID) Mental Development Index (MDI). In contrast, the investigators report postnatal exposure associated with improved MDIs in this cohort. While the reason for this discrepancy is not clear, one hypothesis is that children with higher MDIs may have increased exploratory behavior that influences their postnatal exposure.[17]

In this cohort, increased prenatal DDT was associated with decreases in psychomotor developmental index (PDI) assessed at 6 and 12 months of age but not at 24 months. Prenatal DDE levels were also associated with decreased PDI, but only at 6 months of age. Decreases in MDI at ages 12 and 24 months were related to DDT levels. No significant relationship to DDE was noted on MDI.[27]

Similar studies of neurodevelopment in younger children have been performed in urban birth cohorts. These demonstrate consistent adverse impacts of prenatal chlorpyrifos exposure on neurodevelopmental function in both the motor and mental functional domains at 3 years of life.[28,29]

### *Chronic Effects in School-Age Children*

*[illegible]*

A rapidly increasing body of research associates pesticide exposure with behavioral disorders including Attention Deficit and Hyperactivity Disorder (ADHD) and autism spectrum disorder (ASD), which manifest in preschool and school-age children. A case-control analysis using State of California data on autism diagnosis and a spatial-temporal map of pesticide applications found the risk for ASD was consistently associated with residential proximity to organochlorine pesticide applications occurring around the period of CNS embryogenesis.[30]

Follow-up of the Massachusetts cohort discussed in the prior subsection showed an association between prenatal organochlorine exposure and higher rates of ADHD at school age. This association was consistent for both PCBs and DDE.[31] A cross-sectional analysis from the 2000–2004 National Health and Nutrition Examination Survey (NHANES) linked a representative sample of U.S. children's urinary OP metabolites with diagnoses of ADHD.[32] Increased measures of ADHD behavior using the child behavior checklist (CBCL) in the Salinas Valley cohort found prenatal OP exposure associated with a >70 percentile score on the ADHD confidence index, (OR = 5.1, 95% CI, 1.7-15.7) and the composite ADHD indicator (OR = 3.5, 95% CI, 1.1-10.7). Other measures were also positive but did not reach statistical significance.[33]

The relationship of pesticide exposure on stunting (poor growth) and abnormal neurodevelopment was investigated in school-age Ecuadorian children.[34] Seventy-two children less than 9 years of age in 2$^{nd}$ and 3$^{rd}$ grades were studied via detailed physical exam and neurodevelopmental testing. Prenatal exposure to pesticides was determined by maternal occupational history. Many of these mothers worked in flower production activities leading to extensive pesticide exposure. Concurrent exposure in the children was assessed by measuring red blood cell acetylcholinesterase (AChe) levels and urinary organophosphate metabolites. Many of the children suffered from stunting related to poor nutrition *in utero* and early life. Both stunting and pesticide exposure were associated with decreased performance on the Stanford-Binet copying test, and some of the best scores on the copying test were from children without stunting or pesticide exposure. The independent variables of stunting and pesticide exposure appeared to each contribute independently to the adverse effect. Concurrent exposure to organophosphates affected only simple reaction time. This study provides evidence that poor nutrition, common in many pesticide-exposed children in developing countries or agricultural settings, may increase the adverse effects of pesticide exposure.

Several studies of school-age children with prenatal exposure have been recently published. Though the results are not identical, these studies suggest adverse neurodevelopmental outcomes persist in both urban and agricultural environments in children followed in longitudinal cohorts.[24,25,35] One of these studies reported that adverse neurodevelopmental effects were related to prenatal but not postnatal exposure.[25] The other studies did not differentiate between prenatal and postnatal exposure so it is not clear whether or not postnatal exposure contributes to the neurodevelopmental effects.

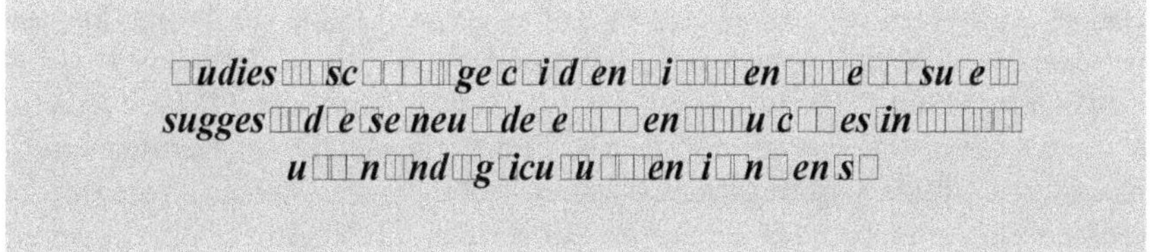

### *Chronic Effects in Adults Following Low-Level Exposure*

In adults, there has been considerable interest in the effects on the nervous system from chronic, low-level exposure to pesticides. A recent review of the evidence regarding the association between pesticide exposure and neurologic dysfunction listed 38 publications reporting various neurologic outcomes of exposure.[36] The studies focused on low-level exposures in adults and can be grouped into two broad categories. The first category is cross-sectional and longitudinal studies of occupationally exposed individuals, observing for a wide range of outcomes. The second focuses on similar populations in relationship to specific disease outcomes, most notably in the neurological area, Parkinson's disease (PD).

Among the first category of studies, an older cross-sectional study evaluated neurotoxicity in pesticide applicators exposed to organophosphates. In this study an increase in vibration sense threshold was thought to represent a loss of peripheral nerve function.[37] A more recent study of termiticide applicators exposed to chlorpyrifos suggested some degree of adverse effects on neurologic function in a subset of tests, specifically pegboard turning and postural sway tests. There was a significant increase in self-reported symptoms in the exposed group. These differences were more marked in the individuals with longer exposure, suggesting a long-term cumulative effect.[38]

A study of farmworkers, as a proxy for pesticide exposure, observed that farmworkers had poorer performance on several neurobehavioral tests compared to nonfarmworker controls. Most notably, farmworkers performed worse with tapping (coefficient [linear regression] = 4.13, 95% CI, 0.0-8.27) and postural sway (coefficient = 4.74, 95% CI, -2.2-11.7). These effects were strongly related to duration of exposure, whether observed in current or former workers.[39] A 5-year prospective longitudinal study of licensed pesticide applicators, the Agricultural Health Study (AHS), reported symptoms at enrollment were strongly correlated with prior pesticide exposure. The symptoms reported, such as headache, fatigue, insomnia, tension, irritability, dizziness, depression and numbness in the hands and feet, were related to duration of exposure to pesticides prior to enrollment. The relationship was still observed after excluding individuals with histories of acute pesticide poisoning or other isolated events with high personal exposure. The strongest associations were with fumigants, organophosphates and organochlorines.[40] This same study reported an association between physician-diagnosed depression and three patterns of exposure to pesticides: acute physician-diagnosed poisoning (OR 2.57 95% CI 1.74-3.79), a high pesticide exposure event (OR 1.65 95% CI 1.33-2.05), and high cumulative exposure (OR 1.54 95% CI 1.16-2.04). It is of interest that the two latter patterns were documented by the study in the absence of a physician-diagnosed acute poisoning episode.[41]

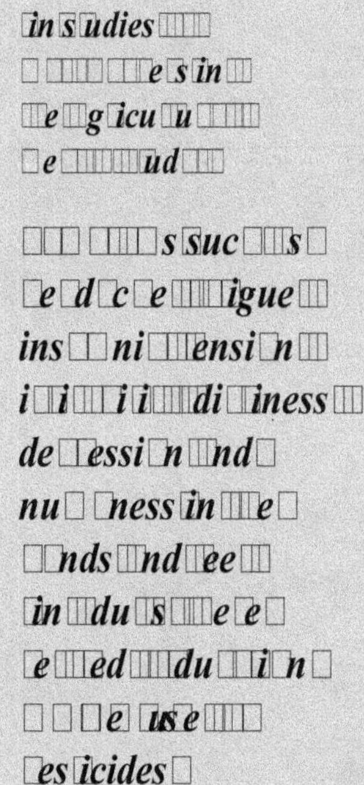

Several studies have examined the potential association between pesticide exposures and Parkinson's disease (PD). A cohort of 238 persons in Washington State who were occupationally exposed to pesticide was compared to 72 non-exposed individuals. In this study, an association between pesticide exposure and PD was reported in the highest tertile of years of exposure (prevalence ratio 2.0, 95% CI, 1.0-4.2).[42] The AHS evaluated the association between the prevalence of physician-diagnosed PD and pesticide usage. The cases were compared to a cohort who did not report PD. Associations were noted with cumulative days of exposure at initial enrollment, frequent personal use of pesticides, and with a few specific pesticides (dieldrin, maneb, paraquat and rotenone).[43] Several postmortem studies of persons with PD have reported a positive association between tissue levels of dieldrin and PD. An evaluation of the biological plausibility of causation concluded that there was sufficient evidence for causation to warrant further evaluation and specific mechanistic studies in animal models.[44] Systematic reviews of the evidence for the association between PD and pesticide exposure have generally concluded that there is evidence for an association. However, at the present time, there is insufficient evidence to conclude that specific pesticide exposures are causative of PD.[45,46]

Following these systematic reviews, several additional studies have evaluated the relationship between PD and pesticides. A multi-country case-control study of pesticide exposure and PD observed a significant exposure between interviewer-administered questionnaire results showing high pesticide exposure and PD (OR = 1.41, 95% CI, 1.06-1.88).[47] Another multi-site case-control study evaluated the risk for PD in various occupations where exposure to toxicants, including pesticides, may occur. Risk of PD was associated with any pesticide use (OR = 1.99, 95% CI, 1.12-3.21). Higher risk was also noted for any of 8 other pesticides selected a priori that have a mechanism that may be associated with PD (OR = 2.2, 95% CI, 1.02-4.75) and for the herbicide 2,4-D (OR = 2.59, 95% CI, 1.03-6.48).[48] To address the difficulty of exposure assessment, a case-control study was conducted using a geographic information system (GIS) that integrated past subject addresses and California pesticide agricultural spray records to characterize exposure. They found that subjects who lived within 500 meters of a field sprayed with paraquat and maneb during the period 1974–1989 were four times more likely to have Parkinson's disease than the control group (OR = 4.17, 95% CI, 1.15-15.16).[49]

Animal studies offer evidence for the basis of a mechanistic association with some pesticides and the development of PD or Parkinsonian features. Two fungicides, mancozeb and maneb, have dose-dependent toxicity on dopaminergic cells in rats. Both the organic component of the fungicide as well as the manganese ion contributed to the toxicity.[50] Additional pesticides of high interest in the relationship with PD include 2,4-D, paraquat, diquat, permethrin, dieldrin and rotenone.[48,51]

## CANCER

Epidemiological data support associations for both adult and childhood cancer,[2,3,52,53] with occupational exposure playing a role in cancer development for both adults and children. However, the most common types of cancer vary for children and adults, and as such, associations between pesticides and cancer are treated separately in this section. As noted at the beginning of this chapter, one common problem in evaluating cancer and pesticide relationships, particularly in children, is the relative rarity of cancer diagnoses.[3,53]

Several meta-analyses and systematic reviews have been published on the association between pesticide exposure and cancer. In most instances, these analyses and reviews serve as the primary source of information for the sections below on childhood and adult cancers.

### Classification Systems for Carcinogenicity in Humans

All active ingredients in pesticides are required to be tested in animals or using *in vitro* tests for their likelihood of causing cancer. The Health Effects Division of the EPA's Pesticide Program performs an independent review of all the available evidence to classify active ingredients according to their potential to cause cancer. The classification systems have changed in the past 30 years from using a letter grade system originally issued in 1986 to a method that uses descriptive phrases based on the weight of evidence. Under the older letter grade system, a grade of "B" was a "probable carcinogen," "C" was equivalent to being classified as "possibly carcinogenic," "D" was "Not classifiable as to human carcinogenicity" and "E" was classified as having "Evidence for non-carcinogenicity for humans."

***Discuss associations between occupational pesticide exposure and cancers in adults and children***

The current system was proposed in 1996, revised in 1999, and released as a final report, *Guidelines for Carcinogen Risk Assessment* in 2005 by the EPA. The report uses one of five specific phrases to designate carcinogenicity: "carcinogenic to humans," "likely to be carcinogenic to humans," "suggestive evidence of carcino-

**CARCINOGEN CLASSIFICATION SYSTEMS AT A GLANCE**

**1986 EPA Classification System**
- Group B: Probable human carcinogen
- Group C: Possible human carcinogen
- Group D: Not classifiable as to human carcinogenicity
- Group E: Evidence of non-carcinogenicity for humans

**2005 EPA Classification System**
- Carcinogenic: Carcinogenic to humans
- Likely: Likely to be carcinogenic to humans
- Suggestive: Suggestive evidence of carcinogenic potential
- Inadequate: Inadequate information to assess carcinogenic potential
- Not Likely: Not likely to be carcinogenic to humans

**IARC Classification System**
- Group 1: Carcinogenic to humans
- Group 2A: Probably carcinogenic to humans
- Group 2B: Possibly carcinogenic to humans
- Group 3: Not classifiable as to its carcinogenicity to humans
- Group 4: Probably not carcinogenic to humans

genic potential," "inadequate information to assess carcinogenic potential," and "not likely to be carcinogenic to humans." This information is available only via an emailed report from the EPA website *http://www.epa.gov/pesticides/carlist*. Although the new guidelines have been in place since 2005, not all pesticides have been evaluated under the 2005 cancer guidelines. Active ingredients in pesticides classified using the older letter designation could be reevaluated on a case-by-case basis.

Another classification system for potentially carcinogenic chemicals was established by the International Agency for Research on Cancer (IARC). This system classifies chemicals using a 1-4 grading system. A classification of 1 indicates the chemical is carcinogenic to humans. A category of 2 is split between 2A (probably carcinogenic to humans) and 2B (possibly carcinogenic to humans). A category of 3 indicates the chemical is not classifiable as to its carcinogenic potential. Generally, this category is used when there is inadequate evidence in humans or animals to establish a cancer-causing relationship. Group 4 indicates that the chemical is probably not carcinogenic to humans.

*The table at the end of this chapter lists selected pesticides and their classification of carcinogenicity.*

**The table at the end of this chapter lists selected pesticides and their classification of carcinogenicity.** The list is not meant to be all inclusive, but an attempt to list agents that are more commonly used or have a higher likelihood of being carcinogenic in humans. It includes a number of chemicals that were classified under both the newer and older EPA systems. The list includes some pyrethroid insecticides, the residential use of which has increased as many of the organophosphates have been phased out.

## Associations between Childhood Cancer and Pesticides

Relationships between childhood cancers and pesticides were summarized in two review articles, the first by Zahm and Ward in 1998, and an update published in 2007 by Infante-Rivard. The pediatric cancer types with the most compelling evidence for an association with pesticides are leukemia and brain tumors. Of note, in most of the studies reviewed, all forms of leukemia were considered in one group because of insufficient numbers of certain types of leukemia – *e.g.*, acute lymphocytic leukemia (ALL) or acute myelocytic leukemia (AML). There were a few studies of sufficient size that were able to evaluate ALL separately. Brain tumors are also reported as a group rather than by individual tumor types as they are even rarer than childhood leukemia.[3,53]

***The pediatric cancer types with the most compelling evidence for an association with pesticides are leukemia and brain tumors.***

### *Childhood Leukemia*

Thirteen of the 18 studies reviewed in the 1998 Zahm and Ward article found an increased risk of leukemia following pesticide exposure. The most common reported exposure was not related to agricultural production but rather household insecticide use during pregnancy or during the preconception period. As mentioned above, mixing leukemia types and recall bias were among the limitations of these earlier studies.[53]

Infante-Rivard reviewed 12 more recent studies in 2007.[3] Most of these studies were larger and used higher-quality exposure assessment methodologies. Five found statistically significant associations between leukemia and pesticide exposure.[54,55,56,57,58]

Two included a detailed exposure assessment and were able to demonstrate a dose-response effect.[56,58] The largest study included 491 subjects and limited the outcome to acute lymphocytic leukemia. In this study, maternal residential use during pregnancy of herbicides (OR = 1.84, 95% CI, 1.32, 2.57), plant insecticides (OR = 1.97, 95% CI, 1.32-2.94), and "pesticides for trees" (OR = 1.70, 95% CI, 1.12-2.59) were all associated with ALL. Childhood exposure (from birth to diagnosis of ALL) to plant insecticides (OR = 1.41, 95% CI, 1.06-1.86) and herbicides (OR = 1.82, 95% CI, 1.31-2.52) were also significantly associated.[56] Two studies by the same author did not find an association between child's residence near agriculture-related pesticide application and childhood leukemia,[59] nor maternal residence near agricultural pesticide application at the time of their child's birth and childhood leukemia.[60]

Two additional meta-analyses have been conducted that further explore associations between pesticides and leukemia and support the previously described associations. The first meta-analysis examined parental occupational exposure to pesticides and leukemia and the second focused on studies of pesticides in the home and garden.[61,62] In the first study, maternal occupational exposure was found to be associated with leukemia, the reported ORs were 2.09, 95% CI, 1.51-2.88 for overall pesticide exposure; 2.38, 95% CI, 1.56-3.62 for insecticide exposure; and 3.62, 95% CI, 1.28-10.3 for herbicide exposure. No associations were found for paternal occupational exposure.[62] In the meta-analysis focused on exposure through home and garden uses of pesticides, 15 studies were included and exposure during pregnancy to unspecified pesticides, insecticides and herbicides were all associated with leukemia (OR = 1.54, 95% CI,1.13-2.11; OR = 2.05, 95% CI, 1.80-2.32; and OR = 1.61, 95% CI, 1.2-2.16, respectively).[61]

*There is evidence for increased risk of developing some types of childhood cancers following preconception and/or prenatal exposure to pesticides.*

### Childhood Brain Tumors

In the 1998 Zahm and Ward review, 12 of the 16 studies presented evidence of an association between pesticide exposure and childhood brain tumors, and seven of these reached statistical significance. Similar to the findings with leukemia, household use by the parent (home and garden and on household pets) were the most commonly associated exposures. The number of children with brain tumors is even fewer than that of leukemia, so all types of brain tumors were used to define "cases."[53]

As noted with leukemia, the body of evidence estimating an association between brain tumors and pesticides since 1998 is more robust, with larger studies and improved exposure assessment. Nine of 10 studies in the 2007 Infante-Rivard review demonstrated an increased risk of brain tumors following maternal and/or paternal exposure, with three of the studies reaching statistical significance.[63,64,65] For all studies, it appeared that prenatal exposure to insecticides, particularly in the household, as well as both maternal and paternal occupational exposure before conception though birth represented the most consistent risk factors.[63,64,65,66,67,68,69,70,71] The largest case/control study (321 cases) limited the case definition to astrocytomas and noted an OR of 1.9, 95% CI, 1.1-3.3, following maternal preconceptual/prenatal exposure to insecticides.[65] One cohort study followed 235,635 children and found an association between all brain tumors and paternal exposure to pesticides immediately before conception (RR = 2.36, 95% CI, 1.27-4.39.[63]

In summary, there is relatively consistent evidence for an increased risk of developing some types of childhood cancers following preconception and/or prenatal exposure to pesticides. The strongest evidence appears to be for ALL, the most common form of childhood leukemia. Maternal exposure to insecticides and paternal occupational exposure appear to carry the greatest risk.

## Associations between Pesticides and Cancer in Adults

Bassil et al. conducted a systematic review of cancer and pesticides, which included studies of children and of adults. Each study was evaluated for methodological quality by two trained reviewers using a standardized assessment tool with a high inter-rater reliability. Only studies with a global rating of 4 or higher were included in the review.[2]

Many of the studies evaluating relationships between cancers in adults and pesticides are conducted in the occupational setting. Associations between pesticide exposure and the development of leukemia and non-Hodgkin lymphoma were noted in most studies. Solid tumors of the prostate, pancreas, kidney and breast were among the more consistently reported findings in studies of adults. As was noted in numerous studies of childhood outcomes, ascertainment of whether exposure actually occurred and the amount of exposure are recurring weaknesses in adult studies.

### *Non-Hodgkin Lymphoma and Other Hematopoietic Cancers*

***Tumors of the prostate, pancreas, kidney and breast have been among the more consistently reported findings.***

Of the 27 studies on non-Hodgkin lymphoma (NHL) that met quality criteria in the Bassil review, 23 found positive associations. Almost half of these studies were conducted in adult cohorts of various occupational groups including farmers, pesticide applicators, landscapers and those who worked in pesticide manufacturing. Ten of the 12 cohort studies reported a positive association, with four reaching statistical significance. One of the larger cohort studies demonstrated a relative risk RR of 2.1, 95% CI, 1.1-3.9. Eleven of the 13 case-control studies (excludes one positive study in children) also demonstrated an association between occupational exposure and NHL, with 7 reaching statistical significance. Multiple classes of pesticides were implicated.[2]

A separate meta-analysis of case-control studies examining the relationship between pesticide exposure and hematopoietic cancers was published in 2007. The authors reviewed 36 case-control studies. After excluding studies with methodological flaws or data concerns, a study that included non-hematopoietic cancers and a study written in Italian, 13 studies remained for analysis. The cancers assessed in the meta-analysis were NHL, leukemia and multiple myeloma.[72] The overall meta-OR for NHL was 1.35, 95% CI, 1.2-1.5. An increased risk for leukemia and multiple myeloma was also demonstrated, though both were just short of reaching statistical significance (OR = 1.35, 95% CI, 0.9-1.2 and OR = 1.16, 95% CI, 0.99-1.36). The authors also conducted a meta-regression to account for the heterogeneity among the studies. They found that exposure for longer than 10 years increased the risk for all hematopoietic cancers (mOR = 2.18, 95% CI, 1.43-3.35) and for NHL (mOR = 1.65, 95% CI, 1.08-2.51).[72]

As with other cancer epidemiologic studies discussed above, the major limitation was the lack of sufficient exposure information in many of the studies. Additionally, the cohort studies in the above meta-analysis only listed the class of pesticide and the corresponding OR (herbicides or insecticides) rather than the individual pesticide.[72] Other individual studies have demonstrated risks from certain specific pesticides. One well-designed cohort study reported risks associated with mecoprop, a chlorophenoxy herbicide.[73] Another study demonstrated risks from another chlorophenoxy herbicide – methyl phenoxyacetic acid (MCPA) – and from glyphosate.[74] Another study demonstrated a significant increased risk of NHL for subjects exposed to 2,4-D.[75] The Agricultural Health Study demonstrated a risk of developing leukemia following exposure to diazinon.[76]

### *Prostate Cancer*

It has been suspected that pesticide exposure may be associated with prostate cancer. This association may be related to hormonally active pesticides, known as endocrine disruptors.[77] Of the eight studies included in the Bassil review, all showed positive associations between pesticide exposure and prostate cancer.[77,78,79,80,81,82,83,84] A particularly well-designed study from the Agriculture Health Cohort included 55,000 men in Iowa and North Carolina. The authors found that farmers who applied pesticides had a small but significant increase in prostate cancer compared to the general male population in Iowa and North Carolina (standardized prostate cancer incidence ratio of 1.14 (1.05-1.24)). The study also evaluated risk to specific pesticides by inquiring about 50 different pesticides to which the farmer was "ever exposed" and found positive associations with carbofuran, permethrin, aldrin and DDT. Each OR was in the range of 1.25 to 1.38, all with statistically significant 95% CIs. However, among those who were in the "highest exposure category," a risk estimate of 3.47, 95% CI, 1.37-8.76, was noted for the fumigant methyl bromide. In addition, six pesticides (chlorpyrifos, fonofos, coumaphos, phorate, permethrin and butylate) were positively associated with prostate cancer in men with a family history of prostate cancer.[83]

Around the same time as Bassil's review was published, Mink et al. conducted a separate review article on prostate cancer. The two authors reviewed and independently assessed each study for inclusion or exclusion, and discrepancies were reconciled. The authors included 13 studies (8 cohort, 5 case-control) in their final review; however, they did not report the total number of studies reviewed and excluded. Despite some scattered positive findings in some of the studies they reviewed, the authors concluded there was no causal link between pesticides and prostate cancer.[52]

Two case-control studies by Settimi et al. evaluated prostate cancer among agricultural workers and included a comprehensive questionnaire to evaluate exposures as well as potential confounders. The first study evaluated numerous types of cancers and demonstrated an excess risk of prostate cancer among farmers and farmworkers (OR = 1.4, 95% CI, 1.0-2.1). When the analysis was limited to those who applied pesticides, the OR = 1.7, 95% CI, 1.2-2.6.[85] Assessment of pesticide classes and individual pesticides within classes demonstrated risk specificity for organochlorine insecticides. Elevated ORs for prostate cancer were found for "ever being exposed" to all organochlorines, DDT and dicofol and tetradifon. All ORs were statistically significant, and were slightly higher for those who reported greater than 15 years of exposure compared to "ever exposed."[78]

Another case-control study included data on exposure, diet, lifestyle and occupational factors. A positive association was found for exposure to pesticides, but the 95% CIs were wide. This may have been attributable to the small size of the study – 40 cases – and fewer reporting exposure to pesticides.[86] Two other case-control studies found no association with prostate cancer and pesticide use.[87,88]

### Tumors of the Kidney

A recent review article evaluated renal cancer in adults (primarily renal cell carcinoma) following occupational exposure to pesticides. This review included four studies, each of which observed positive associations between pesticides and renal cancer.[89,90,91,92]

### Other Associations between Human Cancer and Pesticides

Several different agents used as wood preservatives are currently classified as probable carcinogens. Pentachlorophenol (PCP) has been classified as a B2 (probable human carcinogen). In humans, it has been associated with soft tissue sarcoma and kidney and GI tract cancers; however, a causal link has not been established.[89,93] In animal data submitted to the U.S. EPA in support of re-registration of PCP liver tumors, pheochromocytomas and hemangiosarcomas were noted, supporting the B2 classification.[94]

Arsenic is well established as a human carcinogen. Studies show that arsenic exposure can result in epigenetic dysregulation including DNA methylation, histone modification and microRNA expression. These alterations may play a mechanistic role in cancer development, but long-term studies have not yet confirmed this.[95] Primary cancers caused by arsenic include tumors of the lung, bladder and skin. On occasion, the hyperkeratotic papules described above have undergone malignant transformation. Years after exposure, dermatologic findings include squamous cell and basal cell carcinoma, often in sun-protected areas.[96]

A recent review of lung cancer and arsenic evaluated nine cross-sectional studies, six cohort studies, and two case-control studies. Despite the limitations of some of the study designs, the risk ratios and standardized mortality ratios were consistently high on nearly all of the studies. The evidence was most consistent at high exposure levels. The evidence was weak or lacking for developing cancer from exposure to lower levels of arsenic via contaminated drinking water (<100 μg/L).[97]

## ENVIRONMENTAL ENDOCRINE DISRUPTOR EFFECTS

*Data relating human endocrine disruption has become progressively stronger in supporting a role of pesticides. Extensive research continues in this area of investigation.*

Over the last 15 years there has been increasing interest in the ability of environmental chemicals to disrupt endocrine systems. Many pesticides, pesticide vehicles and contaminants have endocrine-disrupting properties based on *in vitro* and animal studies. While data on human effects remain somewhat fragmentary and inconclusive, the weight of evidence from multiple lines of investigation appears to support the concern for human effects. These effects are discussed briefly below, along with the literature that supports these assertions.

The cellular biology of endocrine disruption is very complex and has been extensively reviewed. While the details are beyond the scope of this manual, the reader is directed to one of several reviews for more specific information.[98,99,100] As a group, exogenous agents including pesticides that affect the endocrine system have been labeled endocrine disruptive chemicals (EDCs). Several basic mechanisms have been identified, including direct interaction with nuclear receptors (NR), disturbance of NR signaling and changes in hormone availability. *In vitro* evidence of the latter exists for several pesticides, by alteration of P450 enzyme activity that influences the availability of steroid hormones either by increasing or decreasing the rates of metabolism. For instance, methoxychlor has been shown to interfere with 5'deiodinase in the liver.[101]

### Animal Toxicology

Animal studies conducted in the laboratory suggest that some pesticides may disrupt the endocrine systems of a variety of animals. Vinclozolin, a fungicide with low acute toxicity, has been shown to be strong antiandrogen in rats when exposure occurs *in utero*.[102] Exposure of female rats to DDT has been shown to lead to precocious puberty.[103] Lindane has been shown to affect adrenal steroid synthesis.[104] There is considerable evidence that a variety of chemicals, including some pesticides, affect thyroid function in animals.[105,106]

Further support for effects comes from observations in wildlife. These studies represent the most robust evidence base for various endocrine effects from many different pesticide classes. Only a few examples are mentioned because of space constraints. A strong antiandrogen effect was shown in alligators in a lake in Florida in response to heavy contamination with pesticides including dicofol, DDT and DDE.[107,108] Likewise, a relatively strong association has been shown between the biocide tributyltin (TBT) and pseudohermaphroditism in 150 species of snails.[109] Marine mammals have been noted to have high levels of contamination with a variety of chemicals including pesticides such as DDT, DDE, mirex, dieldrin and chlordane metabolites.[110] These contaminants have been potentially linked to reproductive failure and other effects due to their endocrine action. For example, PCBs in seals and polar bears have been shown to affect thyroid function. Interestingly, levels of PCBs and organochlorine pesticides are negatively correlated with testosterone levels in male polar bears, but PCBs are positively correlated with testosterone in female polar bears. Each of these testosterone alterations may contribute to reproductive changes.[111]

## Evidence of Human Effects of Endocrine Disruption

A systematic review by the Endocrine Society has led to a scientific statement on endocrine disrupting environmental toxicants and notes potential for a variety of human effects, including alteration in mammary gland development and possible carcinogenesis, alteration in male fertility and testicular cancer, male urogenital malformations, prostate cancer, thyroid disruption and obesity.[112] This is a rapidly evolving field of investigation.

### *Human Outcomes Related to Pesticides*

***Precocious Puberty.*** DDE has been linked to precocious puberty in one study of immigrant females in Belgium.[113] Though estrogenic pesticides have been proposed as a contributor to premature thelarche, the evidence to date is not conclusive.

***Altered Lactation.*** A negative correlation has been shown in several cohorts between DDE and duration of lactation.[114]

***Breast Cancer.*** There is considerable interest in this outcome because of animal studies and the estrogenic activities of pesticides such as DDT, DDE, endosulfan and atrazine. Though atrazine is not a direct mimicker of estrogen, in some models it induces aromatase formation, which converts testosterone to estradiol.[115] This effect is not consistent in all cell lines or animal models. Despite the evidence that estrogen is a promoter of breast cancer, the role of these pesticides in breast cancer remains unclear at this time. A U.S. EPA review in 1998 concluded that the association between organochlorines and PCBs was not sufficient to conclude that they were likely causes of breast cancer.[116] A review by The Endocrine Society in 2009 concluded there was sufficient evidence that endocrine disruptors altered mammary gland morphogenesis in humans, making them more prone to neoplastic development.[112]

***Female Fertility.*** There is limited evidence that female fertility may be decreased in women occupationally exposed to pesticides.[117] However, this evidence has not been linked to specific pesticide exposures.

***Semen Quality.*** Decreased semen quality has been noted in individuals exposed to dioxins and PCBs, which are persistent organic compounds considered related to organochlorine pesticides.[112] Two agents, chlordecone and DBPC (dibromochloropropane), have been shown to affect male fertility by direct testicular toxicity at high levels of exposure.[118,119] However, there is not strong evidence for a relationship between organochlorine pesticides and semen quality. On the other hand, there is significant evidence from epidemiology that non-persistent pesticides may alter semen quality. This has been documented by the relationship between pesticide metabolites measured in men and their semen quality. Among the compounds implicated, some with stronger evidence than others, are alachlor mercapturate, atrazine mercapturate and some metabolites of diazinon, chlorpyrifos and carbaryl.[112,120,121,122,123,124,125,126,127,128]

***Male Urogenital Tract Malformations.*** There is limited evidence that exposure to chemicals, including DDT, is associated with increased rates of cryptorchidism and hypospadias. In some studies there appears to be a weak association between these entities and maternal serum concentrations of these chemicals. There is also epidemiological evidence suggesting a relationship between parental or community exposure to pesticides and these malformations without clear evidence for which pesticides are responsible.[112]

***Prostate Cancer and Prostatic Hyperplasia.*** It is well accepted that endocrine status strongly affects the development of both prostate cancer and prostatic hyperplasia. Both androgens and estrogen have been shown to promote cancer and hyperplasia of the prostate. Likewise, antiandrogens and surgical castration can arrest or regress prostate cancer. It seems reasonable then that endocrine-active pesticides would play a role in recent increases in the rates of these problems. Epidemiologic studies have shown increased rates of prostate cancer in farmworkers. A direct link has been shown between methyl bromide exposure and prostate cancer in farmworkers.[83,129] In addition, though arsenical pesticides are in limited use today, arsenic has been associated with prostate cancer.[129] See the *Cancer* subsection of this chapter for additional information.

***Antiandrogens.*** The active ingredients vinclozolin and DDT, along with DDE (the primary metabolite of DDT), are known to be antiandrogens. The effect of DDE and DDT on hypospadias and cryptorchidism is described above, but other antiandrogenic effects of these agents in humans are unclear at this time.

***Reproductive Neuroendocrine Systems.*** There is a considerable amount of evidence in laboratory animals that pesticides may disrupt reproductive systems and affect sexual behavior. As noted above, vinclozolin has been shown to alter sexual behavior in rats. However, there are limited human data to support such effects in children or adults.[112]

***Thyroid Function.*** In the Agricultural Health Study, an association was shown between pesticide exposure and thyroid disease in female spouses of farmworkers. Increased odds ratios ranging from 1.2-1.5 for hypothyroidism were seen with organochlorines including aldrin, DDT, heptachlor, lindane and chlordane, although only chlordane (OR = 1.3) was statistically significant. Benomyl (3.1) and paraquat (1.8) also had significantly elevated rates of hypothyroidism. Interestingly, maneb/mancozeb appeared to be related to both hypothyroidism and hyperthyroidism.[130] In a study of Inuit adults, negative associations were observed between some organochlorine pesticides and thyroid hormone levels.[131]

The science is rapidly advancing in this field, as most studies in human populations have been published relatively recently. Endocrine disruption continues to be the subject of intense research at a pace suggesting significant discovery in the coming decade.

## ASTHMA

The role of pesticides in the development of and/or exacerbation of asthma has been hypothesized and is under investigation. Pyrethrins have some potential as an allergic sensitizing agent, with reports of contact dermatitis, asthma and anaphylactic reactions occurring following exposure.[132,133,134] Organophosphates appear to have mechanisms that could impact the development or exacerbation of asthma. Toxicological studies demonstrated that subcutaneous injection of the organophosphates chlorpyrifos, diazinon and parathion caused airway hyper reactivity in guinea pigs via inhibition of M2 muscarinic receptors.[135,136] Additional studies suggest an organophosphate exposure may induce lipid peroxidation, which will result in oxidative stress.[137,138] Organophosphates may also play a role in the immunological sensitization of individuals to asthma. In a cohort of women farmworkers and their infants, maternal agricultural work was associated with a 26% increase in proportion of T-helper 2 (TH-2) cells, the phenotype associated with atopic disease, in their 24-month-old infants' blood samples. The percentage of TH-2 cells was associated with both physician-diagnosed asthma and maternal report of wheeze in these infants.[139]

## Pesticides and Asthma in Adults

Some epidemiological evidence supports an association between occupational exposure in adults and asthma. A case-control study conducted in Lebanon evaluated 407 subjects with asthma. Those with any exposure to pesticides exhibited an association with asthma (OR = 2.11, 95% CI, 1.47-3.02). Occupational use resulted in an even higher association (OR = 4.98, 95% CI, 1.07, 23.28), although as noted, the intervals were wide, but significant.[140]

Several examinations of the Agricultural Health Study (AHS, discussed in the previous subsection on cancer) evaluated the relationship of asthma to various exposures occurring in farming occupations.[141,142,143,144] In one of these AHS studies, organophosphate insecticides including chlorpyrifos, malathion and parathion were all positively associated with wheeze in farmers. Chlorpyrifos, dichlorvos and phorate were associated with wheeze in the commercial applicators. Chlorpyrifos had the strongest associations in both groups, with OR = 1.48, 95% CI, 1.00-2.19 for farmers and OR = 1.96, 95% CI, 1.05-3.66 for commercial applicators.[143] The same group of authors identified in an earlier paper that driving diesel tractors was also associated with wheezing (OR = 1.31, 95% CI, 1.13-1.52).[145] In order to control for such exposures unique to farmers, a second paper from the Agricultural Health Study cohort limited analysis to 2,255 commercial pesticide applicators. The authors continued to observe associations with organophosphates including chlorpyrifos (≥40 days per year; OR = 2.4, 95% CI, 1.24-4.65) and dichlorvos (OR = 2.48, 95% CI, 1.08-5.66). The herbicide chlorimuron ethyl was also associated with asthma (OR = 1.62, 95% CI, 1.25-2.1).[144] A third analysis evaluated risk factors for women who lived and grew up on a farm. In general, growing up on a farm was found to be protective for having atopic or non-atopic asthma (defined as "doctor diagnosed, after 19 years of age"). However, any use of pesticides was associated with atopic asthma (OR = 1.46, 95% CI, 1.14-1.87). Those women who grew up on a farm but did not apply pesticides had the greatest protection from asthma (OR = 0.41, 95% CI, 0.27-0.62).[141] As with most epidemiological studies, there were some limitations of exposure assessment, including self-reported behaviors and exposures and misclassification.

*The role of pesticides including pyrethrins and organophosphates with respect to asthma is under investigation.*

Other studies have not found an association between pesticide exposure and asthma. One case-control study evaluated exposure of aerial pesticide applicators and community controls. Self-reported asthma rates were similar in the two groups. There was a slight decrease in lung function among aerial applicators, forced expiratory volume in 1 second (FEV1) <80% predicted (8% v. 2%, p = .02), but otherwise there was no difference between cases and controls of other measures of asthma or asthma severity.[146] Two studies assessed emergency department visits for asthma and hospital admissions following insecticide application to control for mosquitoes potentially carrying West Nile virus (WNV) in New York City. One study evaluated visits at a single hospital in the South Bronx after malathion and resmethrin application during a 4-day period. Using the previous year as a reference point, there was no increase in the rate of ED visits or in the severity of asthma presentations.[147] Another study evaluated the rates of ED visits in all public NYC hospitals during the 14-month period of October 1999 to November 2000. The authors looked at asthma visits in a 3-day period before and after spraying events took place, but did not find an increase in daily ED visit rates that corresponded to pesticide spraying.[148] A multicenter prospective study in Europe did not find any association with asthma and exposure to the fungicide ethylene bis dithiocarbamate.[149]

## Pesticide Exposure and Asthma in Children

The few epidemiologic studies on the association between pesticide exposure and respiratory health in children have reported mixed results. In a cohort of rural Iowan children, multiple farm-related exposures were studied for any associations with several asthma-related outcomes ranging from doctor-diagnosed asthma to cough with exercise. Any pesticide use in the previous year was not significantly associated with asthma symptoms and prevalence.[150]

A cross-sectional study of Lebanese children was conducted using a randomly selected sample from public schools. The authors found increased risks of chronic respiratory symptoms, including wheeze, among children with any pesticide exposure in the home, exposure related to parent's occupation, and use outside the home. For any exposure to pesticides, they found an association with asthma (OR = 1.73, 95% CI, 1.07-2.90). Residential exposure, defined as having regional exposure or living near a treated field, had a stronger association, OR = 2.47, 95% CI, 1.52-4.01. Finally, occupational use of pesticides by a family member had the strongest association, OR = 2.98, 95% CI, 1.58-5.56. In the researchers' multivariable model, parental exposure persisted as a risk factor (OR = 4.61, 95% CI, 2.06-10.29). However, within the study population of 3,291, 407 had chronic respiratory disease. Of those, only 84 had medically confirmed asthma.[151] Main shortcomings include the cross-sectional design and self-reported symptoms versus more objective outcome assessment.

A nested-case control study of the Southern California Children's Health Study was conducted to evaluate the relationship between multiple environmental exposures, early life experiences and the occurrence of asthma. Among environmental exposures in the first year of life, "herbicides" and "pesticides" both had a strong association with asthma diagnosis before age 5 years (OR = 4.58, 95% CI, 1.36-15.43 and OR = 2.39, 95% CI, 1.17-4.89, respectively). Of note, cockroach exposure in the first year and later was also associated with having any type of asthma (OR= 2.03, 95% CI, 1.03-4.02). There were also elevated ORs for cockroach exposure in the first year of life with early persistent asthma and late onset asthma; however, the findings did not reach statistical significance.[152] This relationship is potentially important, since cockroaches are known to exacerbate asthma, and pesticides are likely to be used in homes with cockroach infestation.

Similar studies addressing the respiratory health implications for children for specific pesticide chemical types or groups are rare. However, some evidence is emerging for a link between metabolites of DDT and asthma risk.[153,154] One study of 343 children in Germany found an association between DDE levels and asthma (OR= 3.71, 95% CI, 1.10-12.56) as well as DDE levels and IgE levels >200 kU/l (OR = 2.28, 95% CI, 1.2-4.31).[153] In a prospective cohort study of children in Spain, wheezing at 4 years of age increased with increasing levels of DDE at birth. The adjusted RR for the children with exposure in the highest quartile was 2.63, 95% CI, 1.19–4.69. The use of doctor-diagnosed asthma (occurring in 1.9% of children) instead of wheezing as the outcome variable also resulted in a positive association, although it was not statistically significant.[154]

In summary, the available data regarding chronic exposure to pesticides and asthma and other respiratory health effects provide some suggestion of effect but are limited in number with highly variant designs for exposure assessment and outcome determination.

| SELECTED PESTICIDES AND THEIR CARCINOGENIC POTENTIAL | | | |
|---|---|---|---|
| **Name of Pesticide** | **EPA Cancer Classification*** | **Notes** | **IARC Classi-fication**** |
| Acephate | Group C | | |
| Alachlor | Carcinogenic (High Doses); Not Likely | | |
| Arsenic | | Not listed by EPA, all pesticide uses canceled | 1 |
| Benomyl | Group C | | |
| Bifenthrin | Group C | | |
| Butachlor | Likely | | |
| Captafol | Group B | | 2-A |
| Carbaryl | Likely | | 3 |
| Chlordane | | | 2-B |
| Chlordecone | | | 2-B |
| Chlordimeform | Group B | | 3 |
| Chloroaniline, p- | Group B | | |
| Chlorophenoxy herbicides | 2,4-D listed as Group D | Several are Group C (*e.g.*, DCPA) or Not Likely (*e.g.*, MCPA) | 2-B |
| Chlorothalonil | Group B | | 2-B |
| Cypermethrin | Group C | | |
| Dichlorvos | Suggestive | | 2-B |
| Diclofop-methyl | Likely | | |
| Diuron | Likely | | |
| Ethoprop | Likely | | |
| Fenoxycarb | Likely | | |
| Ferbam | Likely | | 3 |
| Fipronil | Group C | | |
| Furiazole | Likely | | |
| Heptachlor | | | 2-B |
| Hexachloroethane | | | 2-B |
| Hexythiazox | Likely | | |
| Iprodione | Likely | | |
| Iprovalicarb | Likely | | |
| Mancozeb | Group B | | |
| Maneb | Group B | | 3 |
| Metam sodium | Likely | | |
| Metofluthrin | Likely | | |
| Metolachlor | Group C | | |
| Mirex | | | 2-B |
| Nitrapyrin | Likely | | |
| Oryzalin | Likely | | |

| SELECTED PESTICIDES AND THEIR CARCINOGENIC POTENTIAL, CONT. | | | |
|---|---|---|---|
| **Name of Pesticide** | **EPA Cancer Classification*** | **Notes** | **IARC Classi-fication**** |
| Oxyfluorfen | Likely | | |
| Parathion, ethyl- | Group C | Methyl parathion is "Not Likely" | 3 |
| Pentachlorophenol | Group B | | 2-B |
| Permethrin | Likely | | 3 |
| Piperonyl butoxide | Group C | | 3 |
| Pirimicarb | Likely | | |
| Propachlor | Likely | | |
| Propoxur | Group B | | |
| Resmethrin | Likely | | |
| Thiacloprid | Likely | The most commonly used neonicotinod, imidacloprid, is Group E | |
| Thiodicarb | Group B | | |
| Tolyfluanid | Likely | | |
| Toxaphene | | | 2-B |
| Trifluralin | Group C | | 3 |
| Triphenyltin hydroxide | Group B | | |
| Vinclozolin | Group C | | |

* The most recent EPA classification, whether from the 1986 or the 2005 system

1986 Classification
Group B: Probable human carcinogen
Group C: Possible human carcinogen
Group D: Not classifiable as to human carcinogenicity
Group E: Evidence of non-carcinogenicity for humans

2005 Classification
Carcinogenic: Carcinogenic to humans
Likely: Likely to be carcinogenic to humans
Suggestive: Suggestive evidence of carcinogenic potential
Inadequate: Inadequate information to assess carcinogenic potential
Not Likely: Not likely to be carcinogenic to humans

**IARC Classification

Group 1: Carcinogenic to humans
Group 2A: Probably carcinogenic to humans
Group 2B: Possibly carcinogenic to humans
Group 3: Not classifiable as to its carcinogenicity to humans
Group 4: Probably not carcinogenic to humans

## References

1. Hill AB. The Environment and Disease: Association or Causation? *Proc R Soc Med.* May 1965;58:295-300.

2. Bassil KL, Vakil C, Sanborn M, Cole DC, Kaur JS, Kerr KJ. Cancer health effects of pesticides: systematic review. *Can Fam Physician.* Oct 2007;53(10):1704-1711.

3. Infante-Rivard C, Weichenthal S. Pesticides and childhood cancer: an update of Zahm and Ward's 1998 review. *J Toxicol Environ Health.* Jan-Mar 2007;10(1-2):81-99.

4. Gurunathan S, Robson M, Freeman N, et al. Accumulation of chlorpyrifos on residential surfaces and toys accessible to children. *Environ Health Perspect.* Jan 1998;106(1):9-16.

5. Hore P, Robson M, Freeman N, et al. Chlorpyrifos accumulation patterns for child-accessible surfaces and objects and urinary metabolite excretion by children for 2 weeks after crack-and-crevice application. *Environ Health Perspect.* Feb 2005;113(2):211-219.

6. Lewis RG, Fortune CR, Blanchard FT, Camann DE. Movement and deposition of two organophosphorus pesticides within a residence after interior and exterior applications. *J Air Waste Manag Assoc.* Mar 2001;51(3):339-351.

7. Freeman NC, Hore P, Black K, et al. Contributions of children's activities to pesticide hand loadings following residential pesticide application. *J Expo Anal Environ Epidemiol.* Jan 2005;15(1):81-88.

8. Freeman NC, Jimenez M, Reed KJ, et al. Quantitative analysis of children's microactivity patterns: The Minnesota Children's Pesticide Exposure Study. *J Expo Anal Environ Epidemiol.* Nov-Dec 2001;11(6):501-509.

9. Selevan SG, Kimmel CA, Mendola P. Identifying critical windows of exposure for children's health. *Environ Health Perspect.* Jun 2000;108 Suppl 3:451-455.

10. Rosenstock L, Keifer M, Daniell WE, McConnell R, Claypoole K. Chronic central nervous system effects of acute organophosphate pesticide intoxication. The Pesticide Health Effects Study Group. *Lancet.* Jul 27 1991;338(8761):223-227.

11. Savage EP, Keefe TJ, Mounce LM, Heaton RK, Lewis JA, Burcar PJ. Chronic neurological sequelae of acute organophosphate pesticide poisoning. *Arch Environ Health.* Jan-Feb 1988;43(1):38-45.

12. Steenland K, Jenkins B, Ames RG, O'Malley M, Chrislip D, Russo J. Chronic neurological sequelae to organophosphate pesticide poisoning. *Am J Public Health.* May 1994;84(5):731-736.

13. Abdollahi M, Karami-Mohajeri S. A comprehensive review on experimental and clinical findings in intermediate syndrome caused by organophosphate poisoning. *Toxicol Appl Pharmacol.* Feb 1 2012;258(3):309-314.

14. Navarro B, Sayas MJ, Atienza A, Leon P. An unhappily married man with thick soles. *Lancet.* Jun 8 1996;347(9015):1596.

15. Heyman A, Pfeiffer JB, Jr., Willett RW, Taylor HM. Peripheral neuropathy caused by arsenical intoxication; a study of 41 cases with observations on the effects of BAL (2, 3, dimercapto-propanol). *N Engl J Med.* Mar 1 1956;254(9):401-409.

16. Rahman MM, Naidu R, Bhattacharya P. Arsenic contamination in groundwater in the Southeast Asia region. *Environ Geochem Health.* Apr 2009;31 Suppl 1:9-21.

17. Eskenazi B, Marks AR, Bradman A, et al. Organophosphate pesticide exposure and neurodevelopment in young Mexican-American children. *Environ Health Perspect.* May 2007;115(5):792-798.

18. Rosas LG, Eskenazi B. Pesticides and child neurodevelopment. *Curr Opin Pediatr.* Apr 2008;20(2):191-197.

19. Horton MK, Rundle A, Camann DE, Boyd Barr D, Rauh VA, Whyatt RM. Impact of prenatal exposure to piperonyl butoxide and permethrin on 36-month neurodevelopment. *Pediatrics.* Mar 2011;127(3):e699-706.

20. Young JG, Eskenazi B, Gladstone EA, et al. Association between *in utero* organophosphate pesticide exposure and abnormal reflexes in neonates. *Neurotoxicology*. Mar 2005;26(2):199-209.

21. Engel SM, Berkowitz GS, Barr DB, et al. Prenatal organophosphate metabolite and organochlorine levels and performance on the Brazelton Neonatal Behavioral Assessment Scale in a multiethnic pregnancy cohort. *Am J Epidemiol.* Jun 15 2007;165(12):1397-1404.

22. Sagiv SK, Nugent JK, Brazelton TB, et al. Prenatal organochlorine exposure and measures of behavior in infancy using the Neonatal Behavioral Assessment Scale (NBAS). *Environ Health Perspect.* May 2008;116(5):666-673.

23. Fenster L, Eskenazi B, Anderson M, et al. Association of *in utero* organochlorine pesticide exposure and fetal growth and length of gestation in an agricultural population. *Environ Health Perspect.* Apr 2006;114(4):597-602.

24. Engel SM, Wetmur J, Chen J, et al. Prenatal exposure to organophosphates, paraoxonase 1, and cognitive development in childhood. *Environ Health Perspect.* Aug 2011;119(8):1182-1188.

25. Bouchard MF, Chevrier J, Harley KG, et al. Prenatal exposure to organophosphate pesticides and IQ in 7-year-old children. *Environ Health Perspect.* Aug 2011;119(8):1189-1195.

26. Ribas-Fito N, Torrent M, Carrizo D, Julvez J, Grimalt JO, Sunyer J. Exposure to hexachlorobenzene during pregnancy and children's social behavior at 4 years of age. *Environ Health Perspect.* Mar 2007;115(3):447-450.

27. Eskenazi B, Marks AR, Bradman A, et al. *In utero* exposure to dichlorodiphenyltrichloroethane (DDT) and dichlorodiphenyldichloroethylene (DDE) and neurodevelopment among young Mexican American children. *Pediatrics.* Jul 2006;118(1):233-241.

28. Rauh VA, Garfinkel R, Perera FP, et al. Impact of prenatal chlorpyrifos exposure on neurodevelopment in the first 3 years of life among inner-city children. *Pediatrics.* Dec 2006;118(6):e1845-1859.

29. Lovasi GS, Quinn JW, Rauh VA, et al. Chlorpyrifos Exposure and Urban Residential Environment Characteristics as Determinants of Early Childhood Neurodevelopment. *Am J Public Health.* Mar 18 2010.

30. Roberts EM, English PB, Grether JK, Windham GC, Somberg L, Wolff C. Maternal residence near agricultural pesticide applications and autism spectrum disorders among children in the California Central Valley. *Environ Health Perspect.* Oct 2007;115(10):1482-1489.

31. Sagiv SK, Thurston SW, Bellinger DC, Tolbert PE, Altshul LM, Korrick SA. Prenatal organochlorine exposure and behaviors associated with attention deficit hyperactivity disorder in school-aged children. *Am J Epidemiol.* Mar 1 2010;171(5):593-601.

32. Bouchard MF, Bellinger DC, Wright RO, Weisskopf MG. Attention-deficit/hyperactivity disorder and urinary metabolites of organophosphate pesticides. *Pediatrics.* Jun 2010;125(6):e1270-1277.

33. Marks AR, Harley K, Bradman A, et al. Organophosphate pesticide exposure and attention in young Mexican-American children: the CHAMACOS study. *Environ Health Perspect.* Dec 2010;118(12):1768-1774.

34. Grandjean P, Harari R, Barr DB, Debes F. Pesticide exposure and stunting as independent predictors of neurobehavioral deficits in Ecuadorian school children. *Pediatrics.* Mar 2006;117(3):e546-556.

35. Rauh V, Arunajadai S, Horton M, et al. Seven-year neurodevelopmental scores and prenatal exposure to chlorpyrifos, a common agricultural pesticide. *Environ Health Perspect.* Aug 2011;119(8):1196-1201.

36. Kamel F, Hoppin JA. Association of pesticide exposure with neurologic dysfunction and disease. *Environ Health Perspect.* Jun 2004;112(9):950-958.

37. Stokes L, Stark A, Marshall E, Narang A. Neurotoxicity among pesticide applicators exposed to organophosphates. *Occup Environ Med.* Oct 1995;52(10):648-653.

38. Steenland K, Dick RB, Howell RJ, et al. Neurologic function among termiticide applicators exposed to chlorpyrifos. *Environ Health Perspect.* Apr 2000;108(4):293-300.

39. Kamel F, Rowland AS, Park LP, et al. Neurobehavioral performance and work experience in Florida farmworkers. *Environ Health Perspect.* Nov 2003;111(14):1765-1772.

40. Kamel F, Engel LS, Gladen BC, Hoppin JA, Alavanja MC, Sandler DP. Neurologic symptoms in licensed private pesticide applicators in the agricultural health study. *Environ Health Perspect.* Jul 2005;113(7):877-882.

41. Beseler CL, Stallones L, Hoppin JA, et al. Depression and pesticide exposures among private pesticide applicators enrolled in the Agricultural Health Study. *Environ Health Perspect.* Dec 2008;116(12):1713-1719.

42. Engel LS, Checkoway H, Keifer MC, et al. Parkinsonism and occupational exposure to pesticides. *Occup Environ Med.* Sep 2001;58(9):582-589.

43. Kamel F, Tanner C, Umbach D, et al. Pesticide exposure and self-reported Parkinson's disease in the agricultural health study. *Am J Epidemiol.* Feb 15 2007;165(4):364-374.

44. Kanthasamy AG, Kitazawa M, Kanthasamy A, Anantharam V. Dieldrin-induced neurotoxicity: relevance to Parkinson's disease pathogenesis. *Neurotoxicology.* Aug 2005;26(4):701-719.

45. Dick FD. Parkinson's disease and pesticide exposures. *Br Med Bull.* 2006;79-80:219-231.

46. Brown TP, Rumsby PC, Capleton AC, Rushton L, Levy LS. Pesticides and Parkinson's disease--is there a link? *Environ Health Perspect.* Feb 2006;114(2):156-164.

47. Dick FD, De Palma G, Ahmadi A, et al. Environmental risk factors for Parkinson's disease and parkinsonism: the Geoparkinson study. *Occup Environ Med.* Oct 2007;64(10):666-672.

48. Tanner CM, Ross GW, Jewell SA, et al. Occupation and risk of parkinsonism: a multicenter case-control study. *Arch Neurol.* Sep 2009;66(9):1106-1113.

49. Costello S, Cockburn M, Bronstein J, Zhang X, Ritz B. Parkinson's disease and residential exposure to maneb and paraquat from agricultural applications in the central valley of California. *Am J Epidemiol.* Apr 15 2009;169(8):919-926.

50. Domico LM, Zeevalk GD, Bernard LP, Cooper KR. Acute neurotoxic effects of mancozeb and maneb in mesencephalic neuronal cultures are associated with mitochondrial dysfunction. *Neurotoxicology.* Sep 2006;27(5):816-825.

51. Costa LG, Giordano G, Guizzetti M, Vitalone A. Neurotoxicity of pesticides: a brief review. *Front Biosci.* 2008;13:1240-1249.

52. Mink PJ, Adami HO, Trichopoulos D, Britton NL, Mandel JS. Pesticides and prostate cancer: a review of epidemiologic studies with specific agricultural exposure information. *Eur J Cancer Prev.* Apr 2008;17(2):97-110.

53. Zahm SH, Ward MH. Pesticides and childhood cancer. *Environ Health Perspect.* Jun 1998;106 Suppl 3:893-908.

54. Alexander FE, Patheal SL, Biondi A, et al. Transplacental chemical exposure and risk of infant leukemia with MLL gene fusion. *Cancer Res.* Mar 15 2001;61(6):2542-2546.

55. Meinert R, Schuz J, Kaletsch U, Kaatsch P, Michaelis J. Leukemia and non-Hodgkin's lymphoma in childhood and exposure to pesticides: results of a register-based case-control study in Germany. *Am J Epidemiol.* Apr 1 2000;151(7):639-646; discussion 647-650.

56. Infante-Rivard C, Labuda D, Krajinovic M, Sinnett D. Risk of childhood leukemia associated with exposure to pesticides and with gene polymorphisms. *Epidemiology.* Sep 1999;10(5):481-487.

57. Infante-Rivard C, Sinnett D. Preconceptional paternal exposure to pesticides and increased risk of childhood leukaemia. *Lancet.* Nov 20 1999;354(9192):1819.

58. Ma X, Buffler PA, Gunier RB, et al. Critical windows of exposure to household pesticides and risk of childhood leukemia. *Environ Health Perspect.* Sep 2002;110(9):955-960.

59. Reynolds P, Von Behren J, Gunier RB, Goldberg DE, Harnly M, Hertz A. Agricultural pesticide use and childhood cancer in California. *Epidemiology.* Jan 2005;16(1):93-100.

60. Reynolds P, Von Behren J, Gunier RB, Goldberg DE, Hertz A, Harnly ME. Childhood cancer and agricultural pesticide use: an ecologic study in California. *Environ Health Perspect.* Mar 2002;110(3):319-324.

61. Turner MC, Wigle DT, Krewski D. Residential pesticides and childhood leukemia: a systematic review and meta-analysis. *Environ Health Perspect.* Jan 2010;118(1):33-41.

62. Wigle DT, Turner MC, Krewski D. A systematic review and meta-analysis of childhood leukemia and parental occupational pesticide exposure. *Environ Health Perspect.* Oct 2009;117(10):1505-1513.

63. Feychting M, Plato N, Nise G, Ahlbom A. Paternal occupational exposures and childhood cancer. *Environ Health Perspect.* Feb 2001;109(2):193-196.

64. Schuz J, Kaletsch U, Kaatsch P, Meinert R, Michaelis J. Risk factors for pediatric tumors of the central nervous system: results from a German population-based case-control study. *Med Pediatr Oncol.* Feb 2001;36(2):274-282.

65. van Wijngaarden E, Stewart PA, Olshan AF, Savitz DA, Bunin GR. Parental occupational exposure to pesticides and childhood brain cancer. *Am J Epidemiol.* Jun 1 2003;157(11):989-997.

66. Schreinemachers DM. Cancer mortality in four northern wheat-producing states. *Environ Health Perspect.* Sep 2000;108(9):873-881.

67. Rodvall Y, Dich J, Wiklund K. Cancer risk in offspring of male pesticide applicators in agriculture in Sweden. *Occup Environ Med.* Oct 2003;60(10):798-801.

68. Heacock H, Hertzman C, Demers PA, et al. Childhood cancer in the offspring of male sawmill workers occupationally exposed to chlorophenate fungicides. *Environ Health Perspect.* Jun 2000;108(6):499-503.

69. Flower KB, Hoppin JA, Lynch CF, et al. Cancer risk and parental pesticide application in children of Agricultural Health Study participants. *Environ Health Perspect.* Apr 2004;112(5):631-635.

70. McKinney PA, Fear NT, Stockton D. Parental occupation at periconception: findings from the United Kingdom Childhood Cancer Study. *Occup Environ Med.* Dec 2003;60(12):901-909.

71. Cordier S, Iglesias MJ, Le Goaster C, Guyot MM, Mandereau L, Hemon D. Incidence and risk factors for childhood brain tumors in the Ile de France. *Int J Cancer.* Dec 15 1994;59(6):776-782.

72. Merhi M, Raynal H, Cahuzac E, Vinson F, Cravedi JP, Gamet-Payrastre L. Occupational exposure to pesticides and risk of hematopoietic cancers: meta-analysis of case-control studies. *Cancer Causes Control.* Dec 2007;18(10):1209-1226.

73. McDuffie HH, Pahwa P, Robson D, et al. Insect repellents, phenoxyherbicide exposure, and non-Hodgkin's lymphoma. *J Occup Environ Med.* Aug 2005;47(8):806-816.

74. Hardell L, Eriksson M, Nordstrom M. Exposure to pesticides as risk factor for non-Hodgkin's lymphoma and hairy cell leukemia: pooled analysis of two Swedish case-control studies. *Leuk Lymphoma.* May 2002;43(5):1043-1049.

75. Miligi L, Costantini AS, Veraldi A, Benvenuti A, Vineis P. Cancer and pesticides: an overview and some results of the Italian multicenter case-control study on hematolymphopoietic malignancies. *Ann N Y Acad Sci.* Sep 2006;1076:366-377.

76. Beane Freeman LE, Bonner MR, Blair A, et al. Cancer incidence among male pesticide applicators in the Agricultural Health Study cohort exposed to diazinon. *Am J Epidemiol.* Dec 1 2005;162(11):1070-1079.

77. Sharma-Wagner S, Chokkalingam AP, Malker HS, Stone BJ, McLaughlin JK, Hsing AW. Occupation and prostate cancer risk in Sweden. *J Occup Environ Med.* May 2000;42(5):517-525.

78. Settimi L, Masina A, Andrion A, Axelson O. Prostate cancer and exposure to pesticides in agricultural settings. *Int J Cancer.* Apr 20 2003;104(4):458-461.

79. Mills PK, Yang R. Prostate cancer risk in California farmworkers. *J Occup Environ Med.* Mar 2003;45(3):249-258.

80. MacLennan PA, Delzell E, Sathiakumar N, et al. Cancer incidence among triazine herbicide manufacturing workers. *J Occup Environ Med.* Nov 2002;44(11):1048-1058.

81. Fleming LE, Bean JA, Rudolph M, Hamilton K. Cancer incidence in a cohort of licensed pesticide applicators in Florida. *J Occup Environ Med.* Apr 1999;41(4):279-288.

82. Dich J, Wiklund K. Prostate cancer in pesticide applicators in Swedish agriculture. *Prostate.* Feb 1 1998;34(2):100-112.

83. Alavanja MC, Samanic C, Dosemeci M, et al. Use of agricultural pesticides and prostate cancer risk in the Agricultural Health Study cohort. *Am J Epidemiol.* May 1 2003;157(9):800-814.

84. Kross BC, Burmeister LF, Ogilvie LK, Fuortes LJ, Fu CM. Proportionate mortality study of golf course superintendents. *Am J Ind Med.* May 1996;29(5):501-506.

85. Settimi L, Comba P, Bosia S, et al. Cancer risk among male farmers: a multi-site case-control study. *Int J Occup Med Environ Health.* 2001;14(4):339-347.

86. Checkoway H, DiFerdinando G, Hulka BS, Mickey DD. Medical, life-style, and occupational risk factors for prostate cancer. *Prostate.* 1987;10(1):79-88.

87. van der Gulden JW, Kolk JJ, Verbeek AL. Work environment and prostate cancer risk. *Prostate.* Nov 1995;27(5):250-257.

88. Ewings P, Bowie C. A case-control study of cancer of the prostate in Somerset and east Devon. *Br J Cancer.* Aug 1996;74(4):661-666.

89. Ramlow JM, Spadacene NW, Hoag SR, Stafford BA, Cartmill JB, Lerner PJ. Mortality in a cohort of pentachlorophenol manufacturing workers, 1940-1989. *Am J Ind Med.* Aug 1996;30(2):180-194.

90. Mellemgaard A, Engholm G, McLaughlin JK, Olsen JH. Occupational risk factors for renal-cell carcinoma in Denmark. *Scand J Work Environ Health.* Jun 1994;20(3):160-165.

91. Hu J, Mao Y, White K. Renal cell carcinoma and occupational exposure to chemicals in Canada. *Occup Med (Lond).* May 2002;52(3):157-164.

92. Buzio L, Tondel M, De Palma G, et al. Occupational risk factors for renal cell cancer. An Italian case-control study. *Med Lav.* Jul-Aug 2002;93(4):303-309.

93. Proudfoot AT. Pentachlorophenol poisoning. *Toxicol Rev.* 2003;22(1):3-11.

94. Agency USEP. *Reregistration Eligibility Decision (RED) for Pentachlorophenol.* 2008. EPA 739-R-08-008.

95. Ren X, McHale CM, Skibola CF, Smith AH, Smith MT, Zhang L. An emerging role for epigenetic dysregulation in arsenic toxicity and carcinogenesis. *Environ Health Perspect.* Jan 2011;119(1):11-19.

96. Maloney ME. Arsenic in Dermatology. *Dermatol Surg.* Mar 1996;22(3):301-304.

97. Celik I, Gallicchio L, Boyd K, et al. Arsenic in drinking water and lung cancer: a systematic review. *Environ Res.* Sep 2008;108(1):48-55.

98. Gronemeyer H, Benhamou B, Berry M, et al. Mechanisms of antihormone action. *J Steroid Biochem Mol Biol.* Mar 1992;41(3-8):217-221.

99. Ruegg J, Penttinen-Damdimopoulou P, Makela S, Pongratz I, Gustafsson JA. Receptors mediating toxicity and their involvement in endocrine disruption. *EXS.* 2009;99:289-323.

100. Swedenborg E, Ruegg J, Makela S, Pongratz I. Endocrine disruptive chemicals: mechanisms of action and involvement in metabolic disorders. *J Mol Endocrinol.* Jul 2009;43(1):1-10.

101. Zhou LX, Dehal SS, Kupfer D, et al. Cytochrome P450 catalyzed covalent binding of methoxychlor to rat hepatic, microsomal iodothyronine 5'-monodeiodinase, type I: does exposure to methoxychlor disrupt thyroid hormone metabolism? *Arch Biochem Biophys.* Oct 1 1995;322(2):390-394.

102. Gray LE, Ostby J, Furr J, et al. Effects of environmental antiandrogens on reproductive development in experimental animals. *Hum Reprod Update.* May-Jun 2001;7(3):248-264.

103. Rasier G, Parent AS, Gerard A, Lebrethon MC, Bourguignon JP. Early maturation of gonadotropin-releasing hormone secretion and sexual precocity after exposure of infant female rats to estradiol or dichlorodiphenyltrichloroethane. *Biol Reprod.* Oct 2007;77(4):734-742.

104. Lahiri P, Sircar S. Suppression of adrenocortical function in female mice by lindane (gamma-HCH). *Toxicology.* Feb 11 1991;66(1):75-79.

105. Boas M, Feldt-Rasmussen U, Skakkebaek NE, Main KM. Environmental chemicals and thyroid function. *Eur J Endocrinol.* May 2006;154(5):599-611.

106. Boas M, Main KM, Feldt-Rasmussen U. Environmental chemicals and thyroid function: an update. *Curr Opin Endocrinol Diabetes Obes.* Oct 2009;16(5):385-391.

107. Guillette LJ, Jr. Endocrine disrupting contaminants--beyond the dogma. *Environ Health Perspect.* Apr 2006;114 Suppl 1:9-12.

108. Guillette LJ, Jr., Gunderson MP. Alterations in development of reproductive and endocrine systems of wildlife populations exposed to endocrine-disrupting contaminants. *Reproduction.* Dec 2001;122(6):857-864.

109. Leung KM, Kwong RP, Ng WC, et al. Ecological risk assessments of endocrine disrupting organotin compounds using marine neogastropods in Hong Kong. *Chemosphere.* Nov 2006;65(6):922-938.

110. Fair PA, Adams J, Mitchum G, et al. Contaminant blubber burdens in Atlantic bottlenose dolphins (Tursiops truncatus) from two southeastern US estuarine areas: concentrations and patterns of PCBs, pesticides, PBDEs, PFCs, and PAHs. *Sci Total Environ.* Mar 1 2010;408(7):1577-1597.

111. Jenssen BM. Endocrine-disrupting chemicals and climate change: A worst-case combination for arctic marine mammals and seabirds? *Environ Health Perspect.* Apr 2006;114 Suppl 1:76-80.

112. Diamanti-Kandarakis E, Bourguignon JP, Giudice LC, et al. Endocrine-disrupting chemicals: an Endocrine Society scientific statement. *Endocr Rev.* Jun 2009;30(4):293-342.

113. Krstevska-Konstantinova M, Charlier C, Craen M, et al. Sexual precocity after immigration from developing countries to Belgium: evidence of previous exposure to organochlorine pesticides. *Hum Reprod.* May 2001;16(5):1020-1026.

114. Rogan WJ, Ragan NB. Some evidence of effects of environmental chemicals on the endocrine system in children. *Int J Hyg Environ Health.* Oct 2007;210(5):659-667.

115. Fan W, Yanase T, Morinaga H, et al. Herbicide atrazine activates SF-1 by direct affinity and concomitant co-activators recruitments to induce aromatase expression via promoter II. *Biochem Biophys Res Commun.* Apr 20 2007;355(4):1012-1018.

116. Crisp TM, Clegg ED, Cooper RL, et al. Environmental endocrine disruption: an effects assessment and analysis. *Environ Health Perspect.* Feb 1998;106 Suppl 1:11-56.

117. Caserta D, Maranghi L, Mantovani A, Marci R, Maranghi F, Moscarini M. Impact of endocrine disruptor chemicals in gynaecology. *Hum Reprod Update.* Jan-Feb 2008;14(1):59-72.

118. Cannon SB, Veazey JM, Jr., Jackson RS, et al. Epidemic kepone poisoning in chemical workers. *Am J Epidemiol.* Jun 1978;107(6):529-537.

119. Slutsky M, Levin JL, Levy BS. Azoospermia and oligospermia among a large cohort of DBCP applicators in 12 countries. *Int J Occup Environ Health.* Apr-Jun 1999;5(2):116-122.

120. Abell A, Ernst E, Bonde JP. Semen quality and sexual hormones in greenhouse workers. *Scand J Work Environ Health.* Dec 2000;26(6):492-500.

121. Hauser R, Chen Z, Pothier L, Ryan L, Altshul L. The relationship between human semen parameters and environmental exposure to polychlorinated biphenyls and p,p'-DDE. *Environ Health Perspect.* Sep 2003;111(12):1505-1511.

122. Juhler RK, Larsen SB, Meyer O, et al. Human semen quality in relation to dietary pesticide exposure and organic diet. *Arch Environ Contam Toxicol.* Oct 1999;37(3):415-423.

123. Kamijima M, Hibi H, Gotoh M, et al. A survey of semen indices in insecticide sprayers. *J Occup Health.* Mar 2004;46(2):109-118.

124. Larsen SB, Giwercman A, Spano M, Bonde JP. A longitudinal study of semen quality in pesticide spraying Danish farmers. The ASCLEPIOS Study Group. *Reprod Toxicol.* Nov-Dec 1998;12(6):581-589.

125. Lifeng T, Shoulin W, Junmin J, et al. Effects of fenvalerate exposure on semen quality among occupational workers. *Contraception.* Jan 2006;73(1):92-96.

126. Padungtod C, Savitz DA, Overstreet JW, Christiani DC, Ryan LM, Xu X. Occupational pesticide exposure and semen quality among Chinese workers. *J Occup Environ Med.* Oct 2000;42(10):982-992.

127. Rignell-Hydbom A, Rylander L, Giwercman A, Jonsson BA, Nilsson-Ehle P, Hagmar L. Exposure to CB-153 and p,p'-DDE and male reproductive function. *Hum Reprod.* Sep 2004;19(9):2066-2075.

128. Whorton MD, Milby TH, Stubbs HA, Avashia BH, Hull EQ. Testicular function among carbaryl-exposed exployees. *J Toxicol Environ Health.* Sep 1979;5(5):929-941.

129. Prins GS. Endocrine disruptors and prostate cancer risk. *Endocr Relat Cancer.* Sep 2008;15(3):649-656.

130. Goldner WS, Sandler DP, Yu F, Hoppin JA, Kamel F, Levan TD. Pesticide use and thyroid disease among women in the Agricultural Health Study. *Am J Epidemiol.* Feb 15 2010;171(4):455-464.

131. Dallaire R, Dewailly E, Pereg D, Dery S, Ayotte P. Thyroid function and plasma concentrations of polyhalogenated compounds in Inuit adults. *Environ Health Perspect.* Sep 2009;117(9):1380-1386.

132. Moretto A. Indoor spraying with the pyrethroid insecticide lambda-cyhalothrin: effects on spraymen and inhabitants of sprayed houses. *Bull World Health Organ.* 1991;69(5):591-594.

133. Newton JG, Breslin AB. Asthmatic reactions to a commonly used aerosol insect killer. Med J Aust. Apr 16 1983;1(8):378-380.

134. Culver CA, Malina JJ, Talbert RL. Probable anaphylactoid reaction to a pyrethrin pediculocide shampoo. *Clin Pharm.* Nov 1988;7(11):846-849.

135. Fryer AD, Lein PJ, Howard AS, Yost BL, Beckles RA, Jett DA. Mechanisms of organophosphate insecticide-induced airway hyperreactivity. *Am J Physiol Lung Cell Mol Physiol.* May 2004;286(5):L963-969.

136. Proskocil BJ, Bruun DA, Thompson CM, Fryer AD, Lein PJ. Organophosphorus pesticides decrease M2 muscarinic receptor function in guinea pig airway nerves via indirect mechanisms. *PLoS One*. 2010;5(5):e10562.

137. Gultekin F, Ozturk M, Akdogan M. The effect of organophosphate insecticide chlorpyrifos-ethyl on lipid peroxidation and antioxidant enzymes (in vitro). *Arch Toxicol.* Nov 2000;74(9):533-538.

138. Ranjbar A, Pasalar P, Abdollahi M. Induction of oxidative stress and acetylcholinesterase inhibition in organophosphorous pesticide manufacturing workers. *Hum Exp Toxicol.* Apr 2002;21(4):179-182.

139. Duramad P, Harley K, Lipsett M, et al. Early environmental exposures and intracellular Thl/Th2 cytokine profiles in 24-month-old children living in an agricultural area. *Environ Health Perspect.* Dec 2006;114(12):1916-1922.

140. Salameh P, Waked M, Baldi I, Brochard P, Saleh BA. Respiratory diseases and pesticide exposure: a case-control study in Lebanon. *J Epidemiol Community Health.* Mar 2006;60(3):256-261.

141. Hoppin JA, Umbach DM, London SJ, et al. Pesticides and atopic and nonatopic asthma among farm women in the Agricultural Health Study. *Am J Respir Crit Care Med.* Jan 1 2008;177(1):11-18.

142. Hoppin JA, Umbach DM, London SJ, et al. Pesticide use and adult-onset asthma among male farmers in the Agricultural Health Study. *Eur Respir J.* Dec 2009;34(6):1296-1303.

143. Hoppin JA, Umbach DM, London SJ, Lynch CF, Alavanja MC, Sandler DP. Pesticides and adult respiratory outcomes in the agricultural health study. *Ann N Y Acad Sci.* Sep 2006;1076:343-354.

144. Hoppin JA, Umbach DM, London SJ, Lynch CF, Alavanja MC, Sandler DP. Pesticides associated with wheeze among commercial pesticide applicators in the Agricultural Health Study. *Am J Epidemiol.* Jun 15 2006;163(12):1129-1137.

145. Hoppin JA, Umbach DM, London SJ, Alavanja MC, Sandler DP. Diesel exhaust, solvents, and other occupational exposures as risk factors for wheeze among farmers. *Am J Respir Crit Care Med.* Jun 15 2004;169(12):1308-1313.

146. Jones SM, Burks AW, Spencer HJ, et al. Occupational asthma symptoms and respiratory function among aerial pesticide applicators. *Am J Ind Med.* Apr 2003;43(4):407-417.

147. O'Sullivan BC, Lafleur J, Fridal K, et al. The effect of pesticide spraying on the rate and severity of ED asthma. *Am J Emerg Med.* Jul 2005;23(4):463-467.

148. Karpati AM, Perrin MC, Matte T, Leighton J, Schwartz J, Barr RG. Pesticide spraying for West Nile virus control and emergency department asthma visits in New York City, 2000. *Environ Health Perspect.* Aug 2004;112(11):1183-1187.

149. Boers D, van Amelsvoort L, Colosio C, et al. Asthmatic symptoms after exposure to ethylenebisdithiocarbamates and other pesticides in the Europit field studies. *Hum Exp Toxicol.* Sep 2008;27(9):721-727.

150. Merchant JA, Naleway AL, Svendsen ER, et al. Asthma and farm exposures in a cohort of rural Iowa children. *Environ Health Perspect.* Mar 2005;113(3):350-356.

151. Salameh PR, Baldi I, Brochard P, Raherison C, Abi Saleh B, Salamon R. Respiratory symptoms in children and exposure to pesticides. *Eur Respir J.* Sep 2003;22(3):507-512.

152. Salam MT, Li YF, Langholz B, Gilliland FD. Early-life environmental risk factors for asthma: findings from the Children's Health Study. *Environ Health Perspect.* May 2004;112(6):760-765.

153. Karmaus W, Kuehr J, Kruse H. Infections and atopic disorders in childhood and organochlorine exposure. *Arch Environ Health.* Nov-Dec 2001;56(6):485-492.

154. Sunyer J, Torrent M, Munoz-Ortiz L, et al. Prenatal dichlorodiphenyldichloroethylene (DDE) and asthma in children. *Environ Health Perspect.* Dec 2005;113(12):1787-1790.

*Section VI*

# APPENDIXES

## Detailed Occupational and Environmental Exposure History Questions

*(Items marked in* **bold type** *are especially important for a pesticide exposure history)*

### ADULT PATIENT

OCCUPATIONAL EXPOSURE

- **What is your occupation?** *(If unemployed, discuss recent employment as appropriate or go to next section)*
- **How long have you been doing this job?**
- **Describe your work and the hazards to which you are exposed.** (*e.g.,* **pesticides**, solvents or other chemicals, dust, fumes, metals, fibers, radiation, biologic agents, noise, heat, cold, vibration)
- **Under what circumstances do you use protective equipment?** (*e.g.*, **work clothes, safety glasses, respirator, gloves**, and hearing protection)
- **Do you smoke or eat at the worksite?**
- **List previous jobs in chronological order, include full and part-time, temporary, second jobs, summer jobs and military experience.** *(Because this question can take a long time to answer, one option is to ask the patient to fill out a form with this question on it prior to the formal history taking by the clinician. Another option is to take a shorter history by asking the patient to list only the prior jobs that involved the agents of interest. For example, one could ask for all current and past jobs involving pesticide exposure.)*

ENVIRONMENTAL EXPOSURE HISTORY

- **Are pesticides** (*e.g.,* **bug or weed killers, flea and tick sprays, collars, powders or shampoos) used in your home or garden or on your pet?**
- **If pesticides are used:**
    - **Who applies the pesticides?**
    - **Is a licensed pesticide applicator involved?**
    - **Where are the pesticides stored?**
- **Do you or any household members have a hobby with exposure to any hazardous materials** (*e.g.,* **pesticides**, paints, ceramics, solvents, metals, glues)?
- **Is food handled properly** (*e.g.,* **washing of raw fruits and vegetables**)?
- **Do you live within 1/4 mile of an agricultural area** (*e.g.,* **field, orchard, greenhouse) where plants, vegetables or fruits are grown?**
- **Did you ever live near a facility which could have contaminated the surrounding area** (*e.g.,* mine, **plant,** smelter, **dump site)?**
- **Have you ever changed your residence because of a health problem?**
- **Does your drinking water come from a private well, city water supply and/or grocery store?**
- Do you work on your car?
- Which of the following do you have in your home: air conditioner/purifier, central heating (gas or oil), gas stove, electric stove, fireplace, wood stove or humidifier?
- Have you recently acquired new furniture or carpet, or remodeled your home?
- Have you weatherized your home recently?
- Approximately what year was your home built?

SYMPTOMS AND MEDICAL CONDITIONS

- **Does the timing of your symptoms have any relationship to your work hours?** *(If unemployed, skip to 3rd bullet)*
- **Has anyone else at work suffered the same or similar problems?**
- **Does the timing of your symptoms have any relationship to environmental activities listed above?**
- **Has any other household member or nearby neighbor suffered similar health problems?**

NON-OCCUPATIONAL EXPOSURES POTENTIALLY RELATED TO ILLNESS OR INJURY

- Do you use tobacco? If yes, in what forms (cigarettes, pipe, cigar, chewing tobacco)? About how many do you smoke or how much tobacco do you use per day? At what age did you start using tobacco? Are there other tobacco smokers in the home?

NON-OCCUPATIONAL ADULT EXPOSURES, CONT.

- Do you drink alcohol? How much per day or week? At what age did you start?
- What medications or drugs are you taking? (Include prescription and non-prescription uses)
- **Has anyone in the family worked with hazardous materials that they might have brought home** *(e.g.,* **pesticides,** asbestos, lead)? *(If yes, inquire about household members potentially exposed.)*

## PEDIATRIC PATIENT

*(questions asked of parent or guardian)*

ENVIRONMENTAL EXPOSURE HISTORY (PESTICIDE-RELATED QUESTIONS)

- **Are pesticides** *(e.g.,* **insect or weed killers, flea and tick sprays, collars, powders, or shampoos) used in your home or garden or on your pet?**
- **Where** *(e.g.,* **out of reach of children) are pesticides stored? What types of containers are pesticides stored in?**
- **Do you or any household member have a hobby with exposure to any hazardous materials** *(e.g.,* **pesticides, paints, ceramics, solvents, metals, glues)?**
- **If pesticides are used:**
  - **Who applies the pesticides?**
  - **Do you use a licensed pesticide applicator for treatments?**
  - **How long do you wait before letting children play on areas recently treated with pesticides?**
  - **Where are the pesticides stored?**
- **Is food handled properly (***e.g.***, washing of raw fruits and vegetables)?**
- **Do you live within 1/4 mile of an agricultural area (***e.g.,* **field, orchard, greenhouse) where plants, vegetables or fruits are grown?**

OCCUPATIONAL EXPOSURE

- **What is your occupation and that of other household members?** *(If no employed individuals, go to next section)*
- **Describe your work and the hazards to which you are exposed.** *(e.g.,* **pesticides**, solvents or other chemicals, dust, fumes, metals, fibers, radiation, biologic agents, noise, heat, cold, vibration)

ADDITIONAL ENVIRONMENTAL EXPOSURE QUESTIONS

- **Has the child ever lived near a facility that could have contaminated the surrounding area** *(e.g.,* mine, **plant,** smelter, **dump site**)?
- **Has the child ever changed residence because of a health problem?**
- **Does the child's drinking water come from a private well, city water supply and/or grocery store?**
- Which of the following are in the child's home: air conditioner/purifier, central heating (gas or oil), gas stove, electric stove, fireplace, wood stove or humidifier?
- Is there recently acquired new furniture or carpet, or recent home remodeling in the patient's home?
- Has the home been weatherized recently?
- Approximately what year was the home built?

SYMPTOMS AND MEDICAL CONDITIONS

- Does the timing of symptoms have any relationship to environmental activities listed above?
- Has any other household member or nearby neighbor suffered similar health problems?

NON-OCCUPATIONAL EXPOSURES POTENTIALLY RELATED TO ILLNESS OR INJURY

- Are there any tobacco users in the home? If yes, in what forms (cigarettes, pipe, cigar, chewing tobacco)?
- What medications or drugs is the child taking? (Include prescription and non-prescription uses)
- **Has anyone in the family worked with hazardous materials that they might have brought home** *(e.g.,* **pesticides,** asbestos, lead)? (If yes, inquire about household members potentially exposed.)

## Key Competencies for Clinicians *from* National Strategies for Healthcare Providers: Pesticides Initiative

EPA, in partnership with several federal agencies and organizations, leads the National Strategies for Health Care Providers: Pesticides Initiative. Established in 1998, the Initiative goal is to improve the recognition, diagnosis, treatment, and prevention of pesticide-related illnesses. As a framework to achieve this goal, the Initiative developed a set of practical skills to guide students, nurses and practicing clinicians in recognizing and managing pesticide-related illnesses. These skills and competencies can be integrated into existing education and training of healthcare providers to facilitate the effective management of patients with suspected pesticide-related illnesses.[1]

### NATIONAL PESTICIDE PRACTICE SKILLS GUIDELINES FOR MEDICAL AND NURSING PRACTICE[2]

PRACTICE SKILL I: TAKING AN ENVIRONMENTAL HISTORY

- Understand the purposes and general principles for taking an occupational and environmental history.
- Incorporate general occupational and environmental screening questions into routine patient histories.
- Be able to take a complete occupational and environmental exposure/health history for adults and children.

PRACTICE SKILL II: AWARENESS OF COMMUNITY AND INDIVIDUAL PESTICIDE RISK FACTORS

- Possess basic awareness of occupational and environmental aspects of communities in which patients live.
- Recognize high-risk occupations for pesticide exposure.
- Develop community resource list.

PRACTICE SKILL III: KNOWLEDGE OF KEY HEALTH PRINCIPLES

- Demonstrate key principles of environmental/occupational health, epidemiology, and population-based health.
- Understand the dose-response relationship.
- Understand measures of morbidity/mortality and study designs.

PRACTICE SKILL IV: CLINICAL MANAGEMENT OF PESTICIDE EXPOSURE

- Know different groups of pesticides, their mechanism of toxicity (pathophysiology) and adverse health effects.
- Recognize the signs and symptoms of pesticide exposures (both acute and chronic).
- Diagnose pesticide-related illness using appropriate testing procedures and treat pesticide over-exposures.
- Treat and manage health conditions associated with pesticide exposure (know anticholinergic agents and dosages, antidote for organophosphates, treatment of seizures) or refer patients to appropriate specialists and resources, and follow up appropriately.

PRACTICE SKILL V: REPORTING PESTICIDE EXPOSURE AND SUPPORTING SURVEILLANCE EFFORTS

- Understand the importance of surveillance and reporting.
- Know the roles of federal and state regulatory agencies with regard to pesticide exposure control.
- Report pesticide exposures as required.

PRACTICE SKILL VI: PROVIDING PREVENTION GUIDANCE AND EDUCATION TO PATIENTS

- Engage in primary prevention strategies to promote health and prevent disease among patients.
- Work proactively with patients and the community to prevent exposure, ensure early detection, and limit effects of illness.

[1]*For more information visit EPA's Web page on the National Strategies for Health Care Providers: Pesticide Initiative at: http://www.epa.gov/oppfead1/safety/healthcare/healthcare.htm*

[2]*Derived from http://www.epa.gov/oppfead1/safety/healthcare/practiceskifinal.pdf*

*Section VII*

# INDEXES

# Index of Signs and Symptoms of Acute Poisoning

*Common symptoms were not included:*

- *malaise*
- *fatigue*
- *dizziness*
- *nausea*
- *vomiting*

This index provides a table of pesticides and their related symptoms and signs and affected organ systems in poisoned individuals. This may be useful in raising the index of suspicion for pesticide toxicity where these signs and symptoms occur. Such suspicion can be evaluated further within the differential diagnosis as appropriate.

It is important to keep in mind that the signs and symptoms listed have multiple causes, pesticidal and nonpesticidal. In addition, no specific symptoms or signs are invariably present in poisonings by particular pesticides. Toxicological presentation may vary based on dosage, route(s) of exposure, life stage of the patient, and patient's genetic vulnerability, co-exposures and/or underlying health status. This complexity may explain why many poisonings are characterized by unexpected or atypical manifestations.

*It is important to keep in mind that the signs and symptoms listed have multiple causes, pesticidal and nonpesticidal.*

The table does not differentiate clinical presentation by route of exposure or dosage. For example, effects of high-dose ingestion are not distinguished from effects of relatively low-dose dermal absorption, nor are topical effects distinguished from systemic dermal manifestations. Such details are addressed more fully in the chapters addressing the specific pesticides, which make up the bulk of this manual. The list of pesticides in this chapter is intended to serve as a clue for the clinician toward further inquiry.

The word "poisoning" is used loosely in these headings to include topical as well as systemic effects that are acute manifestations rather than delayed sequelae or chronic conditions, although in some cases acute effects may persist. For chronic health conditions including long-term sequelae after acute poisoning and chronic conditions associated with lower level, repeat exposures, see **Chapter 2□**, *Chronic Effects*.

Pesticides that are relatively consistent in causing particular manifestations are listed in the column headed "Characteristic of These Poisonings." Agents that are associated with conditions less consistently or are less prominent features of poisoning are listed in the right-hand column, headed "May Occur in These Poisonings." Obviously, the distinction is not always clear cut.

Some symptoms (malaise, fatigue, dizziness, nausea and vomiting) occur so commonly in poisoned individuals that they have little or no value in differential diagnosis, and are therefore not included in these tables.

**SYSTEM: General/Non-Specific**

| SYMPTOMS/ SIGNS/DISEASE CATEGORIES | CHARACTERISTIC OF THESE POISONINGS | MAY OCCUR IN THESE POISONINGS |
|---|---|---|
| Rotten egg odor | Sulfur | |
| Hypothermia | Creosote | |
| Hyperthermia (fever, pyrexia) | Nitrophenols<br>Pentachlorophenol | Borate<br>Thallium<br>Metaldehyde<br>Inorganic arsenicals<br>Chlorophenoxy compounds<br>Cadmium dusts<br>Naphthalene |
| Chills | Phosphine<br>Arsine | |
| Hot sensations | Nitrophenols<br>Chlordimeform | Pentachlorophenol |
| Myalgia | Paraquat<br>Chlorophenoxy compounds | |
| Thirst | Pentachlorophenol<br>Nitrophenols<br>Inorganic arsenicals<br>Phosphorus<br>Phosphides<br>Sodium Fluoride<br>Cholecalciferol<br>Aminopyridine | Borate<br>Endothall |
| Anorexia | Organophosphates<br>N-methyl carbamates<br>Nicotine<br>Pentachlorophenol<br>Hexachlorobenzene<br>Chlordimeform<br>Cholecalciferol | Halocarbon fumigants<br>Nitrophenols<br>Inorganic arsenicals<br>Amino pyridine |
| Alcohol intolerance | Thiram<br>Calcium cyanamide | |
| Sweet taste in the mouth | Chlordimeform | |
| Metallic taste in the mouth | Inorganic arsenicals<br>Organic mercury | |
| Salty, soapy taste<br>In the mouth | Sodium fluoride | |

## General and Non-Specific

## Skin

| SYSTEM | SYMPTOMS/ SIGNS/DISEASE CATEGORIES | CHARACTERISTIC OF THESE POISONINGS | MAY OCCUR IN THESE POISONINGS |
|---|---|---|---|
| SYSTEM: Skin | Irritation, rash, blistering, or erosion (without sensitization) | Copper, organotin, and cadmium compounds<br>Metam sodium<br>Paraquat<br>Diquat<br>Sodium chlorate<br>Phosphorus<br>Sulfur<br>Thiram<br>Chlordimeform<br>Cationic detergents<br>Hexachlorophene<br>Ethylene oxide<br>Formaldehyde<br>Acrolein<br>Methyl bromide<br>Ethylene dibromide<br>Dibromochlorpropane<br>Dichloropropane<br>Endothall<br>Aliphatic acids | Pentachlorophenol<br>Picloram<br>Chlorophenoxy<br>Captan<br>Rotenone<br>Diethyltoluamide (DEET)<br>Creosote<br>Fungicides<br>Herbicides with irritant properties<br>Petroleum distillate |
| | Contact dermatitis | PCP<br>Paraquat<br>DEET<br>Chlorhexidine<br>Creosote<br>Hexachlorophene<br>Pyrethrins/pyrethroids<br>Chlorothalonil<br>Thiram<br>Thiophthalimides<br>Propachlor<br>Propargite<br>Ethylene oxide | Barban<br>Captafol<br>Formaldehyde |
| | Flushing | Cyanamide<br>Nitrophenol | Thiram plus alcohol |
| | Beefy red palms, soles | Borate | |
| | Urticaria | Chlorhexidine<br>PCP<br>DEET | Fluoride<br>Pentachlorophenol |
| | Bullae | Liquid fumigants | Hexachlorobenzene |
| | Pallor | Organochlorines<br>Fumigants<br>Sodium fluoride<br>Creosote | Coumarins<br>Indandiones |

Skin

## SYSTEM: Skin, cont.

| SYMPTOMS/ SIGNS/DISEASE CATEGORIES | CHARACTERISTIC OF THESE POISONINGS | MAY OCCUR IN THESE POISONINGS |
|---|---|---|
| Cyanosis | Sodium chlorate<br>Paraquat<br>Cadmium dusts<br>Sodium fluoroacetate<br>Strychnine<br>Crimidine<br>Nicotine<br>Organochlorines | Organophosphates<br>N-methyl carbamates |
| Yellow stain | Nitrophenols | |
| Keratoses, brown discoloration | Inorganic arsenicals | |
| Ecchymoses | Coumarins<br>Indandiones | Phosphorus<br>Phosphides |
| Jaundice | Carbon tetrachloride<br>Chloroform<br>Phosphorus<br>Phosphides<br>Phosphine<br>Paraquat<br>Sodium chlorate | Inorganic arsenicals<br>Diquat<br>Copper compounds |
| Excessive hair growth | | Hexachlorobenzene |
| Loss of hair | Thallium | Inorganic arsenicals |
| Loss of fingernails | | Paraquat<br>Inorganic arsenicals |
| Brittle nails, white striations | Inorganic arsenicals | Thallium |
| Sweating, diaphoresis | Organophosphates<br>N-methyl carbamates<br>Nicotine<br>Pentachlorophenol<br>Naphthalene<br>Aminopyridine | Copper compounds |

Eye

| | SYMPTOMS/ SIGNS/DISEASE CATEGORIES | CHARACTERISTIC OF THESE POISONINGS | MAY OCCUR IN THESE POISONINGS |
|---|---|---|---|
| SYSTEM: Eye | Conjunctivitis (irritation of mucous membranes, tearing) | Chloropicrin<br>Acrolein<br>Copper compounds<br>Organotin compounds<br>Cadmium compounds<br>Metam sodium<br>Paraquat<br>Diquat<br>Acrolein<br>Chloropicrin<br>Sulfur dioxide<br>Naphthalene<br>Formaldehyde<br>Ethylene oxide<br>Methyl bromide<br>Endothall<br>Toluene<br>Xylene<br>Fipronil | Thiophthalimides<br>Thiram<br>Thiocarbamates<br>Pentachlorophenol<br>Chlorophenoxy compounds<br>Chlorothalonil<br>Picloram<br>Creosote<br>Aliphatic acids<br>Strobilurin fungicides<br>Pyrethrins |
| | Lacrimation (muscarinic) | Organophosphates<br>N-methyl carbamates | |
| | Yellow sclerae | Nitrophenols | Agents that cause jaundice *(see section on Skin)* |
| | Keratitis | Paraquat | |
| | Ptosis | Thallium | |
| | Diplopia | Organophosphates<br>N-methyl carbamates<br>Nicotine | |
| | Photophobia | | Organotin compounds |
| | Constricted visual fields | Organic mercury | |
| | Optic atrophy | | Thallium |
| | Miosis | Organophosphates<br>N-methyl carbamates | Nicotine (early) |
| | Dilated pupils | Cyanide<br>Fluoride | Nicotine (late) |
| | Non-reactive pupils | Cyanide | |

| SYSTEM | SYMPTOMS/ SIGNS/DISEASE CATEGORIES | CHARACTERISTIC OF THESE POISONINGS | MAY OCCUR IN THESE POISONINGS |
|---|---|---|---|
| SYSTEM: Nervous System | Paresthesia | Pyrethroids<br>Organochlorines<br>Inorganic arsenicals<br>Organic mercury<br>Carbon disulfide<br>Pyriminil | Organophosphates<br>Thiabendazole Phosphides<br>Sodium fluoroacetate<br>Thallium |
| | Headache | Organophosphates<br>N-methyl carbamates<br>Nicotine<br>Inorganic arsenicals<br>Organic mercury<br>Cadmium compounds<br>Organotin compounds<br>Copper compounds<br>Thallium<br>Fluoride<br>Borates<br>Naphthalene<br>Phosphine<br>Halocarbon fumigants<br>Creosote<br>Diquat<br>Cholecalciferol<br>Cyanamide<br>Neonicotinoids<br>Fipronil | Organochlorines<br>Nitrophenols<br>Thiram<br>Pentachlorophenol<br>Paraquat<br>DEET |
| | Behavioral – mood disturbances<br>(confusion, excitement, mania, disorientation, emotional lability) | Organic mercury<br>Inorganic arsenicals<br>Organotin compounds<br>Thallium<br>Nicotine<br>Sodium fluoroacetate<br>Diquat<br>Cyanide<br>Nitrophenols<br>Aminopyridine<br>Carbon disulfide<br>Methyl bromide<br>Fipronil | Organophosphates<br>N-methyl carbamates<br>Pentachlorophenol<br>Sodium fluoride<br>DEET<br>Organochlorines<br>Neonicotinoids |

Nervous System

**Nervous System**

**SYSTEM: Nervous System, cont.**

| SYMPTOMS/ SIGNS/DISEASE CATEGORIES | CHARACTERISTIC OF THESE POISONINGS | MAY OCCUR IN THESE POISONINGS |
|---|---|---|
| Depression, stupor, coma | Organophosphates<br>N-methyl carbamates (particularly in children)<br>Sodium fluoride<br>Borate<br>Diquat<br>Fipronil<br>Avermectins | Inorganic arsenicals<br>Metaldehyde<br>Sulfuryl fluoride<br>Halocarbon fumigants<br>Phosphorus<br>Phosphine<br>Paraquat<br>Chlorophenoxy compounds<br>DEET<br>Alkyl phthalates |
| Seizures/convulsions (clonic-tonic) sometimes leading to coma | Organochlorines<br>Strychnine<br>Crimidine<br>Sodium fluoroacetate<br>Nicotine<br>Cyanide<br>Acrylonitrile<br>Metaldehyde<br>Thallium<br>DEET<br>Chlorobenzilate<br>Carbon disulfide<br>Phosphine<br>Povidone-iodine<br>Hexachlorophene<br>Sodium chlorate<br>Creosote<br>Endothall<br>Fluoride | Nitrophenols<br>Pentachlorophenol<br>Inorganic arsenicals<br>Organotin compounds<br>Diquat<br>Borate<br>Sulfuryl fluoride<br>Methyl bromide<br>Chlorophenoxy compounds<br>Organophosphates<br>N-methyl carbamates<br>Aminopyridine<br>Fipronil |
| Muscle twitching/ fasciculation | Organophosphates<br>N-methyl carbamates<br>Nicotine<br>Sulfuryl fluoride<br>Pyrethroids | Organic mercury<br>Chlorophenoxy compounds |
| Myotonia | | Chlorophenoxy compounds |
| Tetany, carpopedal spasms | Fluoride<br>Phosphides<br>Phosphorus | |

| | SYMPTOMS/ SIGNS/DISEASE CATEGORIES | CHARACTERISTIC OF THESE POISONINGS | MAY OCCUR IN THESE POISONINGS |
|---|---|---|---|
| **SYSTEM: Nervous System, cont.** | Tremor | Organic mercury<br>Thallium<br>Organophosphates<br>N-methyl carbamates<br>Nicotine<br>Metaldehyde<br>Borates<br>Neonicotinoids<br>Pyrethroids | Pentachlorophenol<br>Nitrophenols<br>Thiram |
| | Incoordination<br>(including ataxia) | Halocarbon fumigants<br>Organophosphates<br>N-methyl carbamates<br>Carbon disulfide<br>Nicotine<br>Thallium | Organic mercury<br>Organochlorines<br>Chlorobenzilate |
| | Paralysis<br>Paresis, muscle weakness | Inorganic arsenicals<br>Organophosphates<br>N-methyl carbamates<br>Nicotine<br>Neonicotinoids | Organic mercury<br>DEET |
| | Hearing loss | Organic mercury | |

| | | | |
|---|---|---|---|
| **SYSTEM: Cardiovascular** | Hypotension and shock | Phosphorus<br>Phosphides<br>Phosphine<br>Sodium fluoride<br>Sodium chlorate<br>Borate<br>Thallium<br>Copper compounds<br>Endothall<br>Cyanamide | Inorganic arsenicals<br>Nicotine (late)<br>Creosote<br>Alkyl phthalate<br>Cycloheximide<br>Formaldehyde |
| | Hypertension | Thallium (early)<br>Nicotine (early) | Organophosphates |
| | Cardiac arrhythmias | Sodium fluoroacetate<br>Halocarbon fumigants<br>Nicotine<br>Sodium fluoride<br>Ethylene oxide<br>Sodium chlorate<br>Thallium-ventricular<br>Povidone-iodine<br>Veratrum alkaloid (sabadilla)<br>Neonicotinoids | Inorganic arsenicals<br>Phosphorus<br>Phosphides<br>Phosphine<br>Organochlorines<br>Cyanide<br>Acrylonitrile<br>Fluoride |

**Cardiovascular System**

| SYSTEM: Cardio, cont. | SYMPTOMS/ SIGNS/DISEASE CATEGORIES | CHARACTERISTIC OF THESE POISONINGS | MAY OCCUR IN THESE POISONINGS |
|---|---|---|---|
| | Bradycardia (sometimes to asystole) | Cyanide<br>Organophosphates<br>N-methyl carbamates | Nicotine (late) |
| | Tachycardia | Nitrophenols<br>Cyanamide<br>Nicotine (early)<br>Neonicotinoids | Metaldehyde<br>Organophosphates (early, before bradycardia)<br>Pentachlorophenol |

**Respiratory System**

| SYSTEM: Respiratory | SYMPTOMS/ SIGNS/DISEASE CATEGORIES | CHARACTERISTIC OF THESE POISONINGS | MAY OCCUR IN THESE POISONINGS |
|---|---|---|---|
| | Upper respiratory tract irritation, rhinitis, scratchy throat, cough | Naphthalene<br>Paraquat<br>Chloropicrin<br>Acrolein<br>Dichloropropene<br>Ethylene dibromide<br>Sulfur dioxide<br>Sulfuryl fluoride<br>Acrylonitrile<br>Formaldehyde<br>Cadmium dusts<br>Pyrethroids<br>Strobilurin fungicides | Dry formulation of copper, tin, zinc compounds<br>Dusts of thiocarbamate and other organic pesticides<br>Chlorophenoxy compounds<br>Aliphatic acids<br>Rotenone |
| | Sneezing | Sabadilla | |
| | Runny nose | Pyrethrins/ pyrethroids<br>Inorganic arsenicals<br>Organophosphates<br>N-methyl carbamates | Dry formulation of copper, tin, zinc compounds<br>Dusts of thiocarbamate and other organic pesticides<br>Chlorophenoxy compounds<br>Aliphatic acids<br>Rotenone |
| | Pulmonary edema (many chemicals come packaged in a hydrocarbon vehicle, well known to cause pulmonary edema) | Methyl bromide<br>Phosphine<br>Phosphorus<br>Phosphine<br>Ethylene oxide<br>Ethylene dibromide<br>Acrolein<br>Pyrethroids<br>Sulfur dioxide<br>Cationic detergents<br>Creosote<br>Methylisothiocyanate<br>Cadmium | Organophosphates<br>N-methyl carbamates<br>Paraquat<br>Phosphides |

| SYSTEM: Respiratory, cont. | | | |
|---|---|---|---|
| | Pulmonary consolidation | Paraquat<br>Cadmium dusts<br>Methyl bromide | Diquat |
| | Dyspnea | Organophosphates<br>N-methyl carbamates<br>Nicotine<br>Paraquat<br>Cadmium dusts<br>Cyanamide<br>Sulfuryl fluoride<br>Pentachlorophenol<br>Methyl bromide<br>Sulfur dioxide<br>Chloropicrin<br>Neonicotinoids | Nitrophenols<br>Cyanide<br>Creosote<br>Pyrethrins<br>Pyrethroids |

| SYSTEM: Gastrointestinal Tract and Liver | SYMPTOMS/ SIGNS/DISEASE CATEGORIES | CHARACTERISTIC OF THESE POISONINGS | MAY OCCUR IN THESE POISONINGS |
|---|---|---|---|
| | Diarrhea (non-bloody) | Organophosphates<br>N-methyl carbamates<br>Pyrethroids<br>Borates<br>Sulfur<br>Nicotine<br>B. thuringiensis<br>Thiram<br>Cadmium<br>Avermectins | Cationic detergents<br>Cresol<br>Hexachlorophene<br>Chlorophenoxy |
| | Diarrhea (bloody) | Fluoride<br>Paraquat<br>Diquat<br>Thallium<br>Coumarins<br>Indandiones<br>Endothall<br>Arsenicals | Phosphorus<br>Phosphides<br>Cycloheximide |
| | Stomatitis | Inorganic arsenicals<br>Paraquat<br>Diquat<br>Copper compounds | Thallium |
| | Salivation | Organophosphates<br>N-methyl carbamates<br>Pyrethroids<br>Nicotine<br>Aminopyridine<br>Sodium fluoride<br>Cyanide<br>Cadmium compounds | |

## Gastrointestinal Tract and Liver

SYSTEM: GI Tract and Liver, cont.

| | | |
|---|---|---|
| Abdominal pain | Organophosphates<br>N-methyl carbamates<br>Paraquat<br>Diquat<br>Nicotine<br>Metaldehyde<br>Fluoride<br>Borate<br>Phosphorous<br>Phosphides<br>Inorganic arsenicals<br>Cadmium compounds<br>Copper compounds<br>Thallium<br>Organotin compounds<br>Neonicotinoids | Chlorophenoxy compounds<br>Aliphatic acids<br>Sodium chlorate<br>Creosote<br>Endothall<br>Aminopyridine<br>Coumarins<br>Indandiones<br>Fumigants (ingested)<br>Cycloheximide |
| Ileus | Thallium<br>Diquat | Pyriminil |
| Constipation | Pyriminil | |

## Liver

SYSTEM: Liver

| SYMPTOMS/ SIGNS/DISEASE CATEGORIES | CHARACTERISTIC OF THESE POISONINGS | MAY OCCUR IN THESE POISONINGS |
|---|---|---|
| Enlargement | Copper compounds<br>Sodium chlorate<br>Phosphine<br>Carbon tetrachloride<br>Chloroform | Inorganic arsenicals<br>Hexachlorobenzene |
| Jaundice<br>(*see section on Skin*) | | |

## Kidney

SYSTEM: Kidney

| SYMPTOMS/ SIGNS/DISEASE CATEGORIES | CHARACTERISTIC OF THESE POISONINGS | MAY OCCUR IN THESE POISONINGS |
|---|---|---|
| Proteinuria/hematuria and acute renal failure | Inorganic arsenicals<br>Copper compounds<br>Sodium fluoride<br>Naphthalene<br>Borate<br>Nitrophenols<br>Pentachlorophenol<br>Sodium chlorate<br>Sulfuryl fluoride<br>Paraquat<br>Diquat<br>Arsine<br>Ethylene dibromide | Cadmium compounds<br>Phosphorus<br>Phosphides<br>Phosphine<br>Chlorophenoxy compounds<br>Creosote<br>Organotin compounds |

| SYSTEM: Kidney | | | |
|---|---|---|---|
| | Rhabdomyolysis | Neonicotinoids<br>Avermectins | |
| | Dysuria, hematuria, pyuria | Chlordimeform | |
| | Polyuria | Cholecalciferol | Fluoride |
| | Hemoglobinuria | Naphthalene<br>Sodium chlorate<br>Arsine | |
| | Wine-red urine (porphyrinuria) | Hexachlorobenzene | |
| | Smoky urine | Creosote<br>Endothall | |
| | Glycosuria | | Organotin compounds |
| | Ketonuria | | Borate |

| SYSTEM: Blood | SYMPTOMS/ SIGNS/DISEASE CATEGORIES | CHARACTERISTIC OF THESE POISONINGS | MAY OCCUR IN THESE POISONINGS |
|---|---|---|---|
| | Hemolysis | Naphthalene<br>Sodium chlorate<br>Arsine | Copper compounds<br>Cresol |
| | Methemoglobinemia | Sodium chlorate<br>Creosote<br>Propanil<br>Metobromuron | Chlordimeform<br>Cyanide<br>Cresol<br>Copper<br>Arsine<br>Diflubenzuron<br>Nitrophenol<br>Metolachlor |
| | Hypoprothrombinemia | Coumarins<br>Indandiones | Phosphorus<br>Phosphides<br>Carbon tetrachloride |
| | Hyperkalemia | Sodium chlorate<br>Naphthalene<br>Arsine | Sodium fluoride |
| | Hypocalcemia | Fluoride | Thallium<br>Phosphorus<br>Phosphides |
| | Hypercalcemia | Cholecalciferol | |
| | Carboxyhemoglobinemia | | Organotin compounds |
| | Anemia | Naphthalene<br>Sodium chlorate<br>Arsine<br>Inorganic arsenicals | |
| | Leukopenia, Thrombocytopenia | Inorganic arsenicals | |

**Blood**

**SYSTEM: Blood, cont.**

| | | |
|---|---|---|
| Elevated LDH<br>GOT, GPT,<br>alkaline phosphatase,<br>ALT, AST enzymes | Carbon tetrachloride<br>Chloroform<br>Phosphine | Inorganic arsenicals<br>Phosphorus<br>Phosphides<br>Sodium chlorate<br>Nitrophenols<br>Pentachlorophenol<br>Thallium<br>Organochlorines<br>Chlorophenoxy compounds |
| Depressed RBC acetylcholinesterase and plasma pseudocholinesterase | Organophosphates | N-methyl carbamates |

**Reproductive System**

**SYSTEM: Reproductive**

| SYMPTOMS/ SIGNS/DISEASE CATEGORIES | CHARACTERISTIC OF THESE POISONINGS | MAY OCCUR IN THESE POISONINGS |
|---|---|---|
| Low sperm count | Dibromochloropropane | Kepone |

# Index of Pesticide Products

## B

## C

## D

## E

## F

## G



## H

## I



## J



## K



## L



## M

## N



## O



## P

## Q



## R



## S

## T

## U



## V



## W



## X



## Y



## Z

www.ingramcontent.com/pod-product-compliance
Lightning Source LLC
Chambersburg PA
CBHW080635180526
45168CB00008B/3180

* 9 7 8 1 5 0 6 1 4 1 2 0 6 *